Advances in Discrete Dynamical Systems

ADVANCED STUDIES IN PURE MATHEMATICS 53

Chief Editor of the Series: Toshitake Kohno (The University of Tokyo)

Advances in Discrete Dynamical Systems

Edited by

Saber Elaydi (Trinity University)
Kazuo Nishimura (Kyoto University)
Mitsuhiro Shishikura (Kyoto University)
Nobuyuki Tose (Keio University)

Mathematical Society of Japan

This book was typeset by $\mathcal{A}_{\mathcal{M}}\mathcal{S}$-TeX and $\mathcal{A}_{\mathcal{M}}\mathcal{S}$-LaTeX, the TeX macro systems of the American Mathematical Society, together with the style files `aspm.sty` *and* `aspmfm.sty` *for $\mathcal{A}_{\mathcal{M}}\mathcal{S}$-TeX written by Dr. Chiaki Tsukamoto and* `aspmproc.sty` *for $\mathcal{A}_{\mathcal{M}}\mathcal{S}$-LaTeX written by Dr. Akihiro Munemasa and* `aspm.cls` *for $\mathcal{A}_{\mathcal{M}}\mathcal{S}$-LaTeX provided by Tokyo Shoseki Printing Co., Ltd.*

TeX is a trademark of the American Mathematical Society.

Edited by the Mathematical Society of Japan.

Published by the Mathematical Society of Japan.

Distributed by the Mathematical Society of Japan, the American Mathematical Society and the World Scientific Publishing Co., Ltd.
Distributed exclusively in North America by the American Mathematical Society.

Advanced Studies in Pure Mathematics 53
ISBN 978-4-931469-49-5

PRINTED IN JAPAN
by Tokyo Shoseki Printing Co., Ltd.

2000 Mathematics Subject Classification.
Primary 39A05.
Secondary 37N25, 37N40, 39A11.

Preface

The Eleventh International Conference on Difference Equations and Applications (ICDEA 06) was held on the campus of Kyoto University, July 24-28, 2006.

The conference was sponsored by the Mathematical Society of Japan, the International Society of Difference Equations, The Research Institute for Mathematical Sciences, the Department of Mathematics at Kyoto University, The 21st Century COE Program "Interfaces for Advanced Economic Analysis", the Institute of Economic Research at Kyoto University and in cooperation with Keio University COE "Integrative Mathematics", and Okayama University of Science.

These Proceedings include talks given by the main speakers as well as talks given by the participants. The proceedings cover the most updated developments in the areas of difference equations and discrete dynamical systems with applications in many scientific disciplines, including economics, biology, engineering, and physics.

On behalf of the International Society of Difference Equations (ISDE), The Editors wish to express their gratitude to the Mathematical Society of Japan for publishing these proceedings.

The Editors

All papers in this volume have been refereed and are in final form. No version of any of them will be submitted for publication elsewhere.

CONTENTS

Advanced Studies in Pure Mathematics 53, 2009
Advances in Discrete Dynamical Systems
pp. 1–9

On periodic points of 2-periodic dynamical systems

João Ferreira Alves

§1. Introduction and statement of the result

Motivated by a recent extension of Sharkovsky's theorem to periodic difference equations [1] (see also [4]), here we show that kneading theory can be useful in the study of the periodic structure of a 2-periodic nonautonomous dynamical system.

Since the notions of zeta function and kneading determinant will play a central role in this discussion, we start by recalling them.

Let X be a set and $f : X \to X$ a map. For each $n \in \mathbb{Z}^+$, denote by f^n the nth iterate of f, defined inductively by

$$f^1 = f \text{ and } f^{n+1} = f \circ f^n, \text{ for all } n \in \mathbb{Z}^+.$$

In what follows we assume that each iterate of f has finitely many fixed points. The Artin-Mazur zeta function of f is defined in [3] as the invertible formal power series

$$\zeta_f(z) = \exp \sum_{n \geq 1} \frac{\#\mathrm{Fix}(f^n)}{n} z^n,$$

where

$$\mathrm{Fix}(f^n) = \{x \in X : f^n(x) = x\}.$$

Naturally, this definition is a particular case of a more general definition, necessary for our purposes.

Let $f : Y \to X$ be a map, with $Y \subset X$. In this case the nth iterate of f is the map $f^n : Y_n \to X$ defined inductively by:

$$Y_1 = Y, f^1 = f$$

Received November 14, 2006.
Revised October 12, 2007.
Key words and phrases. Zeta functions, periodic points, kneading determinants.
Partially supported by FCT/POCTI/FEDER.

and

$$Y_{n+1} = f^{-n}(Y), f^{n+1} = f \circ f^n, \text{ for all } n \in \mathbb{Z}^+.$$

We define

$$\zeta_f(z) = \exp \sum_{n \geq 1} \frac{\#\mathrm{Fix}(f^n)}{n} z^n,$$

where

$$\mathrm{Fix}(f^n) = \{x \in Y_n : f^n(x) = x\}$$

Problems concerning rationality and analytic continuation of ζ_f are often considered. In some interesting cases ζ_f is a rational function of z. Notice that in such case, there exist $a_1, ..., a_k, b_1, ..., b_k \in \mathbb{C}$ such that

$$\zeta_f(z) = \prod_{i=1}^{k} \frac{1 - b_i z}{1 - a_i z},$$

and consequently

$$(1) \qquad \#\mathrm{Fix}(f^n) = \sum_{i=1}^{k} a_i^n - b_i^n, \text{ for all } n \geq 1.$$

Milnor and Thurston in [5] studied the Artin-Mazur zeta function of a continuous piecewise monotone map $f : [a, b] \to [a, b]$ introducing a so called kneading determinant of f, $\mathbf{D}_f(z)$, the determinant of a finite matrix, $\mathbf{N}_f(z)$, called kneading matrix, with entries in $\mathbb{Z}[[z]]$ and depending upon the orbits of the turning points of f; they established a fundamental relation between $\mathbf{D}_f(z)$ and $\zeta_f(z)$. We illustrate this relation in the two following examples, without going into full details.

Example 1. *Let $I = [a, b] \subset \mathbb{R}$ be a compact interval. A continuous map $f : I \to I$ is called piecewise monotone if there exist points (called* turning points *of f) $a = c_0 < c_1 < \cdots < c_{k-1} < c_k = b$ such that: f is strictly monotone in $[c_i, c_{i+1}]$, and f has a local extrema at c_i.*

As an example let $s \in]1, 2]$, and $f : [-1, 1] \to [-1, 1]$ be the continuos map defined by $f(x) = s - 1 - s\,|x|$. The simplest case occurs when $s = 2$, in this case we do not need kneading theory to conclude that

$$\zeta_f(z) = \frac{1}{1 - 2z} \text{ and } \#\mathrm{Fix}(f^n) = 2^n, \text{ for all } n \geq 1.$$

The situation is much more complex when $s \in]1, 2[$. Following [2] we consider a modified kneading determinant of f given by

$$D_f(z) = (1 - z) \sum_{n \geq 0} k_n z^n,$$

where the sequence $k_n \in \{-1, 0, 1\}$ *is defined by*

$$k_0 = 1 \text{and } k_n = -\text{sign}(f^n(0))k_{n-1}, \text{for } n \geq 1.$$

Thus, $D_f(z)$ *depends upon the orbit of the turning point* 0, *and, as a consequence of the Milnor-Thurston's identity*

$$\zeta_f(z) = \mathbf{D}_f^{-1}(z) = \frac{1}{(1-z)\sum_{n\geq 0} k_n z^n},$$

we may conclude that $\zeta_f(z)$ *is rational if and only if the sequence* k_n *is eventually periodic. For example let* $s = \frac{1+\sqrt{5}}{2}$. *Since* $f(0) > 0$, $f^2(0) < 0$ *and* $f^3(0) = 0$, *we have* $\mathbf{D}_f(z) = 1 - 2z + z^3$ *and*

$$\zeta_f(z) = \frac{1}{z^3 - 2z + 1} = \frac{1}{(1-z)(1-\frac{1-\sqrt{5}}{2}z)(1-\frac{1+\sqrt{5}}{2}z)},$$

and by (1)

$$\#\text{Fix}(f^n) = 1 + \left(\frac{1-\sqrt{5}}{2}\right)^n + \left(\frac{1+\sqrt{5}}{2}\right)^n, \text{ for all } n \geq 1.$$

Example 2. *It is possible to generalize the notion of kneading determinant for a continuous piecewise monotone map*

$$f : [a_1, b_1] \cup \cdots \cup [a_k, b_k] \to \mathbb{R}.$$

As in the previous situation, this determinant depends upon the orbits of the turning points of f *and there exists a fundamental relation between* $\mathbf{D}_f(z)$ *and* $\zeta_f(z)$.

As an example, let $a \in]0, 1[$ *and* $f : [-1, -a] \cup [a, 1] \to \mathbb{R}$ *be the continuos map defined by* $f(x) = 1 - 2|x|$. *A modified kneading determinant of* f *is given by*

$$D_f(z) = (1-z)\sum_{n\geq 0} k_n z^n,$$

where the sequence $k_n \in \{-1, 0, 1\}$ *is defined by*

$$k_0 = 1 \text{ and } k_{n+1} = \epsilon(f^{n+1}(a))k_n, \text{ for } n \geq 0,$$

and $\epsilon : \mathbb{R} \to \{-1, 0, 1\}$ *is the step function defined by*

$$\epsilon(x) = \begin{cases} 1 \text{ if } x \in]-1, -a[\\ -1 \text{ if } x \in]a, 1[\\ 0 \text{ otherwise} \end{cases}.$$

As a consequence of Milnor Thurston's main identity we have

$$\zeta_f(z) = \mathbf{D}_f^{-1}(z) = \frac{1}{(1-z)\sum_{n\geq 0} k_n z^n}.$$

Consider the particular case $a = \frac{1}{8}$. *We have* $f(a) = \frac{3}{4}$, $f^2(a) = -\frac{1}{2} < 0$, $f^3(a) = 0$, *thus* $\mathbf{D}_f(z) = 1 - 2z + z^3$ *and*

$$\zeta_f(z) = \frac{1}{z^3 - 2z + 1} = \frac{1}{(1-z)(1-\frac{1-\sqrt{5}}{2}z)(1-\frac{1+\sqrt{5}}{2}z)},$$

and by (1)

$$\#\mathrm{Fix}(f^n) = 1 + \left(\frac{1-\sqrt{5}}{2}\right)^n + \left(\frac{1+\sqrt{5}}{2}\right)^n.$$

As mentioned above, our goal is to show that kneading theory can be useful to study the periodic structure of a periodic nonautonomous dynamical system. In this paper we shall restrict the discussion to 2-periodic dynamical systems. As in the autonomous case, we shall need a preparation theorem, which is actually a generalization of (1). First, we need to introduce some notation.

In what follows, by a dynamical system on a set X we mean a pair

$$F = \{f_0, f_1\}$$

of self mappings in X. Given $x \in X$, the orbit of x is the sequence $\{x_n\}_{n=0}^{\infty}$ on X defined by

$$x_0 = x, x_1 = f_0(x), x_2 = f_1(f_0(x)), \ldots$$

or more precisely

$$x_0 = x \text{ and } x_{n+1} = \begin{cases} f_0(x_n) \text{ if } n \text{ is even} \\ f_1(x_n) \text{ if } n \text{ is odd} \end{cases}. \tag{2}$$

The point x is called periodic, with period $\mathrm{p}(x) \in \mathbb{Z}^+$, if the orbit of x is a periodic sequence with period $\mathrm{p}(x)$. The set whose elements are the periodic points of F is denoted by Per_F. For each positive integer, n, we also define

$$\mathrm{Per}_F(n) = \{x \in \mathrm{Per}_F : \mathrm{p}(x) \text{ divides } n\}.$$

We will assume that $\mathrm{Per}_F(n)$ is a finite set for all positive integer n.

Observe that, even in the simplest cases, there exists a relevant difference between the numbers

$$\#\mathrm{Per}_F(n) \text{ and } \#\mathrm{Fix}(f^n).$$

Indeed, as the following example shows, even when the set X is finite, we can not guarantee the existence of complex numbers $a_1, ..., a_k, b_1, ..., b_k$ such that

$$\#\mathrm{Per}_F(n) = \sum_{i=1}^{k} a_i^n - b_i^n, \text{ for all } n \geq 1. \tag{3}$$

Example 3. *Let $X = \{0,1\}$ and define the maps $f_0 : \{0,1\} \to \{0,1\}$ and $f_1 : \{0,1\} \to \{0,1\}$ by $f_0(0) = f_0(1) = 0$ and $f_1(0) = f_1(1) = 1$. We have*

$$\#\mathrm{Per}_F(n) = \begin{cases} 0 & \text{if } n \text{ is odd} \\ 1 & \text{if } n \text{ is even} \end{cases}.$$

Consequently, the formal power series

$$\exp \sum_{n\geq 1} \frac{\#\mathrm{Per}_F(n)}{n} z^n = \frac{1}{\sqrt{1-z^2}}$$

is not rational, and therefore do not exist complex numbers satisfying (3).

Nevertheless, it can be shown that, if the set X is finite, then there are complex numbers $a_1, ..., a_k, b_1, ..., b_k, c_1, ..., c_k, d_1, ..., d_k$ such that

$$\#\mathrm{Per}_F(n) = \begin{cases} \sum_{i=1}^{k} a_i^n - b_i^n & \text{if } n \text{ is odd} \\ \sum_{i=1}^{k} c_i^n - d_i^n & \text{if } n \text{ is even} \end{cases}. \tag{4}$$

This fact rises the following problem. If $F = \{f_0, f_1\}$ is a dynamical system on an infinite set X, under which conditions can we guarantee the existence of complex numbers verifying (4)? Our main theorem concerns this problem. For that purpose, we need to introduce the maps

$$\begin{array}{cccc} g_0 : & X_0 \subset X & \to & X \\ & x & \to & f_0(x) \end{array},$$

where

$$X_0 = \{x \in X : f_0(x) = f_1(x)\},$$

and

$$g_1 = f_1 \circ f_0.$$

Theorem 4. *Let $F = \{f_0, f_1\}$ be a dynamical system on X. If $\zeta_{g_0}(z)$ and $\zeta_{g_1}(z)$ are rational, and $a_1, ..., a_k, b_1, ..., b_k, c_1, ..., c_k, d_1, ..., d_k$ are complex numbers such that*

$$\zeta_{g_0}(z) = \prod_{i=1}^{k} \frac{1 - b_i z}{1 - a_i z} \text{ and } \zeta_{g_1}(z) = \prod_{i=1}^{k} \frac{1 - d_i z}{1 - c_i z},$$

then we have

$$\#\mathrm{Per}_F(n) = \begin{cases} \sum_{i=1}^{k} a_i^n - b_i^n & \text{if } n \text{ is odd} \\ \sum_{i=1}^{k} c_i^{\frac{n}{2}} - d_i^{\frac{n}{2}} & \text{if } n \text{ is even} \end{cases}.$$

This general result has a relevant consequence in the context of interval maps. Let f_0 and f_1 be continuous piecewise monotone self-maps of a compact interval $I \subset \mathbb{R}$. Furthermore, assume that the set

$$X_0 = \{x \in I : f_0(x) = f_1(x)\}$$

has finitely many connected components. Notice that under these conditions both maps g_0 and g_1 are continuous piecewise monotone, and thus we can use the old results on kneading theory to study the zeta functions of g_0 and g_1.

Example 5. *Let $f_0 : [-1,1] \to [-1,1]$ be defined by $f_0(x) = 1-2|x|$, and let $f_1 : [-1,1] \to [-1,1]$ be any continuous expanding map such that: f_1 is increasing on $[-1,0]$ and decreasing on $[0,1]$; $f_1(x) = f_0(x)$, for $x \in \{-1,0,1\}$. Under these conditions, it is easy to see that $\zeta_{g_1}(z)$ do not depend upon f_1. As a matter of fact*

$$\zeta_{g_1}(z) = \zeta_{f_1 \circ f_0}(z) = \zeta_{f_0^2}(z) = \frac{1}{1-4z}.$$

The study of $\zeta_{g_0}(z)$ is much more interesting because it depends on X_0. The simplest case occurs when $X_0 = \{-1,0,1\}$. In this case we have

$$\zeta_{g_0}(z) = \frac{1}{1-z},$$

and from Theorem 4

$$\#\mathrm{Per}_F(n) = \begin{cases} 1 & \text{if } n \text{ is odd} \\ 2^n & \text{if } n \text{ is even} \end{cases}.$$

Of course the situation is more complex if the set X_0 is infinite. As an example, assume that $X_0 = \left[-1, -\frac{1}{8}\right] \cup \{0\} \cup \left[\frac{1}{8}, 1\right]$. We have then (see Example 2)

$$\zeta_{g_0}(z) = \frac{1}{(1-z)(1-\frac{1-\sqrt{5}}{2}z)(1-\frac{1+\sqrt{5}}{2}z)},$$

and from Theorem 4

$$\#\mathrm{Per}_F(n) = \begin{cases} 1 + \left(\frac{1-\sqrt{5}}{2}\right)^n + \left(\frac{1+\sqrt{5}}{2}\right)^n & \text{if } n \text{ is odd} \\ 2^n & \text{if } n \text{ is even} \end{cases}.$$

§2. Proof of Theorem 4

Let us begin by recalling some facts on generating functions. For any sequence $\{s_n\}_{n=1}^{\infty} \subset \mathbb{C}$ let us define the formal power series

$$S(z) = \exp \sum_{n \geq 1} \frac{s_n}{n} z^n.$$

It is well-known that:
i) The generating function $S(z)$ is a rational function of z if and only if there exist $a_1, ..., a_k; b_1, ..., b_k \in \mathbb{C}$ such that

$$S(z) = \prod_{i=1}^{k} \frac{1-b_i z}{1-a_i z}. \tag{5}$$

ii) For any $a_1, ..., a_k; b_1, ..., b_k \in \mathbb{C}$ the identity (5) holds if and only if

$$s_n = \sum_{i=1}^{k} a_i^n - b_i^n, \text{ for } n \geq 1.$$

So, for any map $f : X \to X$, we may write: $\zeta_f(z)$ is rational if and only if there exist $a_1, ..., a_k; b_1, ..., b_k \in \mathbb{C}$ such that

$$\zeta_f(z) = \prod_{i=1}^{k} \frac{1-b_i z}{1-a_i z}, \tag{6}$$

which is equivalent to

$$\#\mathrm{Fix}(f^n) = \sum_{i=1}^{k} a_i^n - b_i^n, \text{ for } n \geq 1. \tag{7}$$

We can now prove Theorem 4. Let $F = \{f_0, f_1\}$ be a dynamical system on X. If the zeta functions $\zeta_{g_0}(z)$ and $\zeta_{g_1}(z)$ are both rational, then by (6) there are complex numbers $a_1, ..., a_k, b_1, ..., b_k, c_1, ..., c_k, d_1, ..., d_k$ such that

$$\zeta_{g_0}(z) = \prod_{i=1}^{k} \frac{1 - b_i z}{1 - a_i z} \text{ and } \zeta_{g_1}(z) = \prod_{i=1}^{k} \frac{1 - d_i z}{1 - c_i z},$$

and by (7)

$$\#\mathrm{Fix}(g_0^n) = \sum_{i=1}^{k} a_i^n - b_i^n \text{ and } \#\mathrm{Fix}(g_1^n) = \sum_{i=1}^{k} c_i^n - d_i^n,\ n \geq 1.$$

So, the theorem will follow from the identity

$$\mathrm{Per}_F(n) = \begin{cases} \mathrm{Fix}(g_0^n) & \text{if } n \text{ is odd} \\ \mathrm{Fix}(g_1^{n/2}) & \text{if } n \text{ is even} \end{cases}. \tag{8}$$

In order to prove (8), let $x \in X$ and $\{x_i\}_{i=0}^{\infty}$ be the orbit of x. Assume first that n is even. In this case we have by (2)

$$x_0 = x \text{ and } x_{kn} = (f_1 \circ f_0)^{kn/2}(x) = g_1^{kn/2}(x), \text{ for } k \geq 1.$$

So, we can write

$$x_{kn} = x_0, \text{ for all } k \geq 1 \text{ if and only if } g_1^{kn/2}(x) = x, \text{ for all } k \geq 1,$$

and therefore $\mathrm{Per}_F(n) = \mathrm{Fix}(g_1^{n/2})$.

For n odd, if $x \in \mathrm{Per}_F(n)$, then the period $p(x)$ is odd, and:

$$f_0(x_j) = x_{j+1} = x_{1+j+p(x)} = f_1(x_{j+p(x)}) = f_1(x_j), \text{ if } j \text{ is even;}$$

$$f_1(x_j) = x_{j+1} = x_{1+j+p(x)} = f_0(x_{j+p(x)}) = f_0(x_j), \text{ if } j \text{ is odd,}$$

which shows that

$$\{x_i\}_{i=0}^{\infty} \subset X_0 = \{x \in X : f_0(x) = f_1(x)\}.$$

Therefore

$$x_i = f_0^i(x) = g_0^i(x), \text{ for } i \geq 1$$

and consequently

$$g_0^{p(x)}(x) = x_{p(x)} = x_0 = x.$$

This proves the inclusion $\mathrm{Per}_F(n) \subset \mathrm{Fix}(g_0^n)$. Since the inclusion $\mathrm{Fix}(g_0^n) \subset \mathrm{Per}_F(n)$ is immediate it follows $\mathrm{Fix}(g_0^n) = \mathrm{Per}_F(n)$, as requested.

Acknowledgments. The author wishes to thank to Saber Elaydi the helpful discussions on the subject in a early phase of this paper.

References

[1] Z. AlSharawi, J. Angelos, S. Elaydi and L. Rakesh, An extension of Sharkovsky's theorem to periodic difference equations, J. Math. Anal. Appl., **316** (2006), 128–141.

[2] J. F. Alves and J. Sousa Ramos, Kneading theory: a functorial approach, Comm. Math. Phys., **204** (1999), 89–114.

[3] M. Artin and B. Mazur, On periodic points, Ann. of Math. (2), **81** (1965), 82–99.

[4] J. S. Cánovas and A. Linero, Periodic structure of alternating continuous interval maps, J. Difference Equ. Appl., **12** (2006), 847–858.

[5] J. Milnor and W. Thurston, On iterated maps of the interval, In: Dynamical systems, Maryland, 1986-1987, (ed. J. C. Alexander), Lecture Notes in Math., **1342**, Springer-Verlag, 1988, pp. 465–563.

Department of Mathematics
Instituto Superior Técnico
T. U. Lisbon
Av. Rovisco Pais 1
1049-001 Lisboa
Portugal

E-mail address: `jalves@math.ist.utl.pt`

Advanced Studies in Pure Mathematics 53, 2009
Advances in Discrete Dynamical Systems
pp. 11–21

Asymptotic behaviour of a nonlinear stochastic difference equation modelling an inefficient financial market

John A. D. Appleby and Catherine Swords

Abstract.

This note studies the asymptotic behaviour of linear and nonlinear stochastic difference equations whose structure is motivated by a financial market model. The asymptotic results show that the models can produce behaviour consistent with random walk efficient markets as well as bubbles or crashes.

§1. Motivation and background material

In recent years, much attention in financial economics has focussed on the trading strategies of investors. Classical models of financial markets assume that agents are rational, have homogeneous preferences, and do not use historical market data in framing their investment decisions. An important and seminal collection of papers which summarise this position is [5].

Econometric evidence of market returns (see e.g., [9]) and analysis of the behaviour of traders in real markets reveal a more complex picture. Traders often employ rules of thumb which do not conform to notions of rational behaviour based on knowledge of the empirical distribution of returns (see e.g., [8]). Moreover, many traders use past prices as a guide to the evolution of the price in the future (see e.g., [10]). Linear continuous–time stochastic models of markets which involve agents using past prices to determine their demand, but in which the traders discount past returns with a simple type of exponentially fading memory, include [1] and [7].

In this paper, we present a stochastic difference equation model of an inefficient financial market. The model is informationally inefficient,

Received November 2, 2006.
Key words and phrases. Volterra difference equation, stochastic Volterra difference equation, resolvent, inefficient market, simple random walk.

in the sense that past movements of the stock price have an influence on future movements. We assume that there is trading at intervals of one time unit, with prices fixed in the intervening period. The inefficiency stems from the presence of trend–following speculators, whose demand for the asset depends on the difference between a weighted average over the last N periods of the cumulative return on the stock and the current cumulative return. More precisely, if $X(n)$ is the cumulative return up to time n, the planned excess demand just before trading at time $n+1$ is $g(X(n)) - \sum_{j=1}^{N} w(j)g(X(n-j))$ where $\sum_{j=1}^{N} w(j) = 1$ and g is an increasing function. Speculators react to other random stimuli—"news"— which is independent of past returns. This news arrives at time $n+1$, adding a further $\xi(n+1)$ to the traders' excess demand. Prices increase when there is excess demand (resp. fall when there is excess supply), with the rise (resp. fall) being larger the greater the excess demand (resp. supply). Hence, the price adjustment at time $n+1$ is given by

$$(1) \quad X(n+1) = X(n) + g(X(n)) - \sum_{j=1}^{N} w(j)g(X(n-j)) + \xi(n+1).$$

We study the almost sure asymptotic behaviour as $n \to \infty$ of solutions of (1).

This paper shows three things: first, if the trend following speculators do not behave very aggressively to the difference between current returns and historical returns, or do not discount prices quickly, then returns behave very similarly to a simple random walk, in that they have the same size of large fluctuations. This is characteristic of an efficient market. This occurs once g is linear (Theorem 3), or obeys $g(x) \sim \beta x$ as $x \to \infty$ for some $\beta \geq 0$ (Theorem 5). Moreover, in the case when g is linear the returns follow a random walk plus a stationary mean–reverting process (Theorem 3). Also when g is linear, and the trend–following speculators behave aggressively, the returns will tend to plus or minus infinity exponentially fast: this is a mathematical realisation of a stock market bubble (Theorem 4).

The distinction between traders who are "aggressive with long memory" or "less aggressive with short memory" depends on whether $\beta \sum_{j=1}^{N} jw(j)$ is greater than or less than unity. Large values of β correspond to aggressive behaviour; if $g(x) = \beta x$ for example, the planned excess demand of traders is β multiplied by the difference between the current returns and a weighted average of returns. Therefore, for larger β, a smaller signal from the market produces a given response from the traders. The term $M = \sum_{j=1}^{N} jw(j)$ is in $[1, N]$, and the greater weight that traders give to returns further back in time, the larger M becomes. Therefore, M is a measure of the effective length of memory of the traders; indeed, if traders make their decisions based only on a comparison of current returns with returns j periods ago, then $M = j$, $j = 1, \dots, N$. A fuller treatment of the economic interpretation of the results, the correlation of the returns, and the effect on the market of contrarian (or negative feedback traders) will be presented in a later paper. Due to restrictions on space, economic interpretation of the results is restricted to this introduction.

The paper has the following structure. Section 2 gives notation and supporting results. The asymptotic behaviour of the linear equation is presented in Section 3 and the nonlinear equation is considered in Section 4. Results employed from the theory of deterministic difference equations are standard, see e.g., [6]. Definitions and results from discrete–time martingale theory may all be found in [11].

§2. Background material

$\mathbb{N}$ denotes the integers $0, 1, 2, \dots$, and $\mathbb{R}$ the real line. A real sequence $a = \{a(n) : n \in \mathbb{N}\}$ obeys $a \in \ell^1(\mathbb{N}; \mathbb{R})$ if $\sum_{n \in \mathbb{N}} |a(n)| < \infty$. The convolution of $f = \{f(n) : n \in \mathbb{N}\}$ and $g = \{g(n) : n \in \mathbb{N}\}$, $f * g$, is a sequence defined by $(f * g)(n) = \sum_{k=0}^{n} f(n-k)g(k)$, $n \in \mathbb{N}$.

Let $\beta > 0$, $N \in \mathbb{N}$, and suppose $w = \{w(n) : n = 1, \dots, N\}$ obeys

$$w(n) \geq 0, \quad n = 1, \dots, N; \quad \sum_{n=1}^{N} w(n) = 1 \tag{2}$$

The resolvent $r = \{r(n) : n \geq -N\}$ is a scalar sequence defined by

$$r(n+1) = r(n) + \beta \left(r(n) - \sum_{j=1}^{N} w(j) r(n-j) \right), \quad n \in \mathbb{N} \tag{3a}$$

$$r(0) = 1, \quad r(n) = 0, \quad n < 0. \tag{3b}$$

Lemma 1. *Let $\beta > 0$, w obey (2), and r be defined by (3).*

(a) *r is a non–decreasing sequence with $r(n) > 0$ for $n \in \mathbb{N}$.*

(b) *If $\beta \sum_{j=1}^{N} j w(j) < 1$, then*

$$\lim_{n \to \infty} r(n) = \frac{1}{1 - \beta \sum_{n=1}^{N} j w(j)} =: r^*, \tag{4}$$

and $\delta = \{\delta(n) : n \geq -N\}$ defined by $\delta(-N) = 0$ and $\delta(n+1) = r(n+1) - r(n)$ for $n \geq -N$ obeys $\delta \in \ell^1(\mathbb{N}; \mathbb{R}^+)$.

(c) *If* $\beta\sum_{j=1}^{N} jw(j) > 1$, *there exists* $\alpha \in (0,1)$ *defined by*

$$\alpha^{-1} = 1 + \beta\left(1 - \sum_{k=1}^{N} \alpha^k w(k)\right) \tag{5}$$

such that $\lim_{n\to\infty} \alpha^n r(n) = R^*$, *where* $R^* > 0$ *is given by*

$$R^* = \frac{1}{(1-\alpha)(1+\beta\alpha\sum_{j=1}^{N-1} j\alpha^j W(j))}, \tag{6}$$

and W *is defined by* $W(j) = \sum_{k=j+1}^{N} w(k)$, $j = 0, \dots, N-1$.

Proof. If $N = 1$, $r(n+1) = r(n) + \beta(r(n) - r(n-1))$, so $\delta(n) = \beta^n$, and the results are trivial. Assume $N \geq 2$. Part (a) follows by induction. For (b), putting $r(n) = \sum_{j=-N}^{n} \delta(j)$ into (3) and using (2) gives

$$\begin{aligned}\delta(n+1) = \beta\delta(n) &+ \beta \sum_{j=-N}^{n-N} \left(\sum_{k=1}^{N} w(k) - \sum_{k=1}^{(n-j)\wedge N} w(k)\right)\delta(j) \\ &+ \beta \sum_{j=n-N+1}^{n-1} \left(\sum_{k=1}^{N} w(k) - \sum_{k=1}^{(n-j)\wedge N} w(k)\right)\delta(j).\end{aligned}$$

Rearrange the righthand side and set $W(j) = \sum_{k=j+1}^{N} w(k)$, $j = 0, \dots,$ $N-1$ to get

$$\delta(n+1) = \beta\sum_{j=0}^{N-1} W(j)\delta(n-j), \quad n \in \mathbb{N}; \quad \delta(0) = 1, \delta(n) = 0, \quad n < 0. \tag{7}$$

If $\beta\sum_{k=1}^{N} kw(k) < 1$, $\sum_{n=-N}^{\infty} \delta(n) =: r^*$ is finite, and $r^* = 1 + \beta r^* \sum_{j=0}^{N-1} W(j)$. Since $\sum_{j=0}^{N-1} W(j) = \sum_{j=1}^{N} jw(j)$, (4) holds. Thus $r^* - r(n) = \sum_{j=n+1}^{\infty} \delta(j) =: \Delta(n+1)$. Then $\Delta(n) = r^*$ for $n = -N, \dots, 0$, $\Delta(1) = r^* - 1$, and from (7)

$$\Delta(n+1) = \beta\sum_{j=0}^{N-1} W(j)\Delta(n-j), \quad n \in \mathbb{N}. \tag{8}$$

For part (c), multiplying across (7) by α^{n+1}, where α is given by (5), we get $\delta_\alpha(n+1) = \sum_{j=0}^{N-1} \beta\alpha W_\alpha(j)\delta_\alpha(n-j)$ where $\delta_\alpha(n) = \alpha^n\delta(n)$,

$W_\alpha(n) = \alpha^n W(n)$. Then, the discrete time renewal theorem or z–transform techniques imply

$$\lim_{n\to\infty} \alpha^n \delta(n) = \lim_{n\to\infty} \delta_\alpha(n) = \frac{1}{1+\beta\alpha\sum_{j=1}^{N-1} j\alpha^j W(j)} > 0.$$

Therefore $\lim_{n\to\infty} \alpha^n r(n) = R^*$, where $R^* > 0$ is given by (6). Q.E.D.

Next we find the growth rate of a moving average of a sequence.

Lemma 2. *Let γ be positive and increasing with $\gamma(n-N)/\gamma(n) \to 1$, as $n \to \infty$, for all $N \in \mathbb{N}$. If $k = \{k(n) : n \in \mathbb{N}\}$ is non-negative with $\sum_{n=0}^\infty k(n) \in (0,\infty)$, then $\lim_{n\to\infty}(k*\gamma)(n)/\gamma(n) = \sum_{n=0}^\infty k(n)$.*

Proof. Without loss of generality, let $\sum_{n=0}^\infty k(n) = 1$. For every $\varepsilon > 0$ there is $N > 0$ such that $\sum_{j=N+1}^\infty k(j) < \varepsilon/2$. For $n \geq N+1$, we have

$$\begin{aligned}\frac{(k*\gamma)(n)}{\gamma(n)} - \sum_{j=0}^{n} k(j) &= \sum_{j=0}^{N} k(j)\left(\frac{\gamma(n-j)}{\gamma(n)} - 1\right) \\ &\quad + \sum_{j=N+1}^{n} k(j)\left(\frac{\gamma(n-j)}{\gamma(n)} - 1\right),\end{aligned}$$

which, using monotonicity of γ, and the fact that $\sum_{n=0}^\infty k(n) = 1$ gives

$$\left|\sum_{j=0}^{n} \frac{k(j)\gamma(n-j)}{\gamma(n)} - \sum_{j=0}^{n} k(j)\right| \leq \left(1 - \frac{\gamma(n-N)}{\gamma(n)}\right) + 2\sum_{j=N+1}^{\infty} k(j).$$

Using $\gamma(n-N)/\gamma(n) \to 1$, and then letting $\varepsilon \to 0$ yields the result. Q.E.D.

§3. Linear stochastic difference equation

We consider the linear stochastic difference equation

(9a)

$$Y(n+1) = Y(n) + \beta\{Y(n) - \sum_{j=1}^{N} w(j)Y(n-j)\} + \xi(n+1),\ n \geq 0$$

(9b) $$Y(n) = \phi(n), \quad n \leq 0,$$

where $\xi = \{\xi(n) : n \in \mathbb{N}\}$ is a sequence of random variables obeying

(10a) ξ is a sequence of independent, identically distributed r.vs;

(10b) $\mathbb{E}[\xi(n)] = 0, \quad \mathbb{E}[\xi(n)^2] = \sigma^2, \quad$ for some $\sigma > 0$, and all $n \in \mathbb{N}$.

If $\beta \sum_{j=1}^{N} jw(j) < 1$, Y behaves asymptotically as a random walk. For instance, if ξ obeys (10) the process S given by $S(n) = \sum_{j=1}^{n} \xi(j)$ is a random walk and obeys *the Law of the Iterated Logarithm*:

$$\limsup_{n\to\infty} \frac{S(n)}{\sqrt{2n \log\log n}} = -\liminf_{n\to\infty} \frac{S(n)}{\sqrt{2n \log\log n}} = \sigma, \quad \text{a.s.}$$

Theorem 3. *Let w obey (2), $0 < \beta \sum_{n=1}^{N} nw(n) < 1$, ξ obey (10), and Y obey (9). Then*

(11)
$$\limsup_{n\to\infty} \frac{Y(n)}{\sqrt{2n \log\log n}} = -\liminf_{n\to\infty} \frac{Y(n)}{\sqrt{2n \log\log n}} = \frac{|\sigma|}{1 - \beta \sum_{j=1}^{N} jw(j)}.$$

Proof. For $n \geq 1$, $Y(n) = y(n) + \sum_{j=1}^{n} r(n-j)\xi(j)$, where $y(n+1) = y(n) + \beta(y(n) - \sum_{j=1}^{N} w(j)y(n-j))$, $n \geq 0$ and $y(n) = \phi(n)$, $n \leq 0$. With $U(n) = \sum_{j=1}^{n} \Delta(n-j)\xi(j)$, we get $Y(n) = y(n) - U(n) + r^* \sum_{j=1}^{n} \xi(j)$, $n \geq 1$. Since $r(n) \to r^*$, $\lim_{n\to\infty} y(n)$ exists. By the law of the iterated logarithm, we need only show $\lim_{n\to\infty} |U(n)|/\sqrt{2n \log\log n} = 0$ a.s. Let $b(x) = \sqrt{x}$, $x \geq 0$. Then $b : [0,\infty) \to [0,\infty)$ is increasing and $b^{-1}(x) = x^2$. If ξ is a random variable with the same distribution as $\xi(n)$, by Corollary 4.1.3 in [4], we have

$$\sum_{n=1}^{\infty} \mathbb{P}[|\xi(n)| > \sqrt{n}] \leq \mathbb{E}[b^{-1}(|\xi|)] = \mathbb{E}[\xi^2] < \infty.$$

By the Borel–Cantelli lemma, $\limsup_{n\to\infty} |\xi(n)|/\sqrt{n} \leq 1$, a.s. which implies that $\lim_{n\to\infty} |\xi(n)|/\sqrt{2n \log\log n} = 0$ a.s. Thus, there is an a.s. event Ω^* such that for all $\omega \in \Omega^*$, and all $\varepsilon > 0$, there is $C(\varepsilon, \omega) > 0$ such that

$$|\xi(n,\omega)| < C(\varepsilon,\omega) + \varepsilon\sqrt{2n \log\log(n + e^e)} =: \gamma(n,\omega), \quad n \in \mathbb{N}.$$

By (8) and $\beta \sum_{j=1}^{N} jw(j) < 1$, $\Delta \in \ell^1(\mathbb{N};\mathbb{R})$, so by Lemma 2

$$\limsup_{n\to\infty} \frac{|U(n,\omega)|}{\gamma(n,\omega)} \leq \limsup_{n\to\infty} \frac{\sum_{j=1}^{n} |\Delta(n-j)|\gamma(\omega,j)}{\gamma(n,\omega)} = \sum_{j=0}^{\infty} |\Delta(j)|;$$

thus $\limsup_{n\to\infty} |U(n,\omega)|/\sqrt{2n\log\log n} < \varepsilon\sum_{j=0}^{\infty}|\Delta(j)|$, hence the result. Q.E.D.

When $\beta\sum_{j=1}^{N} jw(j) < 1$, then $U(n+1) = \beta\sum_{k=0}^{N-1} W(k)U(n-k) + \zeta(n+1)$, $n \geq N$, where $\zeta(n+1) := \sum_{k=0}^{N-1}\theta(k)\xi(n-k+1)$ and θ is a deterministic sequence which depends on Δ. In this case, $y - U$ is an asymptotically stationary ARMA process, so Y is the sum of an asymptotically stationary process and a random walk. ARMA (autoregressive moving average) processes are used widely in financial econometrics (see e.g., [2]). When $\beta\sum_{j=1}^{N} jw(j) > 1$, we now prove $\alpha^n Y(n) \to Y^*$ as $n\to\infty$ where $\alpha \in (0,1)$ and Y^* is a random variable given explicitly in terms of ξ. Hence $Y(n)$ tends to $\pm\infty$ according to the sign of Y^*.

Theorem 4. *Let w obey (2), $\beta\sum_{j=1}^{N} jw(j) > 1$, ξ obey (10), and Y obeys (9). If $\alpha \in (0,1)$ is given by (5), and R^* by (6), then*

$$\lim_{n\to\infty}\alpha^n Y(n) = R^*\left(\phi(0) - \beta\sum_{j=0}^{N-1}\sum_{k=j+1}^{N}\alpha^{j+1}w(k)\phi(j-k) + \sum_{j=1}^{\infty}\alpha^j\xi(j)\right).$$

Proof. Let $y(n+1) = y(n) + \beta(y(n) - \sum_{k=1}^{N} w(k)y(n-k))$, $n \geq 0$ and $y(n) = \phi(n)$, $n \leq 0$. Then $y(n) = r(n)\phi(0) - \beta(\tilde{\phi} * r)(n-1)$, $n \geq 1$, where

$$\tilde{\phi}(n) = \begin{cases} \sum_{j=n+1}^{N} w(j)\phi(n-j), & n = 0,1,\ldots,N-1 \\ 0, & n = N, N+1, \ldots, \end{cases}$$

so $y(n) = r(n)\phi(0) - \beta\sum_{j=0}^{N-1} r(n-1-j)\tilde{\phi}(j)$, $n \geq N$. As $\lim_{n\to\infty} r(n)\alpha^n = R^*$,

$$\lim_{n\to\infty}\alpha^n y(n) = R^*\left(\phi(0) - \beta\sum_{j=0}^{N-1}\sum_{k=j+1}^{N}\alpha^{j+1}w(k)\phi(j-k)\right).$$

Next, for $n \geq 1$, we have

$$\text{(12)}\quad \alpha^n Y(n) = \alpha^n y(n) + \sum_{j=1}^{n}(\alpha^{n-j}r(n-j) - R^*)\alpha^j\xi(j) + R^*\sum_{j=1}^{n}\alpha^j\xi(j).$$

Let $M(n) = R^*\sum_{j=1}^{n}\alpha^j\xi(j)$, $n \geq 0$. Since $\alpha \in (0,1)$ M is martingale with finite quadratic variation, so by the martingale convergence theorem $\lim_{n\to\infty} M(n)$ is finite a.s. Since $\mathbb{E}[|\xi(j)|]^2 \leq \mathbb{E}[\xi(j)^2] = \sigma^2$, $\mathbb{E}\sum_{j=1}^{n}\alpha^j|\xi(j)| = \sum_{j=1}^{n}\alpha^j\mathbb{E}[|\xi(j)|] \leq \alpha\sigma(1-\alpha)^{-1}$, so $\xi_\alpha(n) := \alpha^n\xi(n) \in \ell^1(\mathbb{N};\mathbb{R})$, a.s. But $r_1(n) := \alpha^n r(n) - R^* \to 0$, so $(r_1 * \xi_\alpha)(n) \to 0$ as $n\to\infty$. We finish by letting $n\to\infty$ in (12). Q.E.D.

§4. Nonlinear stochastic difference equation

We now study the nonlinear stochastic difference equation

(13a)

$$X(n+1) = X(n) + g(X(n)) - \sum_{j=1}^{N} w(j) g(X(n-j)) + \xi(n+1), \quad n \in \mathbb{N},$$

(13b)

$$X(n) = \phi(n), \quad n = -N, -N+1, \dots, 0.$$

$g : \mathbb{R} \to \mathbb{R}$ is presumed to have the following properties

$$g \in C(\mathbb{R};\mathbb{R}), \ \lim_{x\to\infty} \frac{g(x)}{x} = \lim_{x\to-\infty} \frac{g(x)}{x} = \beta \text{ for some } \beta \geq 0. \tag{14}$$

We now show if the conditions of Theorem 3 hold, the a.s. partial extrema of the solution of (13) grow exactly as those of the solution of (9), which are consistent with the extrema of a random walk. The proof of this result is partly inspired by work in [3].

Theorem 5. *Let w obey* (2), $0 < \beta \sum_{n=1}^{N} nw(n) < 1$, *$g$ obey* (14), *ξ obey* (10), *and Y obey* (9). *Then the solution of* (13) *obeys*

$$\lim_{n\to\infty} \frac{|X(n) - Y(n)|}{\sqrt{2n \log\log n}} = 0, \tag{15}$$

(16)

$$\limsup_{n\to\infty} \frac{X(n)}{\sqrt{2n\log\log n}} = -\liminf_{n\to\infty} \frac{X(n)}{\sqrt{2n\log\log n}} = \frac{|\sigma|}{1 - \beta\sum_{j=1}^{N} jw(j)}.$$

Proof. Set $Z(n) = X(n) - Y(n)$, $\gamma(x) = g(x) - \beta x$, and $G(n+1) = \gamma(X(n)) - \sum_{j=1}^{N} w(j)\gamma(X(n-j))$, so $Z(n+1) - Z(n) = G(n+1) + \beta[Z(n) - \sum_{j=1}^{N} w(j) Z(n-j)]$. Therefore $Z(n) = \sum_{j=0}^{n-1} r(n-1-j) G(j+1)$, $n \geq 1$. Let $n \geq 2$, $n \geq N+1$, so

$$\begin{aligned} Z(n) &= \sum_{j=0}^{n-1} r(n-1-j)\gamma(X(j)) - \sum_{k=1}^{N} w(k) \sum_{j=0}^{n-1} r(n-1-j)\gamma(X(j-k)) \\ &= \sum_{j=0}^{n-1} r(n-1-j)\gamma(X(j)) - \sum_{k=1}^{N} w(k) \sum_{l=-k}^{n-k-1} r(n-k-l-1)\gamma(X(l)) \end{aligned}$$

Hence

$$Z(n) = \sum_{j=0}^{n-1} r(n-1-j)\gamma(X(j)) - \sum_{j=-N}^{n-2} \left(\sum_{k=-j\vee 1}^{(n-j-1)\wedge N} w(k)r(n-k-j-1) \right) \gamma(X(j))$$

$$= \sum_{l=1}^{n-1} \left(r(l) - \sum_{k=-(n-l-1)\vee 1}^{l\wedge N} w(k)r(l-k) \right) \gamma(X(n-l-1))$$

$$+ \gamma(X(n-1)) - \sum_{j=-N}^{-1} \left(\sum_{k=-j\vee 1}^{(n-j-1)\wedge N} w(k)r(n-k-j-1) \right) \gamma(X(j)).$$

Thus

$$Z(n) = \gamma(X(n-1)) + \sum_{l=N}^{n-1} \delta(l+1)\frac{1}{\beta}\gamma(X(n-l-1))$$

$$+ \sum_{l=1}^{N-1} \left(r(l) - \sum_{k=1}^{l} w(k)r(l-k) \right) \gamma(X(n-1-l)) - f_1(n),$$

where $f_1(n) = \sum_{j=-N}^{-1} \left(\sum_{k=-j}^{N} w(k)r(n-k-j-1) \right) \gamma(X(j))$. Set $u(0) = 1$, $u(l) = r(l) - \sum_{k=1}^{l} w(k)r(l-k)$, $l = 1,.., N-1$, $f_2(n) := |f_1(n)| + L(\varepsilon)[\sum_{l=0}^{N-1} |u(l)| + \beta^{-1} \sum_{l=N}^{n-1} \delta(l+1)]$. As $\lim_{|x|\to\infty} \gamma(x)/x = 0$, for each $\varepsilon > 0$ there is $L(\varepsilon) > 0$ such that $|\gamma(x)| \le L(\varepsilon) + \varepsilon|x|$, $x \in \mathbb{R}$. As $\lim_{n\to\infty} f_1(n)$ exists,

$$|Z(n)| \le \sum_{l=0}^{N-1} |u(l)| \left(L(\varepsilon) + \varepsilon|Z(n-1-l)| + \varepsilon|Y(n-1-l)| \right)$$

$$+ \frac{1}{\beta} \sum_{j=0}^{n-1-N} \delta(n-j) \left(L(\varepsilon) + \varepsilon|Z(j)| + \varepsilon|Y(j)| \right) + |f_1(n)|$$

and so there is an f_2 tending to a finite limit such that

$$|Z(n)| \le \sum_{l=0}^{N-1} |u(l)|\varepsilon|Y(n-1-l)| + \frac{1}{\beta} \sum_{l=N}^{n-1} \delta(l+1)\varepsilon|Y(n-l-1)|$$

$$+ \sum_{l=0}^{N-1} |u(l)|\varepsilon|Z(n-1-l)| + \frac{1}{\beta} \sum_{l=N}^{n-1} \delta(l+1)\varepsilon|Z(n-l-1)| + f_2(n).$$

Since $\delta \in \ell^1(\mathbb{N};\mathbb{R}^+)$, there is a summable κ such that

$$|Z(n)| \le f_3(n) + \varepsilon \sum_{j=0}^{n-1} \kappa(n-1-j)|Y(j)| + \varepsilon \sum_{j=0}^{n-1} \kappa(n-1-j)|Z(j)|,$$

where f_3, which tends to a finite limit, has been introduced so that this estimate holds for $n \ge 0$ too. Fix $\varepsilon > 0$ so that $\varepsilon \sum_{n=0}^{\infty} \kappa(n) < 1/2$. Define ρ by $\rho(0) = 1$, $\rho(n+1) = \varepsilon \sum_{j=0}^{n} \kappa(n-j)\rho(j)$, $n \in \mathbb{N}$, and z by

$$z(n+1) = f_3(n+1) + \varepsilon \sum_{j=0}^{n} \kappa(n-j)|Y(j)| + \varepsilon \sum_{j=0}^{n} \kappa(n-j)z(j), \quad n \in \mathbb{N},$$

where $z(0) = 0$. Therefore $|Z(n)| \le z(n)$ and

$$z(n) = \sum_{j=1}^{n} \rho(n-j) \left(f_3(j) + \varepsilon \sum_{k=0}^{j-1} \kappa(j-1-k)|Y(k)| \right).$$

As $\rho \in \ell^1(\mathbb{N};(0,\infty))$, there is an f_4 obeying $\lim_{n\to\infty} f_4(n) = 0$ and

$$|Z(n)| \le f_4(n) + \varepsilon \sum_{k=0}^{n-1} (\rho * \kappa)(n-k-1)|Y(k)|.$$

Therefore, from (11) and Lemma 2 it follows that

$$\limsup_{n\to\infty} \frac{|Z(n)|}{\sqrt{2n\log\log n}} \le \varepsilon c' \sum_{k=0}^{\infty} (\rho * \kappa)(k) = \varepsilon c' \sum_{k=0}^{\infty} \kappa(k) \frac{1}{1 - \varepsilon \sum_{k=0}^{\infty} \kappa(k)},$$

where $c' > 0$ is the righthand side of (11). Since ε can be taken as small as required, and the last inequality holds pathwise, we have (15). (16) is an immediate consequence of (15) and Theorem 3. Q.E.D.

Acknowledgment. The research is partially supported by Science Foundation Ireland under the Research Frontiers Programme grant RFP/MAT/0018 "Stochastic Functional Differential Equations with Long Memory". The authors are grateful to the referee for their scrutiny of the manuscript. John Appleby also wishes to thank the organisers of the International Conference on Difference Equations and Applications for the opportunity to present this work at the conference.

References

[1] J.-P. Bouchaud and R. Cont, A Langevin approach to stock market fluctuations and crashes, Eur. Phys. J. B, **6** (1998), 543–550.

[2] P. J. Brockwell and R. A. Davis, Introduction to Time Series and Forecasting, 2nd ed., Springer, New York, 2002.

[3] T. A. Burton, Fixed points and differential equations with asymptotically constant or periodic solutions, Electron. J. Qual. Theory Differ. Equ., **11** (2004), p. 31 (electronic).

[4] Y. S. Chow and H. Teicher, Probability Theory: independence, interchangability, martingales, 2nd ed., Springer, Berlin, 1988.

[5] P. H. Cootner, ed. The Random Character of Stock Market Prices, MIT Press, Cambridge, 1964.

[6] S. Elaydi, Introduction To Difference Equations, 3rd ed., Springer, New York, 2005.

[7] D. Hobson and L. C. G. Rogers, Complete models with stochastic volatility, Math. Finance, **8** (1998), 27–48.

[8] D. Kahneman and M. R. Rieple, Aspects of investor psychology: beliefs, preferences and biases investment advisors should know about, J. Portfolio Management, summer (1998), 52–65.

[9] A. Lo and A. C. MacKinlay, A Non-Random Walk Down Wall Street, Princeton Univ. Press, Princeton, 1999.

[10] S. N. Neftci, Naive trading rules in financial markets and Wiener-Kolmogorov prediction: a study of technical analysis, J. Business, **64** (1991), 549–571.

[11] D. Williams, Probability with Martingales, Cambridge Univ. Press, Cambridge, 1991.

John A. D. Appleby
School of Mathematical Sciences
Dublin City University
Dublin 9
Ireland

Catherine Swords
School of Mathematical Sciences
Dublin City University
Dublin 9
Ireland

E-mail address: `john.appleby@dcu.ie`
`catherine.swords2@mail.dcu.ie`

Advanced Studies in Pure Mathematics 53, 2009
Advances in Discrete Dynamical Systems
pp. 23–32

Asymptotic properties of solutions of the discrete analogue of the Emden-Fowler equation

Josef Diblík and Irena Hlavičková

Abstract.

This contribution investigates the discrete analogue of the Emden-Fowler differential equation. The main attention is paid to the asymptotic behavior of solution of this equation. This work takes up again the research started in paper [4], where the authors have already presented some asymptotic estimates of the solution of the considered equation. The aim of this paper is to improve these estimates.

§1. Introduction

The well-known Emden-Fowler differential equation has been investigated from many points of view, e.g. in the book [1, Chapter VII]. The special case of it is the equation

$$u''(t) - t^{-2}u^n(t) = 0$$

with $n > 1$.

Substitution

$$t = \exp(s), \quad u(s) = Cz^{\alpha}(s),$$

where

$$\alpha = -1/(n-1) \quad \text{and} \quad C^{n-1} = -\alpha,$$

leads (under the supposition $z \neq 0$) to the equation

$$z'' + \frac{(\alpha - 1)(z')^2}{z} - z' + 1 = 0. \tag{1}$$

We will investigate the discrete analogue of equation (1), i.e. the second-order difference equation

$$\Delta^2 v(k) + \frac{(\alpha - 1)(\Delta v(k))^2}{v(k)} - \Delta v(k) + 1 = 0, \tag{2}$$

Received October 12, 2007.

where $\alpha \in \mathbb{R}$, $\alpha < 0$, $k \in N(a) := \{a, a+1, \dots\}$, $a \in \mathbb{N}$, and $\Delta v(k) = v(k+1) - v(k)$.
Equation (2) can be rewritten as a system of two first-order difference equations (for the details, see [4])

$$\begin{aligned} \Delta u_1(k) &= u_1(k) - \frac{\alpha - 1}{k(1+u_2(k))} \cdot (1+u_1(k))^2, \\ \Delta u_2(k) &= \frac{1}{k+1}(-u_2(k) + u_1(k)), \end{aligned} \tag{3}$$

where u_1, u_2, v and Δv satisfy

$$v(k) = k(1+u_2(k)), \tag{4}$$

$$\Delta v(k) = 1 + u_1(k). \tag{5}$$

§2. Summary of the previous results

System (3) can be seen as a special case of the general system of two difference equations

$$\begin{aligned} \Delta u_1(k) &= f_1(k, u_1(k), u_2(k)), \\ \Delta u_2(k) &= f_2(k, u_1(k), u_2(k)) \end{aligned} \tag{6}$$

with $f_1, f_2 : N(a) \times \mathbb{R} \times \mathbb{R} \to \mathbb{R}$.
In paper [4] one can find sufficient conditions guaranteeing the existence of at least one solution $u(k) = (u_1^*(k), u_2^*(k))$, $k \in N(a)$, of system (6) satisfying

$$\begin{aligned} b_1(k) &< u_1^*(k) < c_1(k), \\ b_2(k) &< u_2^*(k) < c_2(k) \end{aligned}$$

where $b_i, c_i : N(a) \to \mathbb{R}$, $i = 1, 2$, are auxiliary functions such that $b_i(k) < c_i(k)$ for every $k \in N(a)$.
The main result of [4] can be summarized in the following theorem.

Theorem 1. *Let $b_i(k), c_i(k), b_i(k) < c_i(k), i = 1, 2$, be real functions defined on $N(a)$ and let $f_i : N(a) \times \mathbb{R} \times \mathbb{R} \to \mathbb{R}, i = 1, 2$ be functions that are continuous with respect to their last two arguments.*
Suppose that for all points (u_1, u_2), such that $b_1(k) \le u_1 \le c_1(k)$ and $b_2(k) \le u_2 \le c_2(k)$ for some $k \in N(a)$, the following four conditions

hold:

(7)
$$u_1 = b_1(k) \quad \Rightarrow \quad f_1(k, u_1, u_2) < b_1(k+1) - b_1(k),$$
(8)
$$u_1 = c_1(k) \quad \Rightarrow \quad f_1(k, u_1, u_2) > c_1(k+1) - c_1(k),$$
(9)
$$u_2 = b_2(k) \quad \Rightarrow \quad b_2(k+1) - b_2(k) < f_2(k, u_1, u_2) < c_2(k+1) - b_2(k),$$
(10)
$$u_2 = c_2(k) \quad \Rightarrow \quad b_2(k+1) - c_2(k) < f_2(k, u_1, u_2) < c_2(k+1) - c_2(k).$$

Let, moreover, function $F(w) = w + f_2(k, u_1, w)$ *be monotone for every fixed* $(k, u_1) \in \{(k, u_1) : k \in N(a), b_1(k) \le u_1 \le c_1(k)\}$ *on the interval* $b_2(k) \le w \le c_2(k)$.

Then there exists a solution $u = (u_1^*(k), u_2^*(k))$ *of system (6) satisfying the inequalities*

$$b_1(k) < u_1^*(k) < c_1(k),$$
$$b_2(k) < u_2^*(k) < c_2(k)$$

for every $k \in N(a)$.

Applying this general result to equation (3), in the same paper it was shown that there exists a solution of system (3) satisfying for k sufficiently large the conditions

$$-\left(\frac{1}{k}\right)^{\nu_i} < u_i(k) < \left(\frac{1}{k}\right)^{\nu_i}$$

for $i = 1, 2$, where $0 < \nu_2 < \nu_1 < 1$. Rewritten in the terms of the second order equation (2) , it gives

(11)
$$|v(k) - k| < k \cdot \left(\frac{1}{k}\right)^{\nu_2}$$

and

(12)
$$|\Delta v(k) - 1| < \left(\frac{1}{k}\right)^{\nu_1} .$$

When this result was presented, the question came from the audience, whether this estimate of the solution could not be improved. Our present contribution gives a partial answer to this question (the investigation of this problem continues).

§3. New asymptotic estimate of the solution

We show that inequalities (11), (12) can be changed with the couple of inequalities (22), (23) below, where (23) substantially improves (12).

Theorem 2. *Let numbers ν_1, ν_2, $1 < \nu_1 < 2$, $0 < \nu_2 < 1$, $1 + \nu_2 > \nu_1$, be given. Then the system of difference equations (3) has for sufficiently large $a \in \mathbb{N}$ a solution $u(k) = (u_1(k), u_2(k))$ such that*

$$\frac{\alpha - 1}{k} - \left(\frac{1}{k}\right)^{\nu_1} < u_1(k) < \frac{\alpha - 1}{k} + \left(\frac{1}{k}\right)^{\nu_1}, \tag{13}$$

$$-\left(\frac{1}{k}\right)^{\nu_2} < u_2(k) < \left(\frac{1}{k}\right)^{\nu_2} \tag{14}$$

for $k \in N(a)$.

Proof. Verify the conditions of Theorem 1 with

$$\begin{aligned} f_1(k, u_1, u_2) &= u_1 - \frac{\alpha - 1}{k(1 + u_2)}(1 + u_1)^2, \\ f_2(k, u_1, u_2) &= \frac{1}{k + 1}(-u_2 + u_1), \end{aligned}$$

$$b_1(k) = \frac{\alpha - 1}{k} - \left(\frac{1}{k}\right)^{\nu_1}, \qquad c_1(k) = \frac{\alpha - 1}{k} + \left(\frac{1}{k}\right)^{\nu_1}$$

and

$$b_2(k) = -\left(\frac{1}{k}\right)^{\nu_2}, \quad c_2(k) = \left(\frac{1}{k}\right)^{\nu_2}.$$

With respect to condition (7) we have to verify that

$$f_1(k, u_1, u_2) < b_1(k + 1) - b_1(k)$$

if

$$u_1 = \frac{\alpha - 1}{k} - \left(\frac{1}{k}\right)^{\nu_1} \text{ and } -\left(\frac{1}{k}\right)^{\nu_2} \le u_2 \le \left(\frac{1}{k}\right)^{\nu_2},$$

which gives

$$\begin{aligned} &\frac{\alpha - 1}{k} - \left(\frac{1}{k}\right)^{\nu_1} - \frac{\alpha - 1}{k(1 + u_2)}\left(1 + \frac{\alpha - 1}{k} - \left(\frac{1}{k}\right)^{\nu_1}\right)^2 \\ &\qquad < \frac{\alpha - 1}{k + 1} - \left(\frac{1}{k + 1}\right)^{\nu_1} - \frac{\alpha - 1}{k} + \left(\frac{1}{k}\right)^{\nu_1}. \end{aligned} \tag{15}$$

Denote $L_{(15)}$ and $R_{(15)}$ the left-hand and the right-hand side of inequality (15), respectively.

In the following estimate of $L_{(15)}$ we will use the fact that $\alpha < 0$, i.e. also $\alpha - 1 < 0$, and the fact that $1/(1-x) < 1 + 2x$ for $x > 0$ sufficiently close to zero.

$$
\begin{aligned}
L_{(15)} &\le \frac{\alpha-1}{k} - \left(\frac{1}{k}\right)^{\nu_1} - \frac{\alpha-1}{k(1+u_2)}\left(1+\frac{\alpha-1}{k}\right)^2 \\
&\le \frac{\alpha-1}{k} - \left(\frac{1}{k}\right)^{\nu_1} - \frac{\alpha-1}{k(1-(\frac{1}{k})^{\nu_2})}\left(1+\frac{\alpha-1}{k}\right)^2 \\
&\le \frac{\alpha-1}{k} - \left(\frac{1}{k}\right)^{\nu_1} \\
&\quad -\frac{\alpha-1}{k}\left(1+2\left(\frac{1}{k}\right)^{\nu_2}\right)\left(1+2\,\frac{\alpha-1}{k}+\frac{(\alpha-1)^2}{k^2}\right) \\
&= \frac{\alpha-1}{k} - \frac{1}{k^{\nu_1}} \\
&\quad -\frac{\alpha-1}{k}\left(1+2\,\frac{\alpha-1}{k}+\frac{(\alpha-1)^2}{k^2}+2\,\frac{1}{k^{\nu_2}}\right. \\
&\qquad\qquad \left. +4\,\frac{\alpha-1}{k^{1+\nu_2}}+2\,\frac{(\alpha-1)^2}{k^{2+\nu_2}}\right) \\
&= -\frac{1}{k^{\nu_1}} - 2\,\frac{(\alpha-1)^2}{k^2} - \frac{(\alpha-1)^3}{k^3} \\
&\quad -2\,\frac{\alpha-1}{k^{1+\nu_2}} - 4\,\frac{(\alpha-1)^2}{k^{2+\nu_2}} - 2\,\frac{(\alpha-1)^3}{k^{3+\nu_2}}\,.
\end{aligned}
$$

As, due to the assumptions of the Theorem, $1 < \nu_1 < 2$, $0 < \nu_2 < 1$ and $1 + \nu_2 > \nu_1$, one can state that $L_{(15)}$ is negative for k sufficiently large.

As for $R_{(15)}$, it can be simplified to

$$R_{(15)} = -\frac{\alpha-1}{k(k+1)} + \left(\frac{1}{k}\right)^{\nu_1} - \left(\frac{1}{k+1}\right)^{\nu_1} > 0.$$

Thus, inequality (15) holds.

Now let us prove inequality (9), i.e.

$$f_1(k, u_1, u_2) > c_1(k+1) - c_1(k)$$

if

$$u_1 = \frac{\alpha-1}{k} + \left(\frac{1}{k}\right)^{\nu_1} \text{ and } -\left(\frac{1}{k}\right)^{\nu_2} \le u_2 \le \left(\frac{1}{k}\right)^{\nu_2},$$

which gives

$$(16)\quad \frac{\alpha-1}{k}+\left(\frac{1}{k}\right)^{\nu_1}-\frac{\alpha-1}{k(1+u_2)}\left(1+\frac{\alpha-1}{k}+\left(\frac{1}{k}\right)^{\nu_1}\right)^2 > \frac{\alpha-1}{k+1}+\left(\frac{1}{k+1}\right)^{\nu_1}-\frac{\alpha-1}{k}-\left(\frac{1}{k}\right)^{\nu_1}.$$

Again, denote $L_{(16)}$ and $R_{(16)}$ the left-hand and the right-hand side of (16), respectively.

This time we will use the fact that $1/(1+x)>1-2x$ for $x>0$.

$$\begin{aligned}
L_{(16)} &\geq \frac{\alpha-1}{k}+\left(\frac{1}{k}\right)^{\nu_1}-\frac{\alpha-1}{k(1+u_2)}\left(1+\frac{\alpha-1}{k}\right)^2\\
&\geq \frac{\alpha-1}{k}+\left(\frac{1}{k}\right)^{\nu_1}-\frac{\alpha-1}{k(1+(\frac{1}{k})^{\nu_2})}\left(1+\frac{\alpha-1}{k}\right)^2\\
&\geq \frac{\alpha-1}{k}+\left(\frac{1}{k}\right)^{\nu_1}\\
&\quad -\frac{\alpha-1}{k}\left(1-2\left(\frac{1}{k}\right)^{\nu_2}\right)\left(1+2\,\frac{\alpha-1}{k}+\frac{(\alpha-1)^2}{k^2}\right)\\
&= \frac{\alpha-1}{k}+\frac{1}{k^{\nu_1}}\\
&\quad -\frac{\alpha-1}{k}\left(1+2\,\frac{\alpha-1}{k}+\frac{(\alpha-1)^2}{k^2}-2\,\frac{1}{k^{\nu_2}}\right.\\
&\qquad\qquad \left.-4\,\frac{\alpha-1}{k^{1+\nu_2}}-2\,\frac{(\alpha-1)^2}{k^{2+\nu_2}}\right)\\
&= \frac{1}{k^{\nu_1}}-2\,\frac{(\alpha-1)^2}{k^2}-\frac{(\alpha-1)^3}{k^3}\\
&\quad +2\,\frac{\alpha-1}{k^{1+\nu_2}}+4\,\frac{(\alpha-1)^2}{k^{2+\nu_2}}+2\,\frac{(\alpha-1)^3}{k^{3+\nu_2}}\\
&= O\left(\frac{1}{k^{\nu_1}}\right).
\end{aligned}$$

For the estimate of $R_{(16)}$, let us use the fact (gained with the help of the Mean Value Theorem) that

$$(17)\quad \left(\frac{1}{k+1}\right)^{\nu_1}-\left(\frac{1}{k}\right)^{\nu_1}=-\nu_1\left(\frac{1}{\xi}\right)^{\nu_1+1},$$

where $k \le \xi \le k+1$. That gives

$$R_{(16)} \le -\frac{\alpha-1}{k(k+1)} - \frac{\nu_1}{(k+1)^{1+\nu_1}} = O\left(\frac{1}{k^2}\right)$$

and inequality (16) is verified.

The proof of inequalities (10) and (11) is easier.

First let us take the left part of inequality (10):

$$b_2(k+1) - b_2(k) < f_2(k, u_1, u_2)$$

if

$$u_2 = -\left(\frac{1}{k}\right)^{\nu_2} \quad \text{and} \quad \frac{\alpha-1}{k} - \left(\frac{1}{k}\right)^{\nu_1} \le u_1 \le \frac{\alpha-1}{k} + \left(\frac{1}{k}\right)^{\nu_1},$$

which gives

$$-\left(\frac{1}{k+1}\right)^{\nu_2} + \left(\frac{1}{k}\right)^{\nu_2} < \frac{1}{k+1}\left(\left(\frac{1}{k}\right)^{\nu_2} + u_1\right). \tag{18}$$

Using a relation similar to (17), the left-hand side of (18) can be estimated as follows

$$-\left(\frac{1}{k+1}\right)^{\nu_2} + \left(\frac{1}{k}\right)^{\nu_2} \le \frac{\nu_2}{k^{1+\nu_2}}.$$

The right-hand side of (18) can be estimated as

$$\begin{aligned}\frac{1}{k+1}\left(\left(\frac{1}{k}\right)^{\nu_2} + u_1\right) \quad &\ge \quad \frac{1}{k+1}\left(\left(\frac{1}{k}\right)^{\nu_2} + \frac{\alpha-1}{k} - \left(\frac{1}{k}\right)^{\nu_1}\right) \\ &= \quad \frac{1}{(k+1)k^{\nu_2}} + O\left(\frac{1}{k^2}\right).\end{aligned}$$

That proves inequality (18) because $\nu_2 < 1$ and thus

$$\frac{\nu_2}{k^{1+\nu_2}} < \frac{1}{(k+1)k^{\nu_2}} + O\left(\frac{1}{k^2}\right)$$

for k sufficiently large.

The right part of inequality (10), i.e.

$$f_2(k, u_1, u_2) < c_2(k+1) - b_2(k)$$

if

$$u_2 = -\left(\frac{1}{k}\right)^{\nu_2} \quad \text{and} \quad \frac{\alpha-1}{k} - \left(\frac{1}{k}\right)^{\nu_1} \le u_1 \le \frac{\alpha-1}{k} + \left(\frac{1}{k}\right)^{\nu_1},$$

in our case gives

$$\frac{1}{k+1}\left(\left(\frac{1}{k}\right)^{\nu_2}+u_1\right)<\left(\frac{1}{k+1}\right)^{\nu_2}+\left(\frac{1}{k}\right)^{\nu_2}. \tag{19}$$

The maximum possible value of the left-hand side of (19) is

$$\frac{1}{k+1}\left(\left(\frac{1}{k}\right)^{\nu_2}+\frac{\alpha-1}{k}+\left(\frac{1}{k}\right)^{\nu_1}\right)=O\left(\frac{1}{k^{1+\nu_2}}\right),$$

meanwhile the right-hand side of (19) is $O\left(k^{-\nu_2}\right)$ and hence inequality (19) holds.

Condition (11) can be proved in a very similar way. First its left part:

$$b_2(k+1)-c_2(k)<f_2(k,u_1,u_2)$$

if

$$u_2=\left(\frac{1}{k}\right)^{\nu_2} \quad \text{and} \quad \frac{\alpha-1}{k}-\left(\frac{1}{k}\right)^{\nu_1}\le u_1\le\frac{\alpha-1}{k}+\left(\frac{1}{k}\right)^{\nu_1}.$$

That gives

$$-\left(\frac{1}{k+1}\right)^{\nu_2}-\left(\frac{1}{k}\right)^{\nu_2}<\frac{1}{k+1}\left(-\left(\frac{1}{k}\right)^{\nu_2}+u_1\right).$$

Multiplying it by -1, we get

$$\left(\frac{1}{k+1}\right)^{\nu_2}+\left(\frac{1}{k}\right)^{\nu_2}>\frac{1}{k+1}\left(\left(\frac{1}{k}\right)^{\nu_2}-u_1\right). \tag{20}$$

As the left-hand side of (20) is $O\left(k^{-\nu_2}\right)$ and the right-hand side is $O\left(k^{-(1+\nu_2)}\right)$, inequality (20) has be fulfilled.

Now prove the right part of condition (11):

$$f_2(k,u_1,u_2)<c_2(k+1)-c_2(k)$$

if

$$u_2=\left(\frac{1}{k}\right)^{\nu_2} \quad \text{and} \quad \frac{\alpha-1}{k}-\left(\frac{1}{k}\right)^{\nu_1}\le u_1\le\frac{\alpha-1}{k}+\left(\frac{1}{k}\right)^{\nu_1},$$

i.e.

$$\frac{1}{k+1}\left(-\left(\frac{1}{k}\right)^{\nu_2}+u_1\right)<\left(\frac{1}{k+1}\right)^{\nu_2}-\left(\frac{1}{k}\right)^{\nu_2}.$$

Again we will change the signes:

$$\frac{1}{k+1}\left(\left(\frac{1}{k}\right)^{\nu_2}-u_1\right)>\left(\frac{1}{k}\right)^{\nu_2}-\left(\frac{1}{k+1}\right)^{\nu_2}. \tag{21}$$

The left-hand side of (21) can be expressed as

$$\frac{1}{(k+1)k^{\nu_2}} + O\left(\frac{1}{k^2}\right),$$

the maximum possible value of the right hand side is $\nu_2/k^{1+\nu_2}$ and thus inequality (21) holds.

Finally, the function

$$F(w) = w + f_2(k, u_1, w) = w + \frac{1}{k+1}(-w + u_1)$$

is monotone for every fixed (k, u_1) such that $k \in N(a), b_1(k) \leq u_1 \leq c_1(k)$ on the interval $b_2(k) \leq w \leq c_2(k)$ since its derivative

$$F'(w) = 1 - \frac{1}{k+1} = \frac{k}{k+1}$$

is positive for $k \in N(a)$. Then, in accordance with the statement of Theorem 1, there exists a solution of system (3) satisfying inequalities (13) and (14). Q.E.D.

Now we are able to formulate the final result.

Theorem 3. *Let numbers ν_1, ν_2, $1 < \nu_1 < 2$, $0 < \nu_2 < 1$, $1 + \nu_2 > \nu_1$, be given. Then there exists a solution $v(k)$ of equation (2), such that*

$$|v(k) - k| < k \cdot \left(\frac{1}{k}\right)^{\nu_2} \tag{22}$$

and

$$\left|\Delta v(k) - 1 - \frac{\alpha - 1}{k}\right| < \left(\frac{1}{k}\right)^{\nu_1} \tag{23}$$

for k sufficiently large.

Proof. The statement is a simple consequence of Theorem 3 since u_1, u_2, v and Δv are connected by formulae (4), (5). Q.E.D.

Remark 1. The proof of Theorem 1 (see [4]) connects the so called retract technique with Lyapunov approach. For further results we refer e.g. to [2, 3, 5, 6].

Acknowledgment. The first author was supported by the Grant 201/07/0145 of the Czech Grant Agency (Prague), by the Councils of Czech Government MSM 0021630503, MSM 0021630519 and MSM 0021630529. The second author was supported by the grant No 1/3238/06 of the Grant Agency of Slovak Republic (VEGA).

References

[1] R. Bellman, Stability theory of differential equations, Internat. Ser. in Pure and Appl. Math., New York-London, McGraw-Hill Book Company, XIII, 1953.

[2] J. Diblík, Anti-Lyapunov method for systems of discrete equations, Nonlinear Anal., **57** (2004), 1043–1057.

[3] J. Diblík, Asymptotic behaviour of solutions of systems of discrete equations via Liapunov type technique, Comput. Math. Appl., **45** (2003), 1041–1057.

[4] J. Diblík and I. Růžičková, Compulsory asymptotic behavior of solutions of two-dimensional systems of difference equations, In: Difference Equations and Discrete Dynamical Systems, Proceedings of the 9th International Conference, University of Southern California, Los Angeles, California, USA, 2–7 August 2004, World Scientific Publ. Co. Pte. Ltd, Singapore, 2005, pp. 35–49.

[5] J. Diblík, I. Růžičková and M. Růžičková, A general version of the retract method for discrete equations, Acta Math. Sin., 2006, DOI 10.1007/s10114-005-0729-8 (online).

[6] J. Diblík, M. Migda and E. Schmeidel, Bounded solutions of nonlinear discrete equations, Nonlinear Anal., **65** (2006), 845–853.

Josef Diblík
Brno University of Technology
Faculty of Civil Engineering
Department of Mathematics and Descriptive Geometry, Brno
Czech Republic

Irena Hlavičková
Brno University of Technology
Faculty of Electrical Engineering and Communication
Department of Mathematics, Brno
Czech Republic

E-mail address: `diblik.j@fce.vutbr.cz`
`hlavicka@feec.vutbr.cz`

Advanced Studies in Pure Mathematics 53, 2009
Advances in Discrete Dynamical Systems
pp. 33–40

Intermediate solutions for nonlinear difference equations with p-Laplacian

Mariella Cecchi, Zuzana Došlá and Mauro Marini

Abstract.

The paper deals with the existence of the so-called intermediate solutions for the nonlinear difference equation with deviating argument. The roles of the nonlinearity and deviating argument are discussed and illustrated by examples.

§1. Introduction

Consider the difference equation

$$\Delta(a_n|\Delta x_n|^\alpha \text{sgn } \Delta x_n) + b_n F(x_{n+p}) = 0 \tag{1}$$

where Δ is the forward difference operator $\Delta x_n = x_{n+1} - x_n$, $\alpha > 0$ is a real number, $a = \{a_n\}$, $b = \{b_n\}$ are positive real sequences, $p \geq 0$ is a fixed integer number and F is a real positive continuous function on $(0, \infty)$.

Equation (1) is the discrete analogue of a nonlinear differential equation with p-Laplacian operator, that appears in studying spherically symmetric solutions for certain nonlinear elliptic systems.

An important special case of (1) is the discrete Emden-Fowler equation

$$\Delta(a_n|\Delta x_n|^\alpha \text{sgn } \Delta x_n) + b_n |x_{n+p}|^\beta \text{ sgn } x_{n+p} = 0, \tag{2}$$

where $\beta > 0$ is a real number. Both equations (1), (2) are widely considered in the literature, see, e.g., [2, 3, 6, 8, 9], the monograph [1] and references therein. In particular, in [2, 3] equation (1) has been

---

Received October 12, 2007.

Supported by the Research Project MŠMT 0021622409 of the Ministry of Education of the Czech Republic and by the Grant Agency of the Czech Republic, grant 201/04/0580

investigated when $b_n \le 0$ for large n. In this paper we continue such a study, by assuming the positiveness of b.

Throughout the paper, for brevity, by solution of (1) we mean a nontrivial sequence satisfying (1) for $n \ge p$. As usual, a solution $x = \{x_n\}$ of (1) is said to be *nonoscillatory* if there exists $n_x \ge 1$ such that $x_n x_{n+1} > 0$ for $n \ge n_x$.

For the sake of simplicity, we restrict our study to nonoscillatory solutions x for which $x_n > 0$ for large n and we denote by $x^{[1]} = \{x_n^{[1]}\}$ its quasidifference, where

$$x_n^{[1]} = a_n |\Delta x_n|^\alpha \operatorname{sgn} \Delta x_n. \tag{3}$$

Clearly, $x^{[1]}$ is decreasing for large n. Then for any eventually positive solution x of (1) we say $x \in \mathbb{M}^+$ or $x \in \mathbb{M}^-$, according to $x_n > 0, \Delta x_n > 0$ for $n \ge n_x \ge 1$ or $x_n > 0, \Delta x_n < 0$ for $n \ge n_x \ge 1$.

We deal with the existence of a particular type of nonoscillatory solutions, namely the so called *intermediate solutions.* Solution x of (1) is said to be *intermediate* if either

$$x \in \mathbb{M}^-, \quad \lim_n x_n = 0, \quad \lim_n x_n^{[1]} = -\infty, \tag{4}$$

or

$$x \in \mathbb{M}^+, \quad \lim_n x_n = \infty, \quad \lim_n x_n^{[1]} = 0. \tag{5}$$

This terminology originates from the corresponding continuous case and very few is known in the literature (see, e.g., [7, Remark 1.1.]). A comparison with some partial results in [8, 9] is also given. A complete study concerning the nonoscillation of (1) will be given in a forthcoming paper [5].

§2. Main results

Put

$$S_a := \sum_{k=1}^{\infty} \frac{1}{(a_k)^{1/\alpha}}, \quad S_b := \sum_{k=1}^{\infty} b_k.$$

Lemma 1. *(i_1) If $S_a < \infty$, then any solution $x \in \mathbb{M}^+$ is bounded.*

(i_2) Assume that F is nondecreasing. If $S_b < \infty$, then for any solution $x \in \mathbb{M}^-$ the quasidifference $x^{[1]}$ is bounded.

Proof. Claim (i_1). Since $x^{[1]}$ is positive decreasing for large n, say $n \ge n_0$, from (3) we have

$$\Delta x_n \le \frac{1}{(a_n)^{1/\alpha}} (\Delta x_{n_0}^{[1]})^{1/\alpha}.$$

The assertion follows by summing this inequality.

Claim (i_2). Since x is positive decreasing for large n, say $n \geq n_0$, from (1) we have

$$\Delta x_n^{[1]} \geq -b_n F(x_{n_0+p})$$

and again the assertion follows by summing this inequality. Q.E.D.

As follows from Lemma 1, if F is nondecreasing, equation (1) does not admit intermediate solutions when $S_a + S_b < \infty$. Thus, two cases (i) $S_a < \infty, S_b = \infty$, (ii) $S_a = \infty, S_b < \infty$ are considered here.

Theorem 1. *Assume that F is nondecreasing on $(0,1]$ and*

$$J_p := \sum_{j=1}^{\infty} b_j F\left(\sum_{i=j+p}^{\infty} \frac{1}{(a_i)^{1/\alpha}}\right) = \infty, \tag{6}$$

$$S_1 := \sum_{j=2}^{\infty} \left(\frac{1}{a_j}\sum_{i=1}^{j-1} b_i\right)^{1/\alpha} < \infty. \tag{7}$$

Then (1) has solutions satisfying (4).

Proof. Clearly, (6) and (7) yield $S_a < \infty, S_b = \infty$. Let n_0 large so that $n_0 \geq 2$, and

$$\sum_{j=n_0}^{\infty} \frac{1}{(a_j)^{1/\alpha}} < 1, \quad F(1)\sum_{i=1}^{n_0-1} b_i \geq 1, \tag{8}$$

$$\sum_{j=n_0+1}^{\infty} \left(\frac{1}{a_j}\sum_{i=1}^{j-1} b_i\right)^{1/\alpha} \leq \frac{1}{(F(1))^{1/\alpha}}. \tag{9}$$

Let $\mathbb{N}_{n_0} = \{n \in \mathbb{N}, n \geq n_0\}$ and denote by $\mathbb{X}$ the Fréchet space of the real sequences defined for $n \in \mathbb{N}_{n_0}$, endowed with the topology of convergence on finite subsets of $\mathbb{N}_{n_0}$. Consider the set $\Omega \subset \mathbb{X}$ defined by

$$\Omega = \left\{v = \{v_n\} \in \mathbb{X} : \sum_{j=n}^{\infty} \frac{1}{(a_j)^{1/\alpha}} \leq v_n \leq 1\right\}.$$

Let $\mathcal{T} : \Omega \to \mathbb{X}$ be the map given by $\mathcal{T}(v) = z = \{z_n\}$, where $z_{n_0} = 1$ and

$$z_n = \sum_{j=n}^{\infty} \frac{1}{(a_j)^{1/\alpha}} \left(1 + \sum_{i=n_0}^{j-1} b_i F\left(v_{i+p}\right)\right)^{1/\alpha} \quad \text{for } n > n_0.$$

Clearly $z_n \geq \sum_{j=n}^{\infty} \frac{1}{(a_j)^{1/\alpha}}$. In view of (8) and (9), we have for $n > n_0$

$$z_n \leq \sum_{j=n}^{\infty} \frac{1}{(a_j)^{1/\alpha}} \left(1 + F(1) \sum_{i=n_0}^{j-1} b_i\right)^{1/\alpha} \leq$$

$$\leq (F(1))^{1/\alpha} \sum_{j=n_0+1}^{\infty} \frac{1}{(a_j)^{1/\alpha}} \left(\sum_{i=1}^{j-1} b_j\right)^{1/\alpha} \leq 1,$$

therefore $\mathcal{T}(\Omega) \subset \Omega$. Let us show that $\mathcal{T}(\Omega)$ is relatively compact and $\mathcal{T}$ is continuous on Ω. In virtue of the Ascoli theorem, any bounded set in $\mathbb{X}$ is relatively compact (see, e.g., [1, Theorem 5.6.1]) and so, because $\mathcal{T}(\Omega)$ is bounded on $\mathbb{X}$, the compactness follows.

Let us prove the continuity of $\mathcal{T}$ on Ω. Let $v^{(k)} = \{v_j^{(k)}\}$ be a sequence in Ω, converging on finite subsets of $\mathbb{N}_{n_0}$ to $v^{(\infty)} = \{v_j^{(\infty)}\} \in \Omega$. From (8) we have

$$A_j^{(k)} := \frac{1}{(a_j)^{1/\alpha}} \left(1 + \sum_{i=n_0}^{j-1} b_i F\left(v_{i+p}^{(k)}\right)\right)^{1/\alpha} \leq F^{1/\alpha}(1) \left(\frac{1}{a_j} \sum_{i=1}^{j-1} b_i\right)^{1/\alpha}.$$

Since F is continuous, the sequence $A^{(k)} = \{A_j^{(k)}\}$ converges, on finite subsets of $\mathbb{N}_{n_0}$, to $A^{(\infty)}$ (with clear meaning of $A^{(\infty)}$) and so $\lim_k A_j^{(k)} = A_j^{(\infty)}$ for any fixed $j \in \mathbb{N}_{n_0}$. Hence, in view of (7), the series $\sum_j A_j^{(k)}$ totally converges. Using the discrete analogue of the Lebesgue dominated convergence theorem, the sequence $\mathcal{T}(v^{(k)})$ converges on finite subsets of $\mathbb{N}_{n_0}$ to $\mathcal{T}(v^{(\infty)})$ and so the continuity of $\mathcal{T}$ is proved.

Applying the Tychonov fixed point theorem, there exists a sequence x such that for $n > n_0$

$$x_n = \sum_{j=n}^{\infty} \frac{1}{(a_j)^{1/\alpha}} \left(1 + \sum_{i=n_0}^{j-1} b_i F(x_{i+p})\right)^{1/\alpha}. \tag{10}$$

Clearly, x is solution of (1) and $x \in \Omega$. From (8) we have

$$x_n \leq (F(1))^{1/\alpha} \sum_{j=n}^{\infty} \left(\frac{1}{a_j} \sum_{i=1}^{j-1} b_j\right)^{1/\alpha}$$

and so, in view of (7), x converges to zero. Since

$$F\left(x_{i+p}\right) \geq F\left(\sum_{m=i+p}^{\infty} \frac{1}{(a_m)^{1/\alpha}}\right),$$

from (10) we obtain

$$x_n^{[1]} \leq -1 - \sum_{i=n_0}^{n-1} b_i F\left(\sum_{m=i+p}^{\infty} \frac{1}{(a_m)^{1/\alpha}}\right)$$

and so, by (6), $\lim_n x_n^{[1]} = -\infty$. Q.E.D.

Theorem 2. *Assume that F is nondecreasing on $[1,\infty)$, and*

$$I_p := \sum_{i=3}^{\infty} b_i F\left(\sum_{j=2}^{i+p-1} \frac{1}{(a_j)^{1/\alpha}}\right) < \infty, \tag{11}$$

$$\sum_{i=1}^{\infty}\left(\frac{1}{a_i}\sum_{j=i}^{\infty} b_j\right)^{1/\alpha} = \infty. \tag{12}$$

Then equation (1) has solutions satisfying (5).

Proof (outline). Clearly, $S_a = \infty$ and $S_b < \infty$. Let n_0 large so that $n_0 \geq 3$ and

$$\sum_{i=n_0}^{\infty} b_i F\left(\sum_{m=2}^{i+p-1} \frac{1}{(a_m)^{1/\alpha}}\right) < 1, \qquad \sum_{m=2}^{n_0-1} \frac{1}{(a_m)^{1/\alpha}} \geq 1. \tag{13}$$

Let $\mathbb{X}$ be defined as in Theorem 1. The assertion follows by applying the fixed point Tychonov fixed point theorem in the set

$$\Omega = \left\{u = \{u_n\} \in \mathbb{X} : 1 \leq u_n \leq \sum_{j=2}^{n-1} \frac{1}{(a_j)^{1/\alpha}}\right\}$$

to the map $\mathcal{T} : \Omega \to \mathbb{X}$ given by $\mathcal{T}(u) = y = \{y_n\}$, where $y_{n_0} = 1$,

$$y_n = 1 + \sum_{j=n_0}^{n-1} \frac{1}{(a_j)^{1/\alpha}}\left(\sum_{i=j}^{\infty} b_i F\left(u_{i+p}\right)\right)^{1/\alpha} \quad \text{for } n > n_0. \tag{14}$$

Finally, the fixed point of $\mathcal{T}$, say x, satisfies $\lim_n x_n^{[1]} = 0$ and it is unbounded in virtue of (12). Q.E.D.

If F is bounded or is bounded away zero near zero, the assumption on monotonicity of F in Theorems 1, 2 can be relaxed, as the following result shows.

Theorem 3. *(i_1) Assume* $0 < \inf_{0<u\leq 1} F(u) \leq \sup_{0<u\leq 1} F(u) < \infty$. *If* $S_a < \infty$ *and (7) holds, then (1) has solutions satisfying (4).*

(i_2) Assume $0 < \inf_{u\geq 1} F(u) \leq \sup_{u\geq 1} F(u) < \infty$. *If* $S_b < \infty$ *and (12) holds, then (1) has solutions satisfying (5).*

The proof is similar to the ones of the above theorems with minor changes.

Remark 1. In [9], a more general equation is considered, but the existence results concern with unbounded and zero-convergent solutions which are not intermediate ones; moreover these results require strong assumptions on nonlinearity which are not satisfied for any $\beta > 0$.

In [8], a second order difference system, including (1) with $p = 0$, is considered. Comparing [8, Th.4] with our Theorem 2, both summation conditions are equivalent, but [8, Th.4] is not applicable to (1), due to different assumptions on nonlinearities. In addition, the proof of [8, Th.4] seems not to be correct because a previous result [8, Th.2], with a different assumption, is applied. Comparing [8, Th.11] with our Theorem 1, one can check that assumptions of [8, Th.11] can hold for (1) only if $\limsup_{u\to\infty} F(u) < \infty$, so this result is not applicable to (2).

§3. The role of F and p

When assumptions of Theorem 3 are satisfied, the existence of intermediate solutions does not depend on p. When F is unbounded, the situation can be different, as the following two examples show.

Example 1. Consider equation (2) with $b_n = 1$, $a_n = n(n+1)$, $\alpha = 2/3$, $\beta = 1/2$. It is easy to verify that $S_1 < \infty$, $J_p = \infty$ for any $p \geq 0$ and so, from Theorem 1, equation (2) has intermediate solutions in the class $\mathbb{M}^-$.

Example 2. Consider the equations

$$\Delta(a_n|\Delta x_n|^4 \mathrm{sgn}\, \Delta x_n) + b_n|x_n|^2 \,\mathrm{sgn}\, x_n = 0, \tag{15}$$

$$\Delta(a_n|\Delta x_n|^4 \mathrm{sgn}\, \Delta x_n) + b_n|x_{n+1}|^2 \,\mathrm{sgn}\, x_{n+1} = 0, \tag{16}$$

where $b_n = e^{n^2}$, $a_n = \left(e^{-n^2/2} - e^{-(n+1)^2/2}\right)^{-4}$. We will show that Theorem 1 is applicable if $p = 0$ and not for $p = 1$. This means that equation (15) has solutions satisfying (5), while the existence of such solutions for (16) is an open problem. We have $\sum_{k=n}^{\infty}(a_k)^{-1/4} = e^{-n^2/2}$ and so $S_a < \infty$. Furthermore,

$$S_1 \le \sum_{i=2}^{\infty} \left(e^{-i^2/2} - e^{-(i+1)^2/2}\right) e^{i^2/4} i^{1/4} \le \sum_{i=2}^{\infty} e^{-i^2/4} i^{1/4} < \infty.$$

Concerning (6), if $p = 0$, we have

$$J_0 = \sum_{i=1}^{\infty} e^{i^2} \left(\sum_{j=i}^{\infty} \frac{1}{(a_k)^{1/4}}\right)^2 = \sum_{i=1}^{\infty} e^{i^2} e^{-i^2} = \infty,$$

and if $p = 1$, we have

$$J_1 = \sum_{i=1}^{\infty} e^{i^2} \left(\sum_{j=i+1}^{\infty} \frac{1}{(a_k)^{1/4}}\right)^2 = \sum_{i=1}^{\infty} e^{i^2} e^{-(i+1)^2} = e^{-1} \sum_{i=1}^{\infty} e^{-2i} < \infty.$$

Concerning intermediate solutions in the class $\mathbb{M}^+$, it is easy to produce an example of equation (2) for which Theorem 2 holds for any $p \ge 0$ and, similarly, an example such that $I_0 < \infty$ and $I_1 = \infty$.

Remark 2. It is possible to show, by means of some recent summation inequalities, see [4, Lemma 2], that the conditions (6) and (7) [similarly, (11) and (12)] are not compatible for $F(u) = u^\beta$ if $\beta > \alpha$ and $p \ge 1$. Thus Theorems 1, 2 can be applied to (2) only when $\beta \le \alpha$. A detailed discussion on this problem is given in [5].

References

[1] R. P. Agarwal and D. O'Regan, Infinite Interval Problems for Differential, Difference and Integral Equations, Kluwer Acad. Publ., 2001.

[2] M. Cecchi, Z. Došlá and M. Marini, Unbounded solutions of quasilinear difference equations, Comput. Math. Appl., **45** (2003), 1113–1123.

[3] M. Cecchi, Z. Došlá, M. Marini and I. Vrkoč, Summation inequalities and half-linear difference equations, J. Math. Anal. Appl., **302** (2005), 1–13.

[4] M. Cecchi, Z. Došlá, M. Marini and I. Vrkoč, Asymptotic properties for half-linear difference equations, to appear in Math. Bohemica, 2006.

[5] M. Cecchi, Z. Došlá and M. Marini, On oscillation and nonoscillation properties of Emden-Fowler difference equations, submitted for publication, 2008.

[6] H. F. Huo and W. T. Li, Oscillation of the Emden-Fowler difference systems, J. Math. Anal. Appl., **256** (2001), 478–485.

[7] W. Jingfa, Oscillation and nonoscillation theorems for a class of second order quasilinear functional differential equations, Hiroshima Math. J., **27** (1997), 449–466.

[8] W. T. Li, Classification Schemes for Nonoscillatory Solutions of Two-Dimensional Nonlinear Difference Systems, Comput. Math. Appl., **42** (2001), 341–355.

[9] G. Zhang, S. S. Cheng and Y. Gao, Classification schemes for positive solutions of a second-order nonlinear difference equation, J. Comput. Appl. Math., **101** (1999), 39–51.

Mariella Cecchi
Depart. of Electronics and Telecommunications
University of Florence
Via S. Marta 3,50139 Florence
Italy

Zuzana Došlá
Depart. of Mathematics, Masaryk University
Janáčkovo nám.
2a, 66295 Brno
Czcch Rcpublic

Mauro Marini
Depart. of Electronics and Telecommunication
University of Florence
Via S. Marta 3, 50139 Florence
Italy

E-mail address: mariella.cecchi@unifi.it
dosla@math.muni.cz
mauro.marini@unifi.it

Advanced Studies in Pure Mathematics 53, 2009
Advances in Discrete Dynamical Systems
pp. 41–50

Oscillation theory of symplectic difference systems

Ondřej Došlý

Abstract.

Recent results in the oscillation theory of symplectic difference systems are presented. A particular attention is devoted to Sturmian and oscillation theorems for these systems and for the so-called reversed symplectic systems. Some open problems and conjectures are formulated.

§1. Introduction

We deal in this contribution with oscillatory properties of solutions of symplectic difference systems (further SDS)

$$z_{k+1} = \mathcal{S}_k z_k, \tag{1}$$

where $z_k \in \mathbb{R}^{2n}$ and the $2n \times 2n$-matrices $\mathcal{S}_k$ are *symplectic*, i.e.,

$$\mathcal{S}_k^T \mathcal{J} \mathcal{S}_k = \mathcal{J} \quad \text{with} \quad \mathcal{J} = \begin{pmatrix} 0 & I \\ -I & 0 \end{pmatrix}.$$

If we split a solution $z \in \mathbb{R}^{2n}$ of (1) into two n-dimensional vectors x, u, i.e. $z = \binom{x}{u}$, and $\mathcal{S}$ is considered in the form $\mathcal{S} = \begin{pmatrix} \mathcal{A} & \mathcal{B} \\ \mathcal{C} & \mathcal{D} \end{pmatrix}$ with $n \times n$ matrices $\mathcal{A}, \mathcal{B}, \mathcal{C}, \mathcal{D}$, then system (1) takes the form

$$x_{k+1} = \mathcal{A}_k x_k + \mathcal{B}_k u_k, \quad u_{k+1} = \mathcal{C}_k x_k + \mathcal{D}_k u_k \tag{2}$$

and symplecticity of $\mathcal{S}$ is equivalent to the identities

$$\mathcal{A}^T \mathcal{C} = \mathcal{C}^T \mathcal{A}, \quad \mathcal{B}^T \mathcal{D} = \mathcal{D}^T \mathcal{B}, \quad \mathcal{A}^T \mathcal{D} - \mathcal{C}^T \mathcal{B} = I. \tag{3}$$

Revised October 12, 2007.

Key words and phrases. Oscillation, Symplectic difference system, Discrete eigenvalue problem, principal solution, focal point, Sturmian theorem, Oscillation theorem.

Oscillatory properties of (1) are closely related to positivity/nonnegativity of the associated discrete quadratic functional

$$(4)\quad \mathcal{F}(z)=\mathcal{F}(x,u)=\sum_{k=0}^{N}\{x_k^T\mathcal{A}_k^T\mathcal{C}_kx_k+2x^T\mathcal{C}_k^T\mathcal{B}_ku_k+u_k^T\mathcal{B}_k^T\mathcal{D}_ku_k\}$$

considered over the class of sequences satisfying the first equation in (2) (the so-called *admissible sequences*, the first equation in (1) is usually called the *equation of motion*) and certain boundary conditions at the endpoints $k=0$ and $k=N+1$. We refer to the paper [6] where basic oscillatory properties of (1) are established and to the subsequent papers [8, 7, 11, 13, 14] (and references given therein), where various aspects of oscillation theory of (1) are investigated, including the problem of positivity and nonnegativity of (4).

Symplectic difference systems (1) cover a large variety of difference equations and systems, among them also the *linear Hamiltonian difference system*

$$(5)\qquad \Delta x_k=A_kx_{k+1}+B_ku_k,\quad \Delta u_k=C_kx_{k+1}-A_k^Tu_k,$$

where the $n\times n$-matrices B_k and C_k are symmetric and $I-A_k$ is nonsingular, as discussed e.g. in the monograph by Ahlbrandt and Peterson [3]. Essentials of the oscillation theory of (5) where the matix B is allowed to be singular are established in the fundamental paper [5].

This means, in turn, that systems (1) also cover the *higher order Sturm-Liouville difference equation*

$$\sum_{\mu=0}^{n}(-\Delta)^\mu\{r_k^{[\mu]}\Delta^\mu y_{k+n-\mu}\}=0\quad\text{with}\quad r_k^{[n]}\neq 0,$$

in particular, its special case, the *Sturm-Liouville second order difference equation*

$$(6)\qquad \Delta(r_k\Delta x_k)+p_kx_{k+1}=0\quad\text{with}\quad r_k\neq 0,$$

which are well studied in the recent literature, see [1, 2, 12, 15].

The aim of this paper is to present some recent results of the oscillation theory of (1) which are complemented here by some new results and observations. A particular attention is devoted to the Sturmian theory for (1) and to the formulation of related open problems and conjectures.

§2. Sturmian and Oscillation theorems for SDS

The symplecticity of the matrices $\mathcal{S}_k$ in (1) means that a fundamental matrix $Z_k\in\mathbb{R}^{2n\times 2n}$ of (1) is symplectic, whenever it is symplectic at one particular index, say $k = 0$. If this is the case and we write Z in the form $Z = \binom{X\ \tilde{X}}{U\ \tilde{U}}$, where $\binom{X}{U}$, $\binom{\tilde{X}}{\tilde{U}}$ are $2n \times n$ solutions of (1), then each solution is called a *conjoined basis* and together form the so-called pair of *normalized* conjoined bases. Consequently, a conjoined basis $\binom{X}{U}$ of (1) is a $2n \times n$ matrix solution satisfying

$$\operatorname{rank}\begin{pmatrix} X \\ U \end{pmatrix} = n, \quad X^T U = U^T X.$$

Oscillatory properties of (1) are defined using the concept of a focal point as follows. We say that a conjoined basis $\binom{X}{U}$ has no *focal point* in the interval $(k, k+1]$ if

$$\text{(7)} \qquad \operatorname{Ker} X_{k+1} \subseteq \operatorname{Ker} X_k \quad \text{and} \quad X_k X_{k+1}^{\dagger} \mathcal{B}_k \geq 0$$

holds. Here † and $\geq$ denote the Moore-Penrose generalized inverse and nonnegative definiteness of the matrix indicated, respectively. Note that if the first condition in (7) holds then the matrix $X_k X_{k+1}^{\dagger} \mathcal{B}_k$ is really symmetric, (cf. [6]), and it equals the matrix P_k given below by (8) since $T_k = I$ in this case.

The following matrices were introduced in [16]:

$$\text{(8)} \qquad M_k = (I - X_{k+1} X_{k+1}^{\dagger})\mathcal{B}_k, \quad T_k = I - M_k^{\dagger} M_k, \quad P_k = T_k^T X_k X_{k+1}^{\dagger} \mathcal{B}_k T_k,$$

for $k \in \{0, \dots, N\}$. Then obviously $M_k T_k = 0$ and it can be shown (see, e.g., [16]) that the matrix P_k is symmetric. The *multiplicity* of a focal point in the interval $(k, k+1]$ is defined as the number

$$\operatorname{rank} M_k + \operatorname{ind} P_k,$$

where ind stands for the index, i.e., the number of negative eigenvalues of the matrix indicated.

The next statement is proved in [9], it is a Sturmian type theorem for (1) and it can be regarded as a discrete version of [17, Lemma 7.1, p. 357] which concerns the linear Hamiltonian differential system

$$\text{(9)} \qquad \dot{x} = A(t)x + B(t)u, \quad \dot{u} = C(t)x - A^T(t)u.$$

Proposition 1. *Suppose that there exists a conjoined basis of* (1) *with no focal point in* $(0, N+1]$. *Then any other conjoined basis of this system has at most* n *focal points in* $(0, N+1]$, *each focal point counted a number of times equal to its multiplicity.*

Proof. We will present a general idea of the proof only, details can be found in [9]. The proof is based on the fact that the existence of a conjoined basis of (1) with no focal point in $(0, N+1]$ implies positivity of the associated quadratic functional (4). Hence, by contradiction, suppose that there exists a conjoined basis $\binom{X}{U}$ of (1) having at least $n+1$ focal points in $(0, N+1]$, including multiplicities. Then it is possible to construct admissible sequences $\{z_k^{[i]}\}_{k=0}^{N+1} = \binom{x_k^{[i]}}{u_k^{[i]}}$, $i = 1, \ldots, n+1$, such that $x_{N+1}^{[i]} = 0$, $\mathcal{F}(z^{[i]}) = (x_0^{[i]})^T u_0 + d^{[i]}$, where $d^{[i]}$ are certain nonpositve numbers (which can be specified explicitly, but this specification is rather technical), and

$$\begin{aligned} \mathcal{F}(z^{[i]}; z^{[j]}) &:= \sum_{k=0}^{N} \Big\{ (x_k^{[i]})^T \mathcal{A}_k^T \mathcal{C}_k x_k^{[j]} + 2 (x_k^{[i]})^T \mathcal{C}_k^T \mathcal{B}_k u_k^{[j]} \\ &\qquad\qquad + (u_k^{[i]})^T \mathcal{B}_k^T \mathcal{D}_k u_k^{[j]} \Big\} \\ &= 0 \end{aligned}$$

for $i \neq j$. Now, there exists a nontrivial linear combination

$$z = \begin{pmatrix} x \\ u \end{pmatrix} := \sum_{i=1}^{n+1} \alpha_i \begin{pmatrix} x^{[i]} \\ u^{[i]} \end{pmatrix}$$

such that $x_0 = 0$, $\{x_k\}_{k=1}^{N} \not\equiv 0$, and $\mathcal{F}(x, u) \leq 0$. This contradicts positivity of the functional $\mathcal{F}$. Q.E.D.

When $n = 1$ and (1) is rewritten Sturm-Liouville second order difference equation (6), the previous statement is the Sturm separation theorem for generalized zeros of solutions of (6).

Now we turn our attention to the so-called *oscillation theorem* for symplectic difference systems. It concerns the eigenvalue problem associated with (1)

$$(10) \qquad \begin{cases} x_{k+1} = & \mathcal{A}_k x_k + \mathcal{B}_k u_k, \\ u_{k+1} = & \mathcal{C}_k x_k + \mathcal{D}_k u_k - \lambda \mathcal{W}_k x_{k+1}, \quad 0 \leq k \leq N \\ & x_0 = 0 = x_{N+1}, \end{cases}$$

where $N \in \mathbb{N}$, $\lambda \in \mathbb{R}$ is the eigenvalue parameter, and where we assume that the matrices $\mathcal{W}$ are symmetric and nonnegative definite. The number λ is an *eigenvalue* of (10) if there exists a nontrivial solution $z = \binom{x}{u} = \binom{x_k}{u_k}_{k=0}^{N+1}$, a corresponding *eigenvector* of (10), i.e., z solves (10) and there exists $k \in \{1, \ldots N\}$ with $(x_k, u_k) \neq (0, 0)$.

In the next statement, usually referred to as the *oscillation theorem*, we will use the following notation:

$$
\begin{aligned}
n_1(\lambda) &:= \text{the number of focal points of a conjoined basis } \begin{pmatrix} X \\ U \end{pmatrix} \\
&\quad\;\; \text{in the interval } (0, N+1], \text{ including multiplicities;} \\
n_2(\lambda) &:= \text{the number of eigenvalues of (10),} \\
&\quad\;\; \text{which are less than or equal to } \lambda.
\end{aligned}
$$

Proposition 2. *Let* $Z = \binom{X}{U} = \binom{X_k(\lambda)}{U_k(\lambda)}_{k\in\mathbb{Z}}$ *be the principal solution at* $k = 0$ *of* (1)*, i.e., the solution given by the initial condition* $X_0 = 0$*,* $U_0 = I$*, and suppose that*

$$\lim_{\lambda\to-\infty} n_1(\lambda) = 0 \quad \textit{and} \quad \lim_{\lambda\to-\infty} n_2(\lambda) = 0. \tag{11}$$

Then

$$n_1(\lambda) = n_2(\lambda) \quad \text{for all} \quad \lambda \in \mathbb{R}. \tag{12}$$

Proof. The proof is technically rather complicated and it is based on the so-called *local oscillation theorem* which deals with the local dependece on λ of the quantities which appeared in the definition of a focal point, and states that

$$
\begin{aligned}
\operatorname{rank} M_k(\lambda+) - \operatorname{rank} M_k(\lambda-) &= \operatorname{ind} P_k(\lambda+) - \operatorname{ind} P_k(\lambda-) \\
&= \operatorname{rank} X_k(\lambda) - \operatorname{rank} X_k(\lambda-) - \operatorname{rank} X_{k+1}(\lambda) + \operatorname{rank} X_k(\lambda+)
\end{aligned}
$$

(the matrices M, P are defined by (8), now these matrices depend on λ) for all $\lambda \in \mathbb{R}$. This local result then leads to the equality

$$n_1(\lambda) - n_1(\lambda-) = n_2(\lambda) - n_2(\lambda-)$$

which togeteher with (11) gives the required result. We refer to [7] and [10] for technical details. Q.E.D.

The previous proposition is a combination of [7, Theorem 2] and [10, Theorem 1]. In Theorem 2 of [7], a certain "exceptional" set of (finitely many) λ's appeared, where (12) may fail to hold since oscillation theorem presented in that paper does not take into consideration the concept of *multiplicity* of a focal point (which did not exist when that paper was prepared). Theorem 1 of [10] already reflects this concept, but assumption (11) is not supposed in that paper and this fact required a substantial modification of the concept of eigenvalue of (10). We prefer here the above given combination of [7] and [10] since it is more understandable for the presentation given in this paper.

§3. Reversed symplectic difference system

System (1) can be written in the equivalent form $z_k = \mathcal{S}_k^{-1} z_{k+1}$ and this system we call the *reversed symplectic difference system.* In components x, u this system takes the form

$$(13) \qquad x_k = \mathcal{D}_k^T x_{k+1} - \mathcal{B}_k^T u_{k+1}, \quad u_k = -\mathcal{C}_k^T x_{k+1} + \mathcal{A}_k^T u_{k+1},$$

here we have used the fact that $S^{-1} = \begin{pmatrix} \mathcal{D}^T & -\mathcal{B}^T \\ -\mathcal{C}^T & \mathcal{A}^T \end{pmatrix}$ which follows e.g. from (3). In accordance with the terminology for "non-reversed" symplectic system (1), we say that a conjoined basis $\binom{X}{U}$ of (1) has no *reversed focal point* (shortly *r-focal point*) in the interval $[k, k+1)$, if

$$(14) \qquad \operatorname{Ker} X_k \subseteq \operatorname{Ker} X_{k+1} \quad \text{and} \quad \hat{P}_k = X_{k+1} X_k^{\dagger} \mathcal{B}_k^T \geq 0.$$

The following statement is a "reversed" analogue of [6, pp. 711–712].

Lemma 1. *Let* $\binom{X}{U}$ *be a conjoined basis of* (1). *First condition in* (14) *holds if and only if* $\operatorname{Ker} X_k^T \subseteq \operatorname{Ker} \mathcal{B}_k$ *and this is equivalent to the identity* $X_k X_k^{\dagger} \mathcal{B}_k^T = \mathcal{B}_k^T$. *Also, if* (14) *holds, the matrix* $\hat{P}_k$ *is symmetric.*

Proof. First of all recall that the standard result of the theory od Moore-Penrose generalized inverses claims that $WV^{\dagger}V = W$ holds for a pair of matrices V, W if and only if $\operatorname{Ker} V \subseteq \operatorname{Ker} W$, see [4]. Hence $\mathcal{B}_k (X_k^{\dagger})^T X_k^T = \mathcal{B}$, i.e. $X_k X_k^{\dagger} \mathcal{B}_k^T = \mathcal{B}_k^T$ if and only if the first condition in (14) holds. Suppose now that this condition in (14) holds. Let $\binom{\tilde{X}}{\tilde{U}}$ be a conjoined basis which together with $\binom{X}{U}$ forms a pair of normalized conjoined bases, i.e., the matrix $Z = \begin{pmatrix} X & \tilde{X} \\ U & \tilde{U} \end{pmatrix}$ is symplectic. Then

$$\mathcal{S}_k = Z_{k+1} Z_k^{-1} = \begin{pmatrix} X_{k+1} & \tilde{X}_{k+1} \\ U_{k+1} & \tilde{U}_{k+1} \end{pmatrix} \begin{pmatrix} \tilde{U}_k^T & -\tilde{X}_k \\ -U_k^T & X_k^T \end{pmatrix}$$

and hence

$$\mathcal{B}_k = \tilde{X}_{k+1} X_k^T - X_{k+1} \tilde{X}_k^T.$$

Now, let $\alpha \in \operatorname{Ker} X_k^T$. Since the identity $Z^T \mathcal{J} Z = \mathcal{J}$ which defines symplecticity of the matrix Z is equivalent to the identity $Z \mathcal{J} Z^T = \mathcal{J}$, we also have $\tilde{X}_k X_k^T = X_k \tilde{X}_k^T$, i.e., $\tilde{X}_k X_k^T \alpha = 0$ implies $X_k \tilde{X}_k^T \alpha = 0$. Hence we have $\tilde{X}_k^T \alpha \in \operatorname{Ker} X_k \subseteq \operatorname{Ker} X_{k+1}$. Then $\mathcal{B}_k \alpha = (\tilde{X}_{k+1} X_k^T - X_{k+1} \tilde{X}_k^T)\alpha = -X_{k+1} \tilde{X}_k \alpha = 0$, which means that $\operatorname{Ker} X_k^T \subseteq \operatorname{Ker} \mathcal{B}_k$. Conversely, let $\operatorname{Ker} X_k^T \subseteq \operatorname{Ker} \mathcal{B}_k$ and $\alpha \in \operatorname{Ker} X_k$. Then

$$\begin{aligned} X_{k+1}\alpha &= (\mathcal{A}_k X_k + \mathcal{B}_k U_k)\alpha = \mathcal{B}_k U_k \alpha = \mathcal{B}_k (X_k^T)^{\dagger} X_k^T U_k \alpha \\ &= \mathcal{B}_k (X_k^T)^{\dagger} U_k^T X_k \alpha = 0, \end{aligned}$$

i.e., $\alpha \in \operatorname{Ker} X_{k+1}$. As for the symmetry of the matrix $\hat{P}$, we have

$$\begin{aligned}\hat{P}_k &= X_{k+1}X_k^\dagger \mathcal{B}_k^T = (\mathcal{A}_k X_k + \mathcal{B}_k U_k)X_k^\dagger \mathcal{B}^T \\ &= \mathcal{A}_k \mathcal{B}_k^T - \mathcal{B}_k (X_k^\dagger)^T X_k^T U_k X_k^\dagger \mathcal{B}_k^T,\end{aligned}$$

here we have used the identity $\mathcal{B}_k (X_k^\dagger)^T X_k^T = \mathcal{B}_k$ and the fact that $\binom{X}{U}$ is a conjoined basis. Q.E.D.

Next, following the previous section, we introduce the notation

(15)
$$\hat{M}_k = (I - X_k X_k^\dagger)\mathcal{B}_k^T, \quad \hat{T}_k = I - \hat{M}_k^\dagger \hat{M}_k, \quad \hat{\mathcal{P}}_k = \hat{T}_k^T X_{k+1} X_k^\dagger \mathcal{B}_k^T \hat{T}_k.$$

Using this notation, a conjoined basis $\binom{X}{U}$ has the r-focal point of *multiplicity* m at an integer k if $m = \operatorname{rank} \hat{M}_k$, and this conjoined basis has the r-focal point of multiplicity p in the open interval $(k, k+1)$ if $p = \operatorname{ind} \hat{\mathcal{P}}_k$. The number $m + p$ defines the multiplicity of the r-focal point of $\binom{X}{U}$ in the interval $[k, k+1)$. Our main result of this section is the following complement of Proposition 1.

Theorem 1. *Suppose that there exists a conjoined basis of* (1) *with no r-focal point in* $[0, N+1)$. *Then any other conjoined basis of* (1) *has at most* n *r-focal points in* $[0, N+1)$, *counting multiplicities.*

Proof. To prove the statement, we proceed similarly as in [6], we "reflect" the interval $[0, N+1]$ by the substitution $k \longmapsto N+1-k$. Let $z = \{z_k\}_{k=0}^{N+1}$ be a solution of (1). Denote

$$\tilde{z}_k = \mathcal{K} z_{N+1-k}, \quad \tilde{\mathcal{S}}_k = \mathcal{K}\mathcal{S}_{N-k}^{-1}\mathcal{K}, \quad \text{where } \mathcal{K} = \begin{pmatrix} I & 0 \\ 0 & -I \end{pmatrix}.$$

We have (taking into account that $\mathcal{K}^{-1} = \mathcal{K}$)

$$\tilde{z}_{k+1} = \mathcal{K} z_{N-k} = \mathcal{K}\mathcal{S}_{N-k}^{-1} z_{N+1-k} = \mathcal{K}\mathcal{S}_{N-k}^{-1}\mathcal{K}\tilde{z}_k = \tilde{\mathcal{S}}_k \tilde{z}_k, \tag{16}$$

where we have denoted

$$\tilde{\mathcal{S}}_k = \begin{pmatrix} \tilde{\mathcal{A}}_k & \tilde{\mathcal{B}}_k \\ \tilde{\mathcal{C}}_k & \tilde{\mathcal{D}}_k \end{pmatrix} := \mathcal{K}\mathcal{S}_{N-k}^{-1}\mathcal{K} = \begin{pmatrix} \mathcal{D}_{N-k}^T & \mathcal{B}_{N-k}^T \\ \mathcal{C}_{N-k}^T & \mathcal{A}_{N-k}^T \end{pmatrix},$$

i.e., in particular, $\tilde{\mathcal{B}}_k = \mathcal{B}_{N-k}^T$. Next we show that a conjoined basis $\binom{X}{U}$ of (1) has an r-focal point of multiplicity $m+p$ at $[N-k, N+1-k)$ for some $k \in \{0 \dots, N\}$ (here $m = \operatorname{rank} \hat{M}_{N-k}$ is the multiplicity at $N-k$ and $p = \operatorname{ind} \hat{P}_{N-k}$ is the multiplicity in the open interval $(N-k, N-k+$

1)), if and only if the solution $\binom{\tilde{X}_k}{\tilde{U}_k} := \mathcal{K}\binom{X_{N+1-k}}{U_{N+1-k}}$ has the ("normal") focal point of multiplicity $m+p$ at $(k, k+1]$. We have

$$\hat{M}_{N-k} = (I - X_{N-k}X^{\dagger}_{N-k})\mathcal{B}^T_{N-k} = (I - \tilde{X}_{k+1}\tilde{X}^{\dagger}_{k+1})\tilde{\mathcal{B}}_k = \tilde{M}_k$$

hence $\hat{T}_{N-k} = \tilde{T}_k$, and

$$\hat{P}_{N-k} = \hat{T}_{N-k}X_{N+1-k}X^{\dagger}_{N-k}\hat{T}_{N-k} = T_k\tilde{X}_k\tilde{X}^{\dagger}_{k+1}\tilde{\mathcal{B}}_kT_k = \tilde{P}_k.$$

Here the matrices $\tilde{M}_k, \tilde{T}_k, \tilde{P}_k$ define the multiplicity of a focal point of $\binom{\tilde{X}}{\tilde{U}}$ in $[k, k+1)$. This implies the claimed statement about multiplicity of the r-focal point of $\binom{X}{U}$ at $[N-k, N+1-k)$ and the "normal" focal point of $\binom{\tilde{X}}{\tilde{U}}$ at $(k, k+1]$. To finish the proof, it suffices to apply Proposition 1. Q.E.D.

Remark 1. (i) We start with the open problem formulated in [9], which concerns the number of focal points of conjoined bases of (1). Based on the statement of Proposition 1 and on a known result (see [17, Chap. VII]) for conjoined bases of the linear Hamiltonian differential system (9) (which is a continuous counterpart of (1)), it is conjectured in [9] that the numbers of focal points (including multiplicities) of any pair of conjoined bases of (1) in a given discrete interval differ by at most n, where n is the dimension of square matrices $\mathcal{A}, \mathcal{B}, \mathcal{C}, \mathcal{D}$ in (1). It seems that the crucial role in the proof of this statement could be played by the relationship between the number of focal point of the principal solution of (1) in the interval $(0, N+1]$ and the index and nullity of the quadratic form defined by the functional (4). We conjecture here that this relationship can be obtained using Proposition 2 and the proof of the above mentioned conjecture of [9] is a subject of the present investigation.

(ii) Another open problem concerns the number and the *location* of focal points and r-focal points of *one* conjoined basis of (1) in a given discrete interval. We conjecture that the numbers of focal and r-focal points of a conjoined basis $\binom{X}{U}$ of (1) in the discrete interval $[0, N+1]$ *are the same* and that the location of focal points and r-focal point is essentially the same. In particular, we conjecture that

$$\operatorname{rank} M_k = \operatorname{rank} \hat{M}_{k+1} \quad \text{and} \quad \operatorname{ind} P_k = \operatorname{ind} \hat{P}_k,$$

where the matrices $M_k, P_k, \hat{M}_k, \hat{P}_k$ are defined by (8) and (15).

Acknowledgement. This research is supported by the grant 201/04/0580 of the Grant Agency of Czech Republic and the Research Project MSM00 22162409 of the Czech Ministry of Education.

References

[1] R. P. Agarwal, Difference Equations and Inequalities, Theory, Methods, and Applications. Second ed., Monogr. Textbooks Pure Appl. Math., **228**, Marcel Dekker, Inc., New York, 2000.

[2] R. P. Agarwal, M. Bohner, S. R. Grace and D. O'Regan, Discrete Oscillation Theory, Hindawi Publ. Corporation, New York, 2005.

[3] C. D. Ahlbrandt and A. C. Peterson, Discrete Hamiltonian Systems: Difference Equations, Continued Fractions, and Riccati Equations, Kluwer Academic Publ., Boston, 1996.

[4] A. Ben-Israel and E. N. Greville, Generalized Inverses: Theory and Applications, Springer-Verlag, New York, 2003.

[5] M. Bohner, Linear Hamiltonian difference systems: disconjugacy and jacoby-type conditions, J. Math. Anal. Appl., **199** (1996), 804–826.

[6] M. Bohner and O. Došlý, Disconjugacy and transfromations for symplectic difference systems, Rocky Mountain J. Math., **27** (1997), 707–743.

[7] M. Bohner, O. Došlý and W. Kratz, An oscillation theorem for discrete eigenvalue problems, Rocky Mountain J. Math., **33** (2003), 1233–1260.

[8] M. Bohner, O. Došlý and W. Kratz, Positive semidefiniteness of discrete quadratic functionals, Proc. Edinburg Math. Soc., **46** (2003), 227–236.

[9] O. Došlý and W. Kratz, A Sturmian separation theorem for symplectic difference systems, J. Math. Anal. Appl., **325** (2007), 333–341.

[10] O. Došlý and W. Kratz, Oscillation theorems for symplectic difference systems, J. Difference Equ. Appl., **13** (2007), 585–605.

[11] O. Došlý, R. Hilscher and V. Zeidan, Nonnegativity of discrete quadratic functionals corresponding to symplectic difference systems, Linear Algebra Appl., **375** (2003), 21–44.

[12] S. Elaydi, An Introduction to Difference Equations. Third ed., Undergrad. Texts Math., Springer, New York, 2005.

[13] R. Hilscher and V. Zeidan, A remark on discrete quadratic functionals with separable endpoints, Rocky Mountain J. Math., **33** (2003), 1337–1351.

[14] R. Hilscher and V. Zeidan, Equivalent conditions to the nonnegativity of a quadratic functional in discrete optimal control, Math. Nachr., **266** (2004), 48–59.

[15] W. G. Kelley and A. C. Peterson, Difference Equations. An Introduction with Applications. Second ed., Harcourt/Academic Press, San Diego, 2001.

[16] W. Kratz, Discrete oscillation, J. Difference Equ. Appl., **9** (2003), 135–147.

[17] W. T. Reid, Ordinary Differential Equations, John Wiley & Sons, New York, 1971.

Masaryk University Brno
Department of Mathematics
Faculty of Science
Kotlářská 2
CZ–61137 Brno
Czech Republic

E-mail address: dosly@math.muni.cz

Advanced Studies in Pure Mathematics 53, 2009
Advances in Discrete Dynamical Systems
pp. 51–57

Periodic solutions of periodic difference equations

Tetsuo Furumochi [1] and Masato Muraoka

Abstract.

In this paper, we discuss the existence of periodic solutions of the periodic difference equation

$$x(n+1) = f(n, x(n)), \quad n \in \mathbf{Z}$$

and the periodic difference equation with finite delay

$$x(n+1) = f(n, x_n), \quad n \in \mathbf{Z},$$

where x and f are d-vectors, and $\mathbf{Z}$ denotes the set of integers. We show the existence of periodic solutions by using Browder's fixed point theorem, and illustrate an example by using a boundedness result due to Shunian Zhang.

§1. Introduction

The existence of periodic solutions of periodic difference equations has been discussed in some books and papers (see [2-5, 7] and their references). In particular, Dannan-Elaydi-Liu [2], Elaydi [4] and Elaydi-Zhang [5] have obtained some existence results of periodic solutions by employing various methods including homotopy techniques. In this paper, we give some new existence results of periodic solutions for periodic difference equations by using Browder's fixed point theorem, Liapunov method and Razumikhin method, and show an example by using a boundedness result due to Shunian Zhang.

Fixed point theorems are very useful tools in obtaining existence theorems for periodic solutions. First we state a fixed point theorem

---

Received October 12, 2007.
2000 *Mathematics Subject Classification.* 39A11.
Key words and phrases. Periodic solutions, difference equations, fixed point theorem.

[1]This paper is in final form and no version of it will be submitted for publication elsewhere, and partly supported in part by Grant-in-Aid for Scientific Research (C), No. 16540141, Japan Society for the Promotion of Science.

due to Browder, which we use later to show the existence of periodic solutions of periodic difference equations.

Theorem 1. *(Browder) Let S and S^1 be open convex subsets of the Banach space A, S^0 a closed convex subset of A, $S^0 \subset S^1 \subset S$, F a compact mapping of S into A. Suppose that for a positive integer m, F^m is well-defined on S^1, $\cup_{0 \le j \le m} F^j(S^0) \subset S^1$, while $F^m(S^1) \subset S^0$. Then F has a fixed point in S^0.*

Here, F is said to be a compact mapping of S, if F is a continuous mapping and $F(S)$ is contained in a compact set of A.

§2. Difference equations without delays

Consider the difference equation

$$x(n+1) = f(n, x(n)), \quad n \in \mathbf{Z}, \tag{1}$$

where x and f are d-vectors, and $\mathbf{Z}$ denotes the set of integers. Suppose that $f : \mathbf{Z} \times \mathbf{R}^d \to \mathbf{R}^d$ is continuous in x for each fixed $n \in \mathbf{Z}$, where $\mathbf{R}$ denotes the set of real numbers. For any $n_0 \in \mathbf{Z}$ and $\xi \in \mathbf{R}^d$, $x(n, n_0, \xi)$ denotes the solution of Eq.(1) with $x(n_0, n_0, \xi) = \xi$. For any $\alpha > 0$, define the set S_α by

$$S_\alpha := \{x \in \mathbf{R}^d : |x| \le \alpha\},$$

where $|\cdot|$ denotes the Euclidean norm for $\mathbf{R}^d$.

First we state the definitions of boundedness of solutions of Eq.(1).

Definition 2.1. *The solutions of* Eq.(1) *are equi-bounded, if for any $n_0 \in \mathbf{Z}$ and $\alpha > 0$, there exists a $\beta = \beta(n_0, \alpha) > 0$ such that if $\xi \in S_\alpha$, then $|x(n, n_0, \xi)| < \beta$ for $n \in \mathbf{Z} \cap [n_0, \infty)$.*

Definition 2.2. *The solutions of* Eq.(1) *are uniformly bounded, if the β in* Definition 1 *is independent of n_0.*

Definition 2.3. *The solutions of* Eq.(1) *are equltimately bounded for bound X, if there exists an $X > 0$ and if corresponding to any $n_0 \in \mathbf{Z}$ and $\alpha > 0$, there exists a $\nu = \nu(n_0, \alpha) \in \mathbf{N}$ such that $\xi \in S_\alpha$ implies that $|x(n, n_0, \xi)| < X$ for $n \in \mathbf{Z} \cap [n_0 + \nu, \infty)$, where $\mathbf{N}$ denotes the set of positive integers.*

Definition 2.4. *The solutions of* Eq.(1) *are uniformly ultimately bounded for bound X, if the ν in* Definition 3 *is independent of n_0.*

By using Theorem 1, we can prove the following theorem.

Theorem 2. *If $f(n,x)$ in* Eq.(1) *is N-periodic in n for an $N \in \mathbf{N}$, and if the solutions of* Eq.(1) *are equiultimately bounded for bound $X > 0$, then* Eq.(1) *has an N-periodic solution $x(n)$ such that $|x(n)| < X$.*

Proof. Since equiultimate boundedness of the solutions of Eq.(1) implies equi-boundedness of the solutions of Eq.(1), and since Eq.(1) is N-periodic, the solutions of Eq.(1) are uniformly bounded. Therefore, there exists a $\beta = \beta(X) > 0$ such that if $n_0 \in \mathbf{Z}$ and $\xi \in S_X$, then $|x(n, n_0, \xi)| < \beta$ for $n \in \mathbf{Z} \cap [n_0, \infty)$. Moreover, there are numbers $\gamma > \beta$ and $\gamma^* > \gamma$ such that if $n_0 \in \mathbf{Z}$ and $\xi \in S_\beta$, then

$$|x(n, n_0, \xi)| < \gamma \quad \text{for} \quad n \in \mathbf{Z} \cap [n_0, \infty)$$

and that $n_0 \in \mathbf{Z}$ and $\xi \in S_\gamma$ imply

$$|x(n, n_0, \xi)| < \gamma^* \quad \text{for} \quad n \in \mathbf{Z} \cap [n_0, \infty).$$

From equiultimate boundedness for bound X of the solutions of Eq.(1), it follows that there exists a $\nu \in \mathbf{N}$ such that if $n \in \mathbf{Z} \cap [\nu, \infty)$ and $\xi \in S_\beta$, then $|x(n, 0, \xi)| < X$, and hence, there exists an $m \in \mathbf{N}$ for which

$$|x(mN, 0, \xi)| < X \quad \text{if} \quad \xi \in S_\beta.$$

Let $\mathbf{R}^d$ and S_X be A and S^0 in Theorem 1 respectively, and let S and S^1 be the sets defined by

$$S := \{x \in \mathbf{R}^d : |x| < \gamma\}$$

and

$$S^1 := \{x \in \mathbf{R}^d : |x| < \beta\}.$$

In addition, let $F : S \to \mathbf{R}^d$ be a mapping defined by

$$F(\xi) := x(N, 0, \xi), \quad \xi \in S.$$

Then, the mapping F is continuous, and $F(S)$ is contained in a compact set S_{γ^*}, which shows that F is a compact mapping. Moreover, the convex sets S, S^0 and S^1 satisfy the assumptions in Theorem 1. Therefore, there exists a fixed point $\xi \in S_X$, which implies the existence of an N-periodic solution $x(n) = x(n, 0, \xi)$ of Eq.(1) with $|x(n)| < X$ for $n \in \mathbf{Z} \cap [0, \infty)$. Q.E.D.

§3. Difference equations with finite delay

For a fixed $\kappa \in \mathbf{N}$, let B be the set of sequences $\phi : \mathbf{Z} \cap [-\kappa, 0] \to \mathbf{R}^d$. For any $\phi \in B$, define $\|\phi\|$ by

$$\|\phi\| := \sup\{|\phi(k)| : k \in \mathbf{Z} \cap [-\kappa, 0]\}.$$

For any $\alpha > 0$, the set B_α defined by

$$B_\alpha := \{\phi \in B : \|\phi\| \leq \alpha\}$$

is compact. For any sequence $x(k) : \mathbf{Z} \to \mathbf{R}^d$ and any fixed $n \in \mathbf{Z}$, the symbol x_n denotes the restriction of $x(k)$ on $\mathbf{Z} \cap [n - \kappa, n]$, that is, x_n is an element of B defined by

$$x_n(k) := x(n + k), \quad k \in \mathbf{Z} \cap [-\kappa, 0].$$

Consider the difference equation with finite delay

$$(2) \qquad x(n + 1) = f(n, x_n), \quad n \in \mathbf{Z},$$

where $f : \mathbf{Z} \times B \to \mathbf{R}^d$ is continuous in ϕ for each fixed $n \in \mathbf{Z}$. For any $n_0 \in \mathbf{Z}$ and any initial sequence $\phi \in B$, there is a unique solution of Eq.(2), denoted by $x(n, n_0, \phi)$, such that it satisfies Eq.(2) for $n \in \mathbf{Z} \cap [n_0, \infty)$ and

$$x(n_0 + k, n_0, \phi) = \phi(k) \text{ for } k \in \mathbf{Z} \cap [-\kappa, 0].$$

The definitions of boundedness of solutions of Eq.(2) are similarly obtained by replacing ξ and S_α in Definitions 1-4 by ϕ and B_α, respectively.

By using Theorem 1, we can prove the following theorem, which is similar to Theorem 2.

Theorem 3. *If $f(n, \phi)$ in* Eq.(2) *is N-periodic in n for an $N \in \mathbf{N}$, and if the solutions of* Eq.(2) *are equiultimately bounded for bound $X > 0$, then* Eq.(2) *has an N-periodic solution $x(n)$ such that $|x(n)| < X$.*

This theorem can be easily proved by replacing ξ, $\mathbf{R}^d$, S_X, S_β, S_γ and $\{x \in \mathbf{R}^d : |x| < \gamma\}$ in the proof of Theorem 2 by ϕ, B, B_X, B_β, B_γ and $S := \{\phi \in B : \|\phi\| < \gamma\}$, respectively, and considering a mapping $F : S \to B$ defined by

$$F(\phi) := x_N(0, \phi), \quad \phi \in S.$$

So, we omit the proof.

§4. Boundedness in difference equations with finite delay

In Section 3, equiultimate boundedness of solutions of Eq.(2) is an important assumption. There are many results concerning boundedness of the solutions of difference equations (see for example [6, 8] and their references). Here we state a boundedness theorem due to Shunian Zhang without a proof. By using this theorem, we show an example later.

Theorem 4. *(Zhang) Suppose that there exists a Liapunov function* $V : \mathbf{Z} \times \mathbf{R}^d \to \mathbf{R}^+ := [0, \infty)$*, which satifies the following conditions;*

(i) $a(|x|) \le V(n, x) \le b(|x|)$*, where* $a,\ b : \mathbf{R}^+ \to \mathbf{R}^+$*,* $a(r)$ *and* $b(r)$ *are continuous, increasing and* $a(r) \to \infty$ *as* $r \to \infty$*,*

(ii) $\Delta V_{(2)}(n, x(n)) := V(n+1, x(n+1)) - V(n, x(n)) \le M - c(|x(n)|)$ *whenever*

$$P\big(V(n+1, x(n+1))\big) > V(k, x(k)) \ \text{ for } \ k \in \mathbf{Z} \cap [n - \kappa, n],$$

where $x(n)$ *is a solution of* Eq.(2)*,* M *is a positive constant,* $c : \mathbf{R}^+ \to \mathbf{R}^+$ *is continuous, increasing and* $c(r) \to \infty$ *as* $r \to \infty$*, and* $P : \mathbf{R}^+ \to \mathbf{R}^+$ *is continuous,* $P(u) > u$ *for* $u > 0$*, and* $\kappa \in \mathbf{N}$*.*

Then the solutions of Eq.(2) *are uniformly ultimately bounded for a bound* $X > 0$*.*

By using Theorem 4, we construct an example for Theorem 3.

Example. Consider the scalar N-periodic difference equation

$$x(n+1) = \sum_{k=0}^{\kappa} f_k(n, x(n-k)) + G \cos \frac{2\pi}{N} n, \quad n \in \mathbf{Z}, \tag{3}$$

where G is a nonzero constant, $\kappa,\ N \in \mathbf{N}$ with $N > 1$, $f_k(n, x) : \mathbf{Z} \times \mathbf{R} \to \mathbf{R}$ is continuous in x for each fixed n, $f_k(n, x)$ is N-periodic in n,

$$|f_k(n, x)| \le c_k |x|$$

and

$$\sum_{k=0}^{\kappa} c_k < 1. \tag{4}$$

From (4), we can choose a constant $\rho > 1$ with

$$c_0 + \rho \sum_{k=1}^{\kappa} c_k < 1.$$

Let $P(u) = \rho r$ and $V(n,x) = |x|$, and let $x(n) = x(n, n_0, \phi)$ be any solution of Eq.(3) with $n_0 \in \mathbf{Z}$ and $\phi \in B$. If we have

$$P\big(V(n+1, x(n+1))\big) = \rho|x(n+1)|$$

$$> |x(n-k)| = V(n-k, x(n-k)), \quad k \in \mathbf{Z} \cap [0, \kappa],$$

then we obtain

$$|x(n+1)| \le |f_0(n, x(n))| + \sum_{k=1}^{\kappa} |f_k(n, x(n-k))| + |G \cos \frac{2\pi}{N} n|$$

$$\le c_0|x(n)| + \sum_{k=1}^{\kappa} c_k|x(n-k)| + |G| \le c_0|x(n)| + \rho \sum_{k=1}^{\kappa} c_k|x(n+1)| + |G|,$$

which implies

$$|x(n+1)| \le \frac{c_0|x(n)|}{1 - \rho\sum_{k=1}^{\kappa} c_k} + \frac{|G|}{1 - \rho\sum_{k=1}^{\kappa} c_k}.$$

Thus, for the number M defined by

$$M := \frac{|G|}{1 - \rho\sum_{k=1}^{\kappa} c_k},$$

we have

$$\Delta V_{(3)}(n, x(n)) = |x(n+1)| - |x(n)| \le M - \frac{1 - c_0 - \rho\sum_{k=1}^{\kappa} c_k}{1 - \rho\sum_{k=1}^{\kappa} c_k}|x(n)|,$$

which shows that all assumptions of Theorem 4 are satisfied with $a(r) = b(r) = r$ and

$$c(r) = \frac{1 - c_0 - \rho\sum_{k=1}^{\kappa} c_k}{1 - \rho\sum_{k=1}^{\kappa} c_k} r.$$

Thus, by Theorems 3 and 4, Eq.(3) has an N-periodic solution.

References

[1] F. E. Browder, On a generalization of the Schauder fixed point theorem, Duke Math. J., **26** (1959), 291–303.

[2] F. Dannan, S. Elaydi and P. Liu, Periodic solutions of difference equations, J. Differ. Equations Appl., **6** (2000), 203–232.

[3] S. Elaydi, An Introduction to Difference Equations 3rd ed., Springer, 2005.

[4] S. Elaydi, Periodicity and stability of linear Volterra difference systems, J. Math. Anal. Appl., **181** (1994), 483–492.

[5] S. Elaydi and S. Zhang, Stability and periodicity of difference equations with finite delay, Funkcial. Ekvac., **37** (1994), 401–413.
[6] T. Furumochi, K. Hishitani and K. Tsuchihashi, Liapunov functions and boundedness in difference equations, Dynamics of Continuous, Discrete and Impulsive Systems, to appear.
[7] W. G. Kelley and A. C. Peterson, Difference Equations, Academic Press, 2001.
[8] S. Zhang, Boundedness of finite delay difference systems, Ann. Differential Equations, **9** (1993), 107–115.

Tetsuo Furumochi
Department of Mathematics
Shimane University
1060 Nishikawatsucho
Matsue 690-8504
Japan

Masato Muraoka
Department of Mathematics
Shimane University
1060 Nishikawatsucho
Matsue 690-8504
Japan

E-mail address: `furumochi@riko.shimane-u.ac.jp`
`muuuuuu@ezweb.ne.jp`

Advanced Studies in Pure Mathematics 53, 2009
Advances in Discrete Dynamical Systems
pp. 59–66

On the existence of almost periodic solutions to a nonlinear Volterra difference equation

Yoshihiro Hamaya

Abstract.

The existence of almost periodic solutions of a nonlinear Volterra difference equation with infinte delay is obtained by using some restrictive conditions of the equation.

§1. Introduction

For ordinary differential equations and integrodifferential equations, the existence of almost periodic solutions of almost periodic systems have been studied by many authors [2,3,6,7]. Hamaya [3] has investigated the existence of almost periodic solutions of functional difference equations with infinite delay by using stability properties of bounded solutions. Y. Song, C.T.H. Baker and H. Tian [6] have studied the existence of periodic and almost periodic solutions for nonlinear Volterra difference equations by means of certain stability conditions. However, it is not easy to know whether or not there exists a bounded solutions for nonlinear Volterra difference equations. In this paper, we discuss the existence of nonlinear Volterra difference equations motivated by C. Feng [1] and S. Kato and M. Imai [5]. Our restrictive conditions to guarantee the existence of almost periodic solutions are weaker than that of Hamaya [3], and Y. Song, C.T.H. Baker and H. Tian [6], and moreover, this paper is based on [4].

We first introduce an almost periodic function $f(n,x) : Z \times D \to R^l$, where D is an open set in R^l, and $f(n,\cdot)$ be continuous for each $n \in Z$.

Definition 1. $f(n,x)$ is said to be almost periodic in n uniformly for $x \in D$, if for any $\epsilon > 0$ and any compact set K in D, there exists a

Received November 14, 2006.
Revised October 12, 2007.
2000 *Mathematics Subject Classification.* 39A10, 39A11.

positive integer $L^*(\epsilon, K)$ such that any interval of length $L^*(\epsilon, K)$ contains an integer τ for which

$$(1) \qquad |f(n+\tau, x) - f(n, x)| \leq \epsilon$$

for all $n \in Z$ and all $x \in K$. Such a number τ in (1) is called an ϵ-translation number of $f(n, x)$.

In order to formulate a property of almost periodic functions, which is equivalent to the above definition, we discuss the concept of the normality of almost periodic functions. Namely, Let $f(n, x)$ be almost periodic in n uniformly for $x \in D$. Then, for any sequence $\{h_k'\} \subset Z$, there exists a subsequence $\{h_k\}$ of $\{h_k'\}$ such that the sequence of functions $\{f(n + h_k, x)\}$ converges uniformly on $Z \times S$, where S is a compact set in D. In other words, for any sequence $\{h_k'\} \subset Z$, there exists a subsequence $\{h_k\}$ of $\{h_k'\}$ and function $g(n, x)$ such that

$$f(n + h_k, x) \to g(n, x)$$

uniformly on $Z \times K$ as $k \to \infty$, where K is any compact set in D. There are many properties of the discrete almost periodic functions [6], which are corresponding properties of the continuous almost periodic function $f(t, x) \in C(R \times D, R^l)$ [cf.2,7].
The following concept of asymptotic almost periodicity was introduced by Frechet in the case of continuous function (cf.[2,7]).

Definition 2. $u(n)$ is said to be asymptotically almost periodic if it is a sum of an almost periodic function $p(n)$ and a function $q(n)$ defined on $I^* = [a, \infty) \subset Z^+$ which tends to zero as $n \to \infty$, that is,

$$u(n) = p(n) + q(n).$$

$u(n)$ is asymptotically almost periodic if and only if for any sequence $\{n_k\}$ such that $n_k \to \infty$ as $k \to \infty$ there exists a subsequence $\{n_{k_j}\}$ for which $u(n + n_{k_j})$ converges uniformly on I^*.

We shall consider the following Volterra difference equation

$$(2) \quad \Delta x(n) = f(n, x(n)) + \sum_{m=-\infty}^{n} F(n, m)x(m) + p(n), \qquad n \geq 0,$$

where $f(n, x)$ is almost periodic in n uniformly for x, $p(n)$ is an almost periodic function and $F(n, m)$ is almost periodic, that for any $\epsilon > 0$

and any compact set K, there exists an integer $L^{**} = L^{**}(\epsilon, K) > 0$ such that any interval of length L^{**} contains an τ for which

$$|F(n+\tau, m+\tau) - F(n,m)| \leq \epsilon.$$

Let R^l be Euclidean space with norm denoted by $\|\cdot\|$. We define the functional $[\cdot,\cdot] : R^l \times R^l \to R$ by

$$[x,y] = h^{-1}(\|x+hy\| - \|x\|) \quad \text{for } h > 0.$$

Let x, y and z be in R^l, from [5], the functional $[\cdot,\cdot]$ has the following properties:

(*i*) $|[x,y]| \leq \|y\|$;

(*ii*) $[x, y+z] \leq [x,y] + [x,z]$;

(*iii*) $\Delta_h^+\|u(n)\| = [u(n), \Delta_h u(n)]$,
where u is a function and $\Delta_h^+\|u(n)\|$ denotes the h-differences of $\|u(n)\|$, that is $\Delta_h^+\|u(n)\| = h^{-1}(\|u(n+h)\| - \|u(n)\|)$.

§2. Main results

We assume the following conditions:

(I) $\|p(n) + f(n,0)\| \leq L$ for all $n \in Z$, where L is a positive constant.

(II) There exists a positive constant γ such that

$$\lim_{n-m\to\infty} \frac{1}{n-m} \sum_{j=m}^{n-1} e(j) = -\gamma < 0,$$

where the function $e(n) : Z \to R$.
(III) for all $(n,x) \in Z \times R^l$,

$$[x-y, f(n,x) - f(n,y) + \sum_{m=-\infty}^{n} F(n,m)x(m)$$
$$- \sum_{m=-\infty}^{n} F(n,m)y(m)] \leq e(n)\|x-y\|.$$

We have two preliminary lemmas.

Lemma 1. Suppose that $e(n)$ is an almost periodic function. Then, for any $b \in Z$, the following equality holds

$$\lim_{n-m\to\infty} \frac{1}{n-m} \sum_{j=m}^{n-1} e(j) = \lim_{n-m\to\infty} \frac{1}{n-m} \sum_{j=m}^{n-1} e(j+b).$$

By discrete analogy of Corollary 3.2 in [2], we can prove this lemma. So, we will omit it.

Lemma 2. Suppose that (II) is satisfied. Then, for any $b \in Z$, the following inequality holds

$$\exp\Big(\sum_{j=m}^{n-1} e(j+b)\Big) \le \beta \exp(-\alpha(n-m)). \tag{3}$$

where $e(n)$ is an almost periodic function, α and β are positive constants independent for b.

Proof. From Lemma 1, for $\epsilon = \frac{\gamma}{4} > 0$, there exists a sufficiently large $T = T(\epsilon)$ such that $|\frac{1}{n-m} \sum_{j=m}^{n-1} e(j+b) - (-\gamma)| < \epsilon$, namely,

$$\sum_{j=m}^{n-1} e(j+b) < -\frac{3\gamma(n-m)}{4}, \quad \text{for } n-m \ge T. \tag{4}$$

Since $e(n)$ is an almost periodic function, there exists a positive constant c such that inequality $|e(j+b)| < c$ holds for any $j, b \in Z$. Thus, when $|n-m| \le T$, we have

$$\Big|\sum_{j=m}^{n-1} e(j+b)\Big| \le \sum_{j=m}^{n-1} |e(j+b)| \le cT. \tag{5}$$

Let $\alpha = \frac{3\gamma}{4}$ and $\beta = \exp(cT + \alpha T)$, inequality (4) and (5) imply that (3) holds. Q.E.D.

Now, we have main result of this paper.

Theorem. Suppose that assumptions (I),(II) and (III) hold, then equation (2) has a unique almost periodic solution.

Proof. We first show that equation (2) has a bounded solution.

Let $x(n)$ be a solution of equation (2), then we have

$$\begin{aligned}
\Delta^+\|x(n)\| &= [x(n), \Delta x(n)] \\
&= [x(n), f(n, x(n)) + \sum_{m=-\infty}^{n} F(n,m)x(m) + p(n)] \\
&\leq [x(n), f(n, x(n)) - f(n,0) + \sum_{m=-\infty}^{n} F(n,m)x(m) - 0] \\
&+ [x(n), f(n,0) + p(n)] \\
&\leq [x(n), f(n, x(n)) - f(n,0) + \sum_{m=-\infty}^{n} F(n,m)x(m)] \\
&+ \|f(n,0) + p(n)\| \\
&\leq e(n)\|x(n)\| + L. \qquad (6)
\end{aligned}$$

Solving this difference inequality, from Lemma 2 and

$$\prod_{i=0}^{n-1}(1 + |e(i)|) \leq \prod_{i=0}^{n-1} \exp(|e(i)|) = \exp(\sum_{i=0}^{n-1} |e(i)|) \leq \beta \exp(-\alpha n),$$

we obtain

$$\|x(n)\| \leq \beta\|x(0)\| + \frac{L\beta}{\alpha^*} \quad n \geq 0, \qquad (7)$$

where $\alpha^* = \dfrac{\exp(\alpha) - 1}{\exp(\alpha)}$.

This implies that $x(n)$ $(n \geq 0)$ is bounded. Since equation (2) is almost periodic equation, from our assumptions, there exists subsequence $\{n_k\}, n_k \to +\infty$ as $k \to +\infty$ such that $x(n + n_k)$ locally uniformly converges to $\hat{x}$ for $n \geq 0$. Moreover, $x(n + n_k)$ is a solution of the following equation

$$\begin{aligned}
\Delta x(n + n_k) &= f(n + n_k, x(n + n_k)) \\
&+ \sum_{m=-\infty}^{n} F(n + n_k, m + n_k)x(m + n_k) + p(n + n_k).
\end{aligned}$$

Let $k \to \infty$, we have

$$\Delta\hat{x}(n) = f(n, \hat{x}(n)) + \sum_{m=-\infty}^{n} F(n,m)\hat{x}(m) + p(n),$$

that is, $\hat{x}(n)$ is a solution of equation (2). On the other hand, $x(n+n_k) \leq \beta\|x(0)\| + \frac{L\beta}{\alpha^*}$ for $n > -n_k$. Let $k \to \infty$ we have $\hat{x}(n) \leq \beta\|x(0)\| + \frac{L\beta}{\alpha^*}$ for $n \in Z$. This means that equation (2) has a bounded solution.
Now, we will show that equation (2) has a unique bounded solution. Suppose that $\hat{x}_1(n)$ and $\hat{x}_2(n)$ for $n \in Z$ are two bounded solutions of equation (2), then we have

$$\begin{aligned}\Delta^+\|\hat{x}_1(n) - \hat{x}_2(n)\| &= [\hat{x}_1(n) - \hat{x}_2(n), \Delta(\hat{x}_1(n) - \hat{x}_2(n))]\\ &\leq e(n)\|\hat{x}_1(n) - \hat{x}_2(n)\|.\end{aligned}$$

Solving this difference inequality, we obtain

$$\|\hat{x}_1(n) - \hat{x}_2(n)\| \leq \|\hat{x}_1(0) - \hat{x}_2(0)\| \exp(\sum_{i=0}^{n-1} e(i)).$$

Then $\|\hat{x}_1(n) - \hat{x}_2(n)\| \to 0$ as $n \to \infty$, this implies that there is a unique bounded solution of equation (2). According to [3,6,7], if there exists an asymptotically almost periodic solution for an almost periodic difference equation, then there exists an almost periodic solution of the equation. Therefore, in order to show that there exists an almost periodic solution of the equation (2), we prove that there exists an asymptotically almost periodic solution of this equation. Suppose that $x(n)$ is a bounded solution of equation (2), then for sequence $\{r_k\}, r_k \to \infty$ as $k \to \infty$, $\{x(r_k)\}$ is a bounded sequence. Thus, there exists a convergent subsequence $\{x(n_k)\}$ of $\{x(r_k)\}$. Consider sequence $\{x(n+n_k)\}$, then we have

$$\begin{aligned}&\Delta^+\|x(n+n_k) - x(n+n_j)\|\\ &\leq [x(n+n_k) - x(n+n_j), \Delta(x(n+n_k) - x(n+n_j))]\\ &\leq [x(n+n_k) - x(n+n_j), f(n+n_k, x(n+n_k))\\ &- f(n+n_j, x(n+n_j)) + \sum_{m=-\infty}^{n} F(n+n_k, m+n_k)x(m+n_k)\\ &- \sum_{m=-\infty}^{n} F(n+n_j, m+n_j)x(m+n_j)] + \|f(n+n_k, 0)\\ &- f(n+n_j, 0)\| + \|p(n+n_k) - p(n+n_j)\|\\ &\leq e(n)\|x(n+n_k) - x(n+n_j)\| + \|f(n+n_k, 0) - f(n+n_j, 0)\|\\ &+ \|p(n+n_k) - p(n+n_j)\|. \qquad (8)\end{aligned}$$

Solving difference inequality (8), we obtain

$$\begin{aligned}&\|x(n+n_k)-x(n+n_j)\|\\ &\quad\leq \beta\|x(n_k)-x(n_j)\|+\frac{\beta}{\alpha^*}\|f(n+n_k,0)-f(n+n_j,0)\|\\ &\quad+ \frac{\beta}{\alpha^*}\|p(n+n_k)-p(n+n_j)\|.\end{aligned}$$

Since sequence $\{x(n_k)\}$ is convergent, so there exists N_1 sufficently large such that for any $\epsilon>0$, if $k,j>N_1$, then

$$\|x(n_k)-x(n_j)\|<\frac{\epsilon}{3\beta}.$$

Note that $f(n,x(n))$ is almost periodic in n uniformly for x, and $p(n)$ is a almost periodic function. Therefore, there exists N_2 sufficently large, such that if $k,j>N_2$, we have $\|f(n+n_k,0)-f(n+n_j,0)\|<\frac{\epsilon\alpha^*}{3\beta}$ and similar, if $k,j>N_3$, we have $\|p(n+n_k)-p(n+n_j)\|<\frac{\epsilon\alpha^*}{3\beta}$. Let $N_0=\max\{N_1,N_2,N_3\}$. Then if $k,j>N_0$, we have

$$\|x(n+n_k)-x(n+n_j)\|<\beta\cdot\frac{\epsilon}{3\beta}+\frac{\beta}{\alpha^*}\cdot\frac{\epsilon\alpha^*}{3\beta}+\frac{\beta}{\alpha^*}\cdot\frac{\epsilon\alpha^*}{3\beta}=\epsilon.$$

This means $x(n)$ is an asymptotically almost periodic solution of equation (2). Note that almost periodic solution is bounded, equation (2) has one and only one bounded solution, thus, there exists a unique almost periodic solution of equation (2). The proof is completed.

Q.E.D.

The next example can be adapt to Theorem by slight modification for equation (2) with term $g(n,x(n-\tau))$ (cf. [4]).

Example. We consider the following almost periodic equation

$$\begin{aligned}\Delta x(n) &= -(\frac{2}{3}+\frac{x(n)\sin n}{\sqrt{1+x^2(n)}})x(n)\\ &+\frac{1}{4}\sum_{m=-\infty}^{n}x(m)e^{-(n-m)}\cos n+\sqrt{|x(n-k_0)|}\sin\sqrt{2}n+2\sin n. \qquad (9)\end{aligned}$$

We do not know whether or not this equation (9) has a bounded solutiona . Even if suppose that equation (9) has a bounded solution, since $\sqrt{|x(n-k_0)|}\sin\sqrt{2}n$ can not be sufficently small as $n\to\infty$, so the Song, Baker and Tian's theorem [6] also fails for equation (9). We

note that $F(n,m) = \frac{1}{4}\exp(-(n-m))\cos n$ and
$\frac{1}{4}\sum_{m=-\infty}^{n} |F(n,m)| < \infty$. Obviously, equation (9) satisfies all assumptions of our theorem, so there exists a unique almost periodic solution of equation (9).

References

[1] C. Feng, On the existence of almost periodic solutions for some nonlinear delayed integrodifferential equations, to appear.

[2] A. M. Fink, Almost Periodic Differential Equations, Lecture Notes in Math., **377**, Springer-Verlag, Berlin, 1974.

[3] Y. Hamaya, Existence of almost periodic solutions in a difference equation with infinite delay, J. Difference Equ. Appl., **9** (2003), 227–237.

[4] Y. Hamaya, On the existence of almost periodic solutions of a nonlinear Volterra difference equation, Int. J. Difference Equ., **2** (2007), 187–196.

[5] S. Kato and M. Imai, Remarks on the existence of almost periodic solutions of systems of nonlinear differential equations, Nonlinear Anal., **25** (1995), 490–415.

[6] Y. Song and H. Tian, Periodic and almost periodic solutions of nonlinear discrete Volterra equations with unbounded delay, J. Comput. Appl. Math., **205** (2007), 859–870.

[7] T. Yoshizawa, Stability theory and the existence of periodic solutions and almost periodic solutions, Applied Mathematical Science, **14**, Springer, 1975.

Department of Information Science
Okayama University of Science
1-1 Ridai-cho, Okayama 700-0005
Japan

E-mail address: `hamaya@mis.ous.ac.jp`

Advanced Studies in Pure Mathematics 53, 2009
Advances in Discrete Dynamical Systems
pp. 67–76

Indeterminacy in a monetary economy with heterogeneous agents

Katsuhiko Hori [1]

§1. Introduction

It is well known that indeterminacy and chaotic behaviour of equilibria can arise in a monetary economy. To the best of our knowledge, the first work analysing the relationship between a monetary economy and the indeterminacy of equilibria is Brock (1974), who shows that there are multiple equilibrium paths in a discrete-time version of a monetary model with a single agent and elastic labour supply. Gray (1984) and Obstfeld (1984) show that indeterminacy of monetary equilibria may arise in a model with a nonseparable utility function in real money holdings and consumption in continuous-time frameworks. In addition, Mino (1984) studies indeterminacy in connection with several endogenized money supply rules. Matsuyama (1991) finds that chaotic behaviour of equilibria also arises in a discrete-time framework. Fukuda (1993) demonstrates that these results also hold in a model with separable utility function. However, all of the above studies mainly concern indeterminacy and chaotic behaviour of equilibria in connection with the monetary policy or preferences of a single agent in an economy. In contrast, this paper focuses on heterogeneity of real asset holdings and its relationship to indeterminacy of monetary equilibria.

The linkage between indeterminacy and heterogeneity of agents has been investigated in several recent studies. Using an overlapping-generations model with heterogeneous agents, Ghiglino and Tvede (1995) show that heterogeneity may generate indeterminacy and cycles. Ghiglino and Olszak-Duquenne (2001) and Ghiglino and Soger (2002) demonstrate that these results also hold in the discrete-time version of a two-sector

---

Received October 12, 2007.

[1] The author is grateful to Takuma Kunieda, John Stachurski, Makoto Yano, and all other participants of the conference, some of whom provided him with helpful comments. Especially, he is grateful to Atsumasa Kondo, Kazuo Nishimura, and Akihisa Shibata for their generous encouragement and suggestions and to anonymous referee for his helpful suggestions.

model with Leontief-type production and in a continuous-time version of a one-sector model with externalities and elastic labour supply[1]. In a similar spirit, this paper investigates indeterminacy of equilibria in connection with wealth distribution in a standard model of money-in-the-utility function.

§2. Model

In the economy, there are J types of household, indexed by $j = 1, \cdots, J$. Each household has additive separable preferences between periods and between goods and money holdings. The households also have the same positive discount factor, denoted by β. In specific terms, the problem to be solved is as follows:

$$\max \quad \sum_{t=0}^{\infty} \beta^t \left(u_j \left(c_{jt}\right) + v_j \left(m_{jt}\right)\right) \qquad j = 1, \cdots, J$$

(1)

$$\text{s.t.} \quad P_t y_j + (1 + i_t) Q_t a_{jt} + M_{jt} + \psi_j X_t = P_t c_{jt} + Q_{t+1} a_{jt+1} + M_{jt+1},$$

where i_t denotes the nominal interest; P_t, the price of goods; Q_t, the price of the capital asset, and X_t, an aggregate nominal transfer to households in period t[2]. Further, ψ_j and y_j denote respectively an exogenous income received by and the share of the nominal transfer to household j, and they are assumed to be independent of periods. Finally, c_{jt} denotes the consumption of goods; M_{jt}, money holdings; m_{jt}, real money holdings, that is, $m_{jt} = M_{jt}/P_t$; and a_{jt}, the non-produced capital asset, such as land, of household j in period t. The capital asset is assumed to be initially supplied to each household at an amount $\theta_j \bar{k}$, or $a_{j0} = \theta_j \bar{k}$, where $\bar{k}$ is the aggregate amount of capital and θ_j is the initial share of the endowment of household j. The capital asset is assumed to yield a fixed outcome of r per unit[3]. This implies that the nominal

[1]See Ghiglino (2005) and Ghiglino and Olszak-Duquenne (2005) for other works studying the linkage between indeterminacy and heterogeneity.

[2]The exsistence of nominal transfers, or negative inflation taxes, will be the source of indeterminacy of equilibria in the model of this paper. In general, the distortion tax, as well as externalities, is known to the one of the sources of indeterminacy.

[3]The assumptions that the exogenous income and the interest rate are ensured by the assumptions that the amounts of labour and the initial endowment of capital are exogenous and they cannot employ other inputs. Suppose that the aggregate production function takes a Cobb=Douglas form: $y = \bar{k}^{\alpha} l^{1-\alpha}$, where

interest rate satisfies $(1+i_t)\,Q_t = P_{t+1}r + Q_{t+1}$. The left-hand side of this equation presents the gross nominal revenue obtained by giving up a unit of the capital asset. The right-hand side of the equation is the gross nominal revenue of a unit of the capital asset since a unit of the capital asset in period t yields r units of outcome in period $t+1$ and its price is Q_{t+1}. It follows that the nominal interest rate can be written as $1+i_t = (r+q_{t+1})\,(1+\pi_t)\,/q_t$, where q_t — the relative price of capital to output — Q_t/P_t and the inflation rate π_t equals $(P_{t+1}-P_t)\,/P_t$. Hence, it follows from (1) that the lifetime budget constraint can be written as

$$(r+q_1)\,P_1\theta_j\bar{k} + \sum_{t=0}^{\infty}\prod_{s=1}^{t}\frac{1}{1+i_s}\,(P_t y_j + M_{jt} + \psi_j X_t)$$

$$(2) \qquad = \sum_{t=0}^{\infty}\prod_{s=1}^{t}\frac{1}{1+i_s}\,(P_t c_{jt} + M_{jt+1})\,,$$

where θ_j denotes the share of household j in the aggregate capital stock. This implies that $\sum_{j=1}^{J} n_j\theta_j = 1$, where n_j is the number of household j. Moreover, we assume that $\prod_{s=1}^{0} 1/(1+i_s) = 1$ for tractability. Therefore, the Lagrangian of this problem can be written as

$$\mathfrak{L}_j = \sum_{t=0}^{\infty}\beta^t\,(u_j\,(c_{jt}) + v_j\,(m_{jt}))$$

(3)

$$+\,\lambda_j\left[(1+q_1)\,P_1\theta_j\bar{k} + \sum_{t=0}^{\infty}\prod_{s=1}^{t}\frac{1}{1+i_s}\,(P_t y_j + M_{jt} + \psi_j X_t - P_t c_{jt} - M_{jt+1})\right].$$

The first-order conditions of this problem are as follows:

$$\beta^t u_j'\,(c_{jt}) = \lambda_j P_t \prod_{s=1}^{t}\frac{1}{1+i_s} \quad \text{and} \quad \beta^t v_j'\,(m_{jt}) = \lambda_j i_{t+1} P_t \prod_{s=1}^{t+1}\frac{1}{1+i_s}.$$

---

y and l are the aggregate amounts of output and labour employed, respectively. In this case, the interest rate, r, and the wage rate, w, are determined independently of periods since the amounts of capital and labour are fixed over periods: $r = \alpha\bar{k}^{\alpha-1}l^{1-\alpha}$ and $w = (1-\alpha)k^{\alpha}l^{-\alpha}$, respectively. The latter of the two equations also implies the income of household j is constant through periods: $y_j = wl_j$, where l_j is the amount of labour supplied by household j.

Moreover, the market-clearing conditions of this economy are

$$(4) \qquad y + r\bar{k} = c_t \qquad \text{and} \qquad M_t = \sum_{j=1}^{J} n_j M_{jt},$$

where y and c_t denote the aggregate amount of income and consumption in period t, respectively; $y = \sum_{j=1}^{J} n_j y_j$ and $c_t = \sum_{j=1}^{J} n_j c_{jt}$. The market-clearing conditions implies Walras's law that yields

$$(5) \qquad X_t = (\mu - 1) M_t,$$

where M_t denotes the aggregate money supply in period t and μ denotes the gross growth rate of money supply; $\mu = M_{t+1}/M_t$. To be well defined the problem, we assume that $\mu > \beta$. This also implies that

$$(6) \qquad \pi_t = m_t / m_{t+1}\mu - 1.$$

We then proceed to consider the steady-state equilibria, the following must hold: $c_j^* = c_{jt}$, $m_j^* = m_{jt}$ and $q^* = q_t$ for all t. It follows from (6) that these conditions imply that $P_{t+1} = \mu P_t$, $Q_{t+1} = \mu Q_t$ and $1 + i^* = (r/q^* + 1)\mu$. Thus, the first-order conditions can be rewritten as

(7)

$$u_j'(c_j^*) = \lambda_j P_0 \left[\frac{1}{\beta\left(\frac{r}{q^*} + 1\right)} \right]^t \quad \text{and} \quad v_j'(m_j^*) = \lambda_j P_0 i^* \left[\frac{1}{\beta\left(\frac{r}{q^*} + 1\right)} \right]^{t+1}.$$

For these two conditions to be well defined, the equality $q^* = r\beta/(1-\beta)$ must hold. Substituting it back into (7), we have

$$(8) \qquad u_j'(c_j^*) = \lambda_j P_0 \qquad \text{and} \qquad v_j'(m_j^*) = \lambda_j P_0 \left(\frac{\mu}{\beta} - 1 \right).$$

Here, these two equations show c_j^* and m_j^* to be decreasing functions of the Lagrange multiplier λ_j. Moreover, it follows from (2) that the consumption of household j in a steady state can be written as

$$(9) \qquad c_j^* = y_j + r\theta_j \bar{k} + (\mu - 1)\left(\psi_j m^* - m_j^*\right),$$

where

$$(10) \qquad m^* = \sum_{j=1}^{J} n_j m_j^*.$$

Moreover, (9) can be rewritten as

$$c_j^* + (\mu - 1)m_j^* = y_j + r\theta_j\bar{k} + (\mu - 1)\psi_j m^*. \tag{11}$$

To ensure the uniqueness of the steady state, we assume that the demand of each household for money holdings is equal to the amount of monetary transfer; thus, $m_j^* = \psi_j m^*$[4]. In this case, (11) is simplified as

$$c_j^* = y_j + r\theta_j\bar{k}. \tag{12}$$

Combining equations (7) and (12), we have

$$\frac{v_j'\left(m_j^*\right)}{u_j'\left(y_j + r\theta_j\bar{k}\right)} = \frac{\mu}{\beta} - 1. \tag{13}$$

This equation implies that the real money holdings of household j are increasing in its income and capital endowment.

§3. Aggregate behaviour

As shown in Negishi (1960) and in Kehoe, Levine, and Romer (1992), the aggregate behaviour of the economy can be characterized by the following problem:

$$\max \sum_{t=0}^{\infty} \beta^t W\left(c_t, m_t; \alpha_1, \cdots, \alpha_J\right)$$

$$\text{s.t.} \quad c_t = y + r\bar{k} + m_t - \left(1 + \pi_t\right) m_{t+1} + \frac{X_t}{P_t},$$

where, $W(\cdot)$ is a Negishi function that is defined as follows:

$$W\left(c_t, m_t; \alpha_1, \cdots, \alpha_J\right) = \max_{\{c_{jt}, m_{jt}\}} \sum_{j=1}^{J} \alpha_j n_j \left(u_j\left(c_{jt}\right) + v_j\left(m_{jt}\right)\right). \tag{14}$$

Here, α_j is the reciprocal of the Lagrange multiplier of household j weighted by its population, $\alpha_j = 1/\left(\lambda_j n_j\right)$. To characterize the Negishi

[4]Equations (10) and (11) determine the value of λ_j, and thus, that of c_j^* and m_j^* from (8). However, these equations imply the possibility of multiple steady states since both hand sides of (10) are decreasing in λ_j from (8) and (10). Although the existence of multiple steady states is an interesting issue, we only address the case of a unique steady state in this paper.

function, we define the Lagrangian of this problem as

$$\mathcal{L}_t = \sum_{j=1}^{J} \alpha_j n_j \left(u_j\left(c_{jt}\right) + v_j\left(m_{jt}\right)\right)$$

$$+ \lambda_{ct}\left(c_t - \sum_{j=1}^{J} n_j c_{jt}\right) + \lambda_{mt}\left(m_t - \sum_{j=1}^{J} n_j m_{jt}\right). \tag{15}$$

The necessary conditions of this problem can be written as

$$\alpha_j n_j u_j'\left(c_{jt}^*\right) = \lambda_{ct} \quad \text{and} \quad \alpha_j n_j v_j'\left(m_{jt}^*\right) = \lambda_{mt} \ \text{ for all } j, \tag{16}$$

where c_{jt}^* and m_{jt}^* denote the optimal consumption and money holdings of household j, respectively[5]. Since, from (16), both c_{jt} and m_{jt} are monotonically decreasing in λ_{ct} and λ_{mt}, respectively, c_{jt} and m_{jt} are uniquely determined if c_t and m_t are given. Therefore, we define Negishi functions for goods and money holdings as $\hat{u}\left(c_t\right) = \sum_{j=1}^{J} \alpha_j n_j u_j\left(c_{jt}^*\right)$ and $\hat{v}\left(m_t\right) = \sum_{j=1}^{J} \alpha_j n_j v_j\left(m_{jt}^*\right)$, respectively. Using these notations, we can express the Bellman equation for the intertemporal problem as[6]

$$\hat{V}_t\left(m_t\right) = \max\ \hat{u}\left(y + r\bar{k} + m_t - \left(1 + \pi_t\right) m_{t+1} + \frac{X_t}{P_t}\right) + \hat{v}\left(m_t\right) + \beta \hat{V}_{t+1}\left(m_{t+1}\right), \tag{17}$$

where $\hat{V}_t$ is the social value function. Therefore, the necessary and envelope conditions for maximization yields

$$\hat{u}'\left(c_t\right) - \beta \frac{P_t}{P_{t+1}}\left(\hat{u}'\left(c_{t+1}\right) + \hat{v}'\left(m_{t+1}\right)\right) = 0. \tag{18}$$

Using (4), (5) and (6), equation (18) can be rewritten as

$$m_t = \frac{\beta}{\mu} m_{t+1}\left(1 + \frac{\hat{v}'\left(m_{t+1}\right)}{\hat{u}'\left(y + r\bar{k}\right)}\right). \tag{19}$$

---

[5]Note that the first-order conditions (16) are the same as the above equations of (4) if $\lambda_{ct} = \beta^{-t} P_t \prod_{s=1}^{t} 1/\left(1 + i_s\right)$ and $\lambda_{mt} = \beta^{-t} P_t i_{t+1} \prod_{s=1}^{t+1} 1/\left(1 + i_s\right)$, which imply that the solution of the problem in this section represents equilibria of the market economy considered in Section 2.

[6]As in Kehoe, Levine, and Romer (1992) and in Ghiglino and Olszak-Duquenne (2001), we can call the solution of this problem to be *pseudo*-Pareto optimum in the sense that it is the solution to the maximization of a Negishi function under given nominal transfers.

The above equation describes the aggregate behavior of the real stock of money. Moreover, in a steady state, since the amount of real money holdings is constant over periods, (19) can be rewritten as follows:

$$\frac{\hat{v}'(m^*)}{\hat{u}'(y + r\bar{k})} = \frac{\mu}{\beta} - 1. \tag{20}$$

This equation determines the aggregate real stock of money in a steady state.

§4. Indeterminacy

This section considers a condition for indeterminacy of equilibria. From (19), we know equilibria are locally indeterminate if the absolute value of the gradient of the right-hand side of (19) with respect to m_{t+1} is greater than 1 around the steady state. Therefore, it follows from (20) that this condition can be written as

$$\hat{\eta}\,(m^*) < \frac{1}{2}\left(1 - \frac{\beta}{\mu}\right), \tag{21}$$

where $\hat{\eta}$ denotes a social intertemporal elasticity of substitution in money holdings; that is, $\hat{\eta}\,(m^*) = -\hat{v}'\,(m^*)\,/\,(\hat{v}''\,(m^*)\,m^*)$. Thus, (21) suggests that a lower social intertemporal elasticity of substitution tends to generate indeterminacy. To investigate this condition in greater detail, we calculate the social intertemporal elasticity of substitution in a manner similar to that of Ghiglino (2005), that yields

$$\hat{\eta}\,(m^*) = \sum_{j=1}^{J} n_j \eta_j\left(m_j^*\right)\frac{m_j^*}{m^*}, \tag{22}$$

where $\eta_j\left(m_j^*\right)$ denotes the intertemporal elasticity of substitution of the individual utility of money holdings: $\eta_j\left(m_j^*\right) = -v_j'\left(m_j^*\right)/\left(m_j^* v_j''\left(m_j^*\right)\right)$. This equation, together with (21), implies that wealth distribution may cause the indeterminacy of monetary equilibria since (13) suggests that m_j^* depends on the distribution of income and the initial shares of the capital asset among agents.

§5. Examples

Here, we present a few examples with specific forms of utility functions. In the following examples, we assume throughout that there are only two types of households with the same population 1/2 and income flow y and that their utility functions with respect to consumption take an identical logarithmic form: $u_j\,(c_{jt}) = \ln c_{jt}$.

■ CIES Utility When $v_j(m_{jt}) = \eta_j / (\eta_j - 1) m_{jt}^{\eta_j/(\eta_j - 1)}$, the social intertemporal elasticity of substitution can be written as

$$\hat{\eta}_{\text{CIES}} = \frac{\sum_{j=1}^{2} \left(\frac{\mu}{\beta} - 1 \right)^{\eta_j - 1} \left(y + r\theta_j \bar{k} \right)^{1-\eta_j} \eta_j}{\sum_{j=1}^{2} \left(\frac{\mu}{\beta} - 1 \right)^{\eta_j - 1} \left(y + r\theta_j \bar{k} \right)^{1-\eta_j}}.$$

The above expression implies that the difference of the individual intertemporal elasticity of substitutions plays an important role in indeterminacy of monetary equilibria. However, the heterogeneneity of the initial share of capital asset holdings plays no role in the occurrence of indeterminacy if the individual intertemporal elasticity of substitution is identical over households.

■ CARA Utility In contrast to the first example, the following two examples are more interesting since wealth distribution has a crucial role in indeterminacy even if the utility functions are identical across households. When $v_j(m_{jt}) = -1/a \exp(-am_{jt})$, it can be written as

$$\hat{\eta}_{\text{CARA}} = \frac{1}{\frac{1}{2} \sum_{j=1}^{2} \ln \left(y + r\theta_j \bar{k} \right) - \ln \left(\frac{\mu}{\beta} - 1 \right)}.$$

In this case, wealth distribution is crucial even if the preferences of agents are identical. Figure 1 illustrates that indeterminacy tends to arise in a highly egalitarian economy.

■ Quadratic Utility When $v_j(m_{jt}) = -b/2 \left(m_{jt} - \bar{m} \right)^2$, it can be written as

$$\hat{\eta}_{\text{QD}} = \frac{1}{\frac{\beta}{\mu - \beta} b\bar{m} \left(\sum_{j=1}^{2} \frac{1}{y_j + r\theta_j k} \right)^{-1} - 1}.$$

This case also derives a result similar to that in the case of CARA utility.

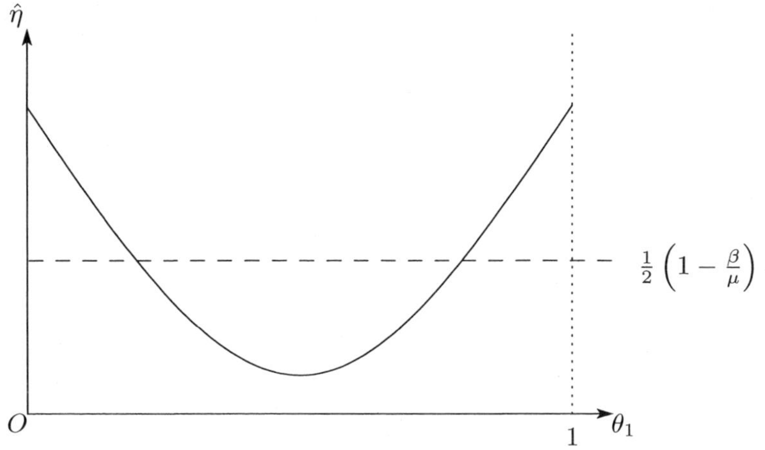

Fig. 1. CARA and Quadratic Cases

References

[1] W. A. Brock, Money and growth: The case of long-run perfect foresight, Internat. Econom. Rev., **15** (1974), 750–777.

[2] S. Fukuda, The emergence of equilibrium cycles in a monetary economy with a separable utility function, J. Monetary Economics, **32** (1993), 321–334.

[3] C. Ghiglino, Wealth inequality and dynamic stability, J. Econom. Theory, **124** (2005), 106–115.

[4] C. Ghiglino and M. Olszak-Duquenne, Inequalities and fluctuations in a dynamic general equilibrium model, Econom. Theory, **17** (2001), 1–24.

[5] C. Ghiglino and M. Olszak-Duquenne, On the impact of heterogeneity on indeterminacy, Internat. Econom. Rev., **46** (2005), 171–188.

[6] C. Ghiglino and G. Sorger, Poverty traps, indeterminacy, and the wealth distribution, J. Econom. Theory, **105** (2003), 120–139.

[7] C. Ghiglino and M. Tvede, Endowments, stability, and fluctuations in OG models, J. Econom. Dynam. Control, **19** (1995), 621–653.

[8] J. A. Gray, Dynamic instability in rational expectation model: An attempt to clarify, Internat. Econom. Rev., **25** (1984), 93–122.

[9] T. J. Kehoe, D. K. Levine and P. M. Romer, On characterizing equilibria of economies with externalities and taxes as solutions to optimization problems, Econom. Theory, **2** (1992), 43–68.

[10] K. Matsuyama, Endogenous price fluctuations in an optimizing model of a monetary economy, Econometrica, **59** (1991), 1617–1631.

[11] K. Mino, Implications of endogenous money supply rules in dynamic models with perfect foresight, J. Macroeconomics, **11** (1989), 181–197.

[12] T. Negishi, Welfare economics and existence of an equilibrium for a competitive economy, Metroeconomica, **12** (1960), 92–97.
[13] M. Obstfeld, Multiple stable equilibria in an optimizing perfect-foresight model, Econometrica, **52** (1984), 223–228.

Institute of Economic Research
Kyoto University
Yoshida Hon-machi, Sakyo-ku
Kyoto, 606-8501
Japan

E-mail address: hori@kier.kyoto-u.ac.jp

Advanced Studies in Pure Mathematics 53, 2009
Advances in Discrete Dynamical Systems
pp. 77–86

Golden optimal path in discrete-time dynamic optimization processes

Seiichi Iwamoto and Masami Yasuda

Abstract.

We are concerned with dynamic optimization processes from a viewpoint of Golden optimality. A path is called Golden if any state moves to the next state repeating the same Golden section in each transition. A policy is called Golden if it, together with a relevant dynamics, yields a Golden path. The probelm is whether an optimal path/policy is Golden or not. This paper minimizes a quadratic criterion and maximizes a square-root criterion over an infinite horizon. We show that a Golden path is optimal in both optimizations. The Golden optimal path is obtained by solving a corresponding Bellman equation for dynamic programming. This in turn admits a Golden optimal policy.

§1. Introduction

Recently the Golden optimal solution, its duality, and its equivalence have been discussed in static optimization problems [4, 5, 6]. In this paper we consider the Golden optimal solution in dynamic optimization problems.

We consider two typical types of criterion — quadratic and square-root — in a deterministic optimization. We minimize quadratic criteria

$$I(x) = \sum_{n=0}^{\infty} \left[x_n^2 + (x_n - x_{n+1})^2\right], \quad J(x) = \sum_{n=0}^{\infty} \left[(x_n - x_{n+1})^2 + x_{n+1}^2\right]$$

and maximize square-root criteria

$$K(x) = \sum_{n=0}^{\infty} \beta^n \left(\sqrt{x_n} + \sqrt{x_n - x_{n+1}}\right),$$

$$L(x) = \sum_{n=0}^{\infty} \beta^n \left(\sqrt{x_n - x_{n+1}} + \sqrt{x_{n+1}}\right),$$

Received October 12, 2007.

respectively. Here β is $0 < \beta < 1$. The differences between I and J and between K and L are

$$J(x) = I(x) - x_0^2$$

$$L(x) = K(x) + \sum_{n=0}^{\infty} \left(\beta^{n-1} - \beta^n\right) \sqrt{x_n} \; (\beta^{-1} = 0).$$

We show that a Golden path is optimal in these four optimization problems. The Golden optimal path is obtained by solving Bellman equation for dynamic programming [3, 7].

§2. Golden paths

A real number

$$\phi = \frac{1+\sqrt{5}}{2} \approx 1.618$$

is called *Golden number* [1, 2, 8]. It is the larger of the two solutions to quadratic equation

$$x^2 - x - 1 = 0. \tag{1}$$

Sometimes (1) is called *Fibonacci quadratic equation.* The Fibonacci quadratic equation has two real solutions: ϕ and its *conjugate* $\overline{\phi} := 1 - \phi$. We note that

$$\phi + \overline{\phi} = 1, \; \phi \cdot \overline{\phi} = -1.$$

Further we have

$$\phi^2 = 1 + \phi, \; \overline{\phi}^2 = 2 - \phi$$
$$\phi^2 + \overline{\phi}^2 = 3, \; \phi^2 \cdot \overline{\phi}^2 = 1.$$

A point $(2-\phi)x$ splits an interval $[0, x]$ into two intervals $[0, (2-\phi)x]$ and $[(2-\phi)x, x]$. A point $(\phi-1)x$ splits the interval into $[0, (\phi-1)x]$ and $[(\phi-1)x, x]$. In either case, the length constitutes the Golden ratio $(2-\phi) : (\phi-1) = 1 : \phi$. Thus both divisions are the *Golden section.*

Definition 2.1. *A sequence* $x : \{0, 1, \ldots\} \to R^1$ *is called Golden if and only if either*

$$\frac{x_{t+1}}{x_t} = \phi - 1 \; or \; \frac{x_{t+1}}{x_t} = 2 - \phi.$$

Lemma 2.1. *A Golden sequence* x *is either*

$$x_t = x_0(\phi-1)^t \; or \; x_t = x_0(2-\phi)^t.$$

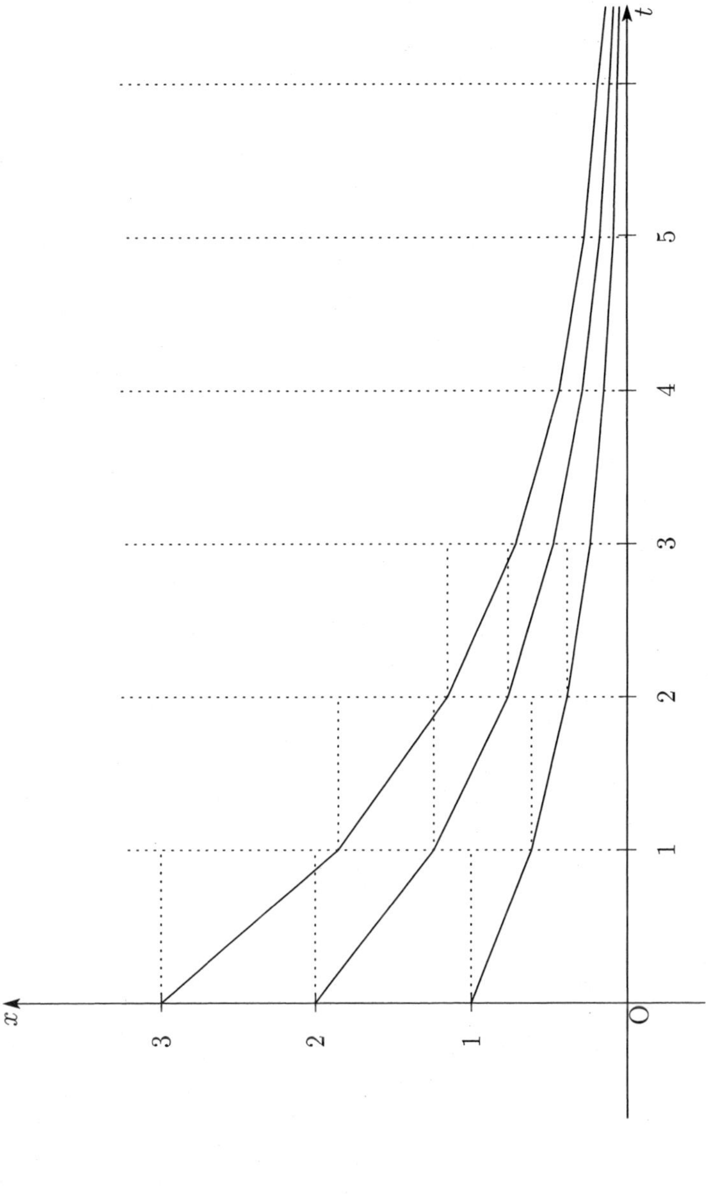

Fig. 1 Golden paths $x = c(\phi - 1)^t \; c = 1, 2, 3$

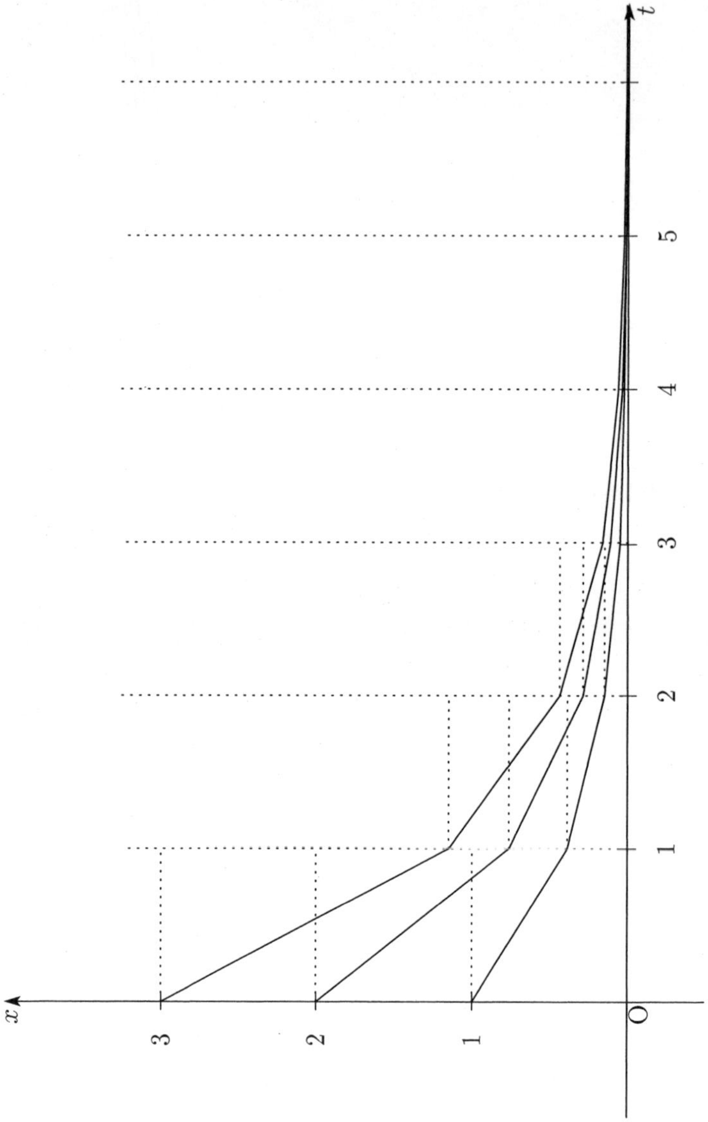

Fig. 2 Golden paths (c) $x = c(2 - \phi)^t$ $c = 1, 2, 3$

We remark that

$$(\phi - 1)^t = \phi^{-t},\ (2 - \phi)^t = (1 + \phi)^{-t}$$

where

$$\phi - 1 = \phi^{-1} \approx 0.618,\ 2 - \phi = (1 + \phi)^{-1} \approx 0.382$$

Let us introduce a controlled linear dynamics with real parameter b as follows.

$$x_{t+1} = bx_t + u_t \ t = 0, 1, \ldots \tag{2}$$

where $u : \{0, 1, \ldots\} \to R^1$ is called *control.* If $u_t = px_t$, the control u is called *proportional,* where p is a real constant, *proportional rate.* A sequence x satisfying (2) is called *path.*

Definition 2.2. *A proportional control u on dynamics (2) is called Golden if and only if it generates a Golden path x.*

Lemma 2.2. *A proportional control $u_t = px_t$ on (2) is Golden if and only if*

$$p = -b + \phi - 1 \ \textit{or} \ p = -b + 2 - \phi. \tag{3}$$

§3. Control processes

This section minimizes two quadratic cost functions

$$\sum_{n=0}^{\infty} \left[x_n^2 + (x_n - x_{n+1})^2\right] \text{ and } \sum_{n=0}^{\infty} \left[(x_n - x_{n+1})^2 + x_{n+1}^2\right].$$

Both problems are solved as a control process with criterion

$$\sum_{n=0}^{\infty} \left(x_n^2 + u_n^2\right) \text{ and } \sum_{n=0}^{\infty} \left(u_n^2 + x_{n+1}^2\right)$$

under a common additive dynamics with a given initial state

$$x_{n+1} = bx_n + u_n,\ x_0 = c$$

where $c \in R^1$.

3.1. Quadratic in current state

Let us now consider a control process with an additive transition $T(x,u) = bx + u$. Here b is a constant, which represents a characteristics of the process :

$$\mathrm{C}(c) \qquad \begin{array}{lll} \text{minimize} & \displaystyle\sum_{n=0}^{\infty} (x_n^2 + u_n^2) & \\ \text{subject to} & \text{(i)} \quad x_{n+1} = bx_n + u_n & \\ & \text{(ii)} \quad -\infty < u_n < \infty & n \geq 0 \\ & \text{(iii)} \quad x_0 = c. & \end{array}$$

Let $v(c)$ be the minimum value of C(c). Then the value function v satisfies Bellman equation [3]:

$$(4) \qquad v(x) = \min_{-\infty<u<\infty} \left[x^2 + u^2 + v(bx + u) \right].$$

Eq. (4) has a quadratic form $v(x) = vx^2$, where $v \in R^1$.

Lemma 3.1. *The control process* C(c) *with characteristic value* b ($\in R^1$) *has a proportional optimal policy* $f^\infty, f(x) = px$, *and a quadratic minimum value function* $v(x) = vx^2$, *where*

$$v = \frac{b^2 + \sqrt{b^4 + 4}}{2}, \quad p = -\frac{v}{1+v} b.$$

The proportional optimal policy f^∞ splits at any time an interval $[0, x]$ into $[0, (b+p)x] = \left[0, \dfrac{bx}{1+v}\right]$ and $\left[\dfrac{bx}{1+v}, x\right]$. In particular, when $b = 1$, the quadratic coefficient v is reduced to the *Golden number*

$$\phi = \frac{1 + \sqrt{5}}{2} \approx 1.618$$

Further the division of $[0, x]$ into $\left[0, \dfrac{x}{1+\phi}\right]$ and $\left[\dfrac{x}{1+\phi}, x\right]$ is Golden. A quadratic function $w(x) = ax^2$ is called *Golden* if $a = \phi$.

Theorem 3.1. *The control process* C(c) *with characteristic value* $b = 1$ *has a Golden optimal policy* $f^\infty, f(x) = (1-\phi)x$, *and the Golden quadratic minimum value function* $v(x) = \phi x^2$.

3.2. Qquadratic in next state

Here we consider the cost function $r : X{\times}U \to R^1$ which is quadratic in current control and next state :

$$r(x,u) \;=\; u^2 + (bx+u)^2.$$

Then a control process is represented by the following sequential minimization problem :

$$\mathrm{C}'(c)\qquad \begin{array}{lll} \text{minimize} & \sum_{n=0}^{\infty}\left(u_n^2 + x_{n+1}^2\right) & \\ \text{subject to} & \text{(i)}\quad x_{n+1} = bx_n + u_n & \\ & \text{(ii)}\quad -\infty < u_n < \infty & n \ge 0 \\ & \text{(iii)}\quad x_0 = c. & \end{array}$$

The value function v satisfies Bellman equation [3]:

$$\text{(5)}\qquad v(x) \;=\; \min_{-\infty<u<\infty}\left[u^2 + (bx+u)^2 + v(bx+u)\right].$$

Eq. (5) has a quadratic solution $v(x) = vx^2$, where $v \in R^1$.

Lemma 3.2. *The control process* $\mathrm{C}'(c)$ *with characteristic value* b *has a proportional optimal policy* $f^\infty, f(x) = px$, *and a quadratic minimum value function* $v(x) = vx^2$, *where*

$$v \;=\; \frac{b^2 - 2 + \sqrt{b^4+4}}{2}, \quad p \;=\; -\frac{1+v}{2+v}b.$$

The policy f^∞ splits an interval $[0,\ x]$ into $\left[0,\ \dfrac{bx}{2+v}\right]$ and $\left[\dfrac{bx}{2+v},\ x\right]$. When $b = 1$, the coefficient v is reduced to the *inverse Golden number*

$$\phi^{-1} = \phi - 1 = \frac{-1+\sqrt{5}}{2} \approx 0.618$$

Further the division of $[0,\ x]$ into $[0,\ (2-\phi)x]$ and $[(2-\phi)x,\ x]$ is Golden. A quadratic function $w(x) = ax^2$ is called *inverse Golden* if $a = \phi^{-1}$.

Theorem 3.2. *The control process* $\mathrm{C}'(c)$ *with characteristic value* $b = 1$ *has a Golden optimal policy* $f^\infty, f(x) = (1-\phi)x$, *and the inverse Golden quadratic minimum value function* $v(x) = (\phi - 1)x^2$.

§4. Allocation processes

This section maximizes two discounted square-root reward functions

$$\sum_{n=0}^{\infty} \beta^n \left(\sqrt{x_n} + \sqrt{x_n - x_{n+1}}\right) \text{ and } \sum_{n=0}^{\infty} \beta^n \left(\sqrt{x_n - x_{n+1}} + \sqrt{x_{n+1}}\right).$$

Both problems are solved as an allocation process with criterion

$$\sum_{n=0}^{\infty} \beta^n \left(\sqrt{x_n} + \sqrt{u_n}\right) \text{ and } \sum_{n=0}^{\infty} \beta^n \left(\sqrt{u_n} + \sqrt{x_{n+1}}\right)$$

under a common subtractive dynamics with a given initial state

$$x_{n+1} = x_n - u_n, \; x_0 = c$$

where $c \geq 0$.

4.1. Square-root in current state

Let us now consider an allocation process with a subtractive transition $T(x, u) = x - u$:

$$\mathrm{A}(c) \quad \begin{array}{ll} \text{Maximize} & \displaystyle\sum_{n=0}^{\infty} \beta^n \left(\sqrt{x_n} + \sqrt{u_n}\right) \\ \text{subject to} & \text{(i)} \quad x_{n+1} = x_n - u_n \\ & \text{(ii)} \quad 0 \leq u_n \leq x_n \\ & \text{(iii)} \quad x_0 = c. \end{array} \quad n \geq 0$$

Let $v(c)$ be the maximum value of $\mathrm{A}(c)$. Then the maximum value function v satisfies the following Bellman equation:

$$(6) \qquad v(x) = \max_{0 \leq u \leq x} \left[\sqrt{x} + \sqrt{u} + \beta v(x - u)\right].$$

Eq. (6) has a square-root form $v(x) = v\sqrt{x}$, where $v \in R^1$.

Let us adopt a proportional policy f^∞ $(f(x) = px)$ with proportional rate p $(0 < p < 1)$. Then state x under the control $u = px$ goes deterministically to the next state $T(x, u) = x - u = x - px = (1 - p)x$. Thus we have $x = (1 - p)x + px$. The state transition of control process $\mathrm{A}(c)$ governed by the proportional policy f^∞ means that the current control $u = px$ splits the state interval $[0, x]$ into two intervals $[0, (1 - p)x]$ and $[(1 - p)x, x]$. When the split yields a *Golden section*, the proportional policy f^∞ $(f(x) = px)$ is called *Golden*.

Lemma 4.1. *The allocation process* A(c) *has a proportional optimal policy* $f^\infty, f(x) = px$, *and a square-root maximum value function* $v(x) = v\sqrt{x}$, *where*

$$v = \frac{2}{1-\beta^2}, \; p = \frac{(1-\beta^2)^2}{(1+\beta^2)^2}.$$

We remark that the coefficient v is the solution to

$$v = 1 + \sqrt{1 + (\beta v)^2}, \; v \geq 2.$$

Let us solve $1 - p = \phi - 1$ or $2 - \phi$. Then we have the following result.

Theorem 4.1. *When* $\beta = \phi(1 - \sqrt{\phi - 1}) \approx 0.346$ *or* $\beta = \sqrt{\phi} - \sqrt{\phi - 1} \approx 0.486$, *the proportional policy* $f^\infty, f(x) = px$, *is Golden optimal.*

4.2. Square-root in next state

Now we consider an allocation process with transition $T(x, u) = x - u$:

$$\mathrm{A}'(c) \qquad \begin{array}{ll} \text{Maximize} & \displaystyle\sum_{n=0}^{\infty} \beta^n \left(\sqrt{u_n} + \sqrt{x_{n+1}}\right) \\ \text{subject to} & \text{(i)} \quad x_{n+1} = x_n - u_n \qquad\qquad n \geq 0 \\ & \text{(ii)} \quad 0 \leq u_n \leq x_n \\ & \text{(iii)} \quad x_0 = c. \end{array}$$

Let $v(c)$ be the maximum value of A$'(c)$. Then the maximum value function v satisfies an optimality equation:

$$v(x) = \max_{0 \leq u \leq x} \left[\sqrt{u} + \sqrt{x-u} + \beta v(x-u)\right]. \tag{7}$$

Eq. (7) has a square-root solution $v(x) = v\sqrt{x}$, where $v \in R^1$.

Let us adopt a proportional policy f^∞ ($f(x) = px$) with p ($0 < p < 1$). Then the current control $u = px$ splits the interval $[0,\ x]$ into $[0,\ (1-p)x]$ and $[(1-p)x,\ x]$.

Lemma 4.2. *The allocation process* A$'(c)$ *has a proportional optimal policy* $f^\infty, f(x) = px$, *and a square-root maximum value function* $v(x) = v\sqrt{x}$, *where*

$$v = \frac{\beta + \sqrt{2-\beta^2}}{1-\beta^2}, \; p = \frac{1 - \beta\sqrt{2-\beta^2}}{2}.$$

Note that the coefficient v is the positive solution to

$$v = \sqrt{1 + (1 + \beta v)^2}.$$

By solving $1 - p = \phi - 1$, we have the following result.

Theorem 4.2. *When* $\beta = \sqrt{1 - 2\sqrt{2\phi - 3}} \approx 0.168$, *the proportional policy* $f^\infty, f(x) = px$, *is Golden optimal.*

References

[1] A. Beutelspacher and B. Petri, Der Goldene Schnitt 2., überarbeitete und erweiterte Auflange, Elsevier GmbH, Spectrum Akademischer Verlag, Heidelberg, 1996.

[2] R. A. Dunlap, The Golden Ratio and Fibonacci Numbers, World Scientific Publ. Co. Pte. Ltd., 1977.

[3] S. Iwamoto, Theory of Dynamic Program: Japanese, Kyushu Univ. Press, Fukuoka, 1987.

[4] S. Iwamoto, Cross dual on the Golden optimum solutions, Math. Econom., Res. Inst. Math. Sci., Kyoto Univ., RIMS Kôkyûroku, **1443**, 2005, pp. 27–43.

[5] S. Iwamoto, The Golden optimum solution in quadratic programming, In: Proceedings of the International Conference on Nonlinear Analysis and Convex Analysis, Okinawa, 2005, (eds. W. Takahashi and T. Tanaka), Yokohama Publ., Yokohama, 2007, pp.109–205.

[6] S. Iwamoto, The Golden trinity — optimility, inequality, identity —, Math. Econom., Res. Inst. Math. Sci., Kyoto Univ., RIMS Kôkyûroku, **1488**, 2006, pp. 1–14.

[7] S. Iwamoto and M. Yasuda, Dynamic programming creates the golden ratio, too, Decision-Making and Mathematical Model under Uncertainty, Res. Inst. Math. Sci., Kyoto Univ., RIMS Kôkyûroku, **1477**, 2006, pp. 136–140.

[8] H. Walser, Der Goldene Schnitt, B. G. Teubner, Leibzig, 1996.

Seiichi Iwamoto
Department of Economic Engineering
Graduate School of Economics
Kyushu University
Fukuoka 812-8581, Japan

Masami Yasuda
Department of Mathematical Science
Faculty of Science, Chiba University
Chiba 263-8522, Japan

E-mail address: `iwamoto@en.kyushu-u.ac.jp`
`yasuda@math.s.chiba-u.ac.jp`

Advanced Studies in Pure Mathematics 53, 2009
Advances in Discrete Dynamical Systems
pp. 87–94

Cannibalism in a discrete predator-prey model with an age structure in the prey

Sophia R. Jang

Abstract.

A discrete-time predator-prey model with an age structure in the prey population in which both the predator and the adult prey can consume juvenile prey population is proposed. The fecundity of the adult prey population is assumed to be a constant and the juvenile survival probability is density dependent on its own population size. It is shown that both populations go to extinction if the inherent net reproductive number of the prey is less than 1. Both populations can survive if the prey's inherent net reproductive number is larger than one and the predator's reproductive number is greater than 1 when the prey population is stabilized in the steady state fashion.

§1. Model derivation and analysis

Cannibalism has been observed in a variety of species, including some fish species such as adult Atlantic cod which eat large numbers of their young [1]. On the other hand, the adult cod is also preyed upon by some fish species such as killer whales and sharks. It has been demonstrated mathematically that cannibalism can have some effects on population dynamics [2, 3, 5, 11, 12]. Kohlmeier and Ebenhoh [10] illustrated that cannibalism in the prey can increase equilibrium population sizes for a predator-prey system modeled by ordinary differential equations. See also [8, 9] and references cited therein for continuous-time predator-prey models with cannibalism.

In this manuscript we propose a simple discrete-time predator-prey model, where each prey individual is classified as either juvenile or adult, and the adults may prey on its own juveniles. Let $x_1(t), x_2(t)$ and $p(t)$ denote the juvenile prey, adult prey, and predator population sizes at

---

Received October 12, 2007.
2000 *Mathematics Subject Classification.* 92D25.
Key words and phrases. Cannibalism, inherent net reproductive number, Liapunov-Schmidt expansion, uniform persistence.

time t, respectively. The probability that a juvenile will escape from being cannibalized is e^{-cx_2} when the adult prey population is of size x_2, where $c > 0$ may be interpreted as the cannibalism rate. It is assumed that only the adult prey population can reproduce with a constant birth rate $b > 0$. The transition probability from juvenile to adult is density dependent on the juvenile prey population and is modeled by a Bervton-Holt type equation.

In addition, the predator also preys on the juvenile prey but not on the adult prey population. Let β be the average number of offspring that a predator can reproduce from a consumed juvenile prey, and the probability that an individual juvenile prey will escape from the predator when the predator population is of size p is e^{-ap}, where the searching efficiency a is a constant. Under the above biological assumptions, our model of predator-prey interaction can be written in terms of the following system of difference equations:

$$
\begin{aligned}
x_1(t+1) &= bx_2(t)e^{-cx_2(t)} \\
x_2(t+1) &= \frac{\alpha x_1(t)}{1+mx_1(t)}e^{-ap(t)} \\
p(t+1) &= \beta x_1(t)(1-e^{-ap(t)}) \\
& x_1(0), x_2(0), p(0) \geq 0,
\end{aligned} \tag{1}
$$

where $b, a, c, m, \beta > 0$ and $0 < \alpha < 1$.

The following result shows that the model is biologically sound and there exists a threshold below which populations become extinct. The proof is straightforward and is therefore omitted.

Lemma 1.1. *Solutions of (1) remain nonnegative and are bounded. Moreover, solutions* $(x_1(t), x_2(t), p(t))$ *of (1) satisfy* $\lim_{t\to\infty}(x_1(t), x_2(t), p(t)) = (0, 0, 0)$ *if* $\alpha b < 1$.

The threshold αb can be interpreted as the inherent net reproductive number of the prey population. It is the average number of offspring that a juvenile prey individual can reproduce during its lifetime when the prey population is very small. Denote $n = \alpha b$. Then both populations go extinct if $n < 1$ by the above lemma.

A simple calculation shows that an x_2-component of a nontrivial steady state $(x_1, x_2, 0)$ with the absence of the predator must satisfy the following equation:

$$\frac{ne^{-cx}}{1+m\frac{n}{\alpha}xe^{-cx}} = 1. \tag{2}$$

Since the left hand side of (2) as a function of x is decreasing with value n at $x = 0$, (2) has a positive solution $\bar{x}_2$ if and only if $n > 1$. In which case the positive solution is unique. Consequently, (1) has a unique biologically feasible steady state $E_1 = (\bar{x}_1, \bar{x}_2, 0)$ if $n > 1$, where $\bar{x}_2$ satisfies (2) and $\bar{x}_1 = b\bar{x}_2 e^{-c\bar{x}_2}$. The linearization of system (1) yields the following Jacobian matrix:

$$(3) \quad J = \begin{pmatrix} 0 & be^{-cx_2} - cbx_2e^{-cx_2} & 0 \\ \dfrac{\alpha}{(1+mx_1)^2} & 0 & \dfrac{-a\alpha x_1}{1+mx_1} \\ 0 & 0 & a\beta x_1 \end{pmatrix}.$$

The local stability of E_1 then depends on the eigenvalues of J evaluated at E_1 [4], $J(E_1)$, which consists of $a\beta\bar{x}_1$ and the eigenvalues of the submatrix

$$(4) \qquad \hat{J} = \begin{pmatrix} 0 & be^{-cx_2} - cbx_2e^{-cx_2} \\ \dfrac{\alpha}{(1+mx_1)^2} & 0 \end{pmatrix}$$

evaluated at $(\bar{x}_1, \bar{y}_1)$. Notice $\hat{J}$ is the Jacobian matrix of the x_1x_2-subsystem of (1):

$$(5) \qquad \begin{aligned} &x_1(t+1) = bx_2(t)e^{-cx_2(t)} \\ &x_2(t+1) = \frac{\alpha x_1(t)}{1+mx_1(t)} \\ &x_1(0), x_2(0) \geq 0, \end{aligned}$$

where $(\bar{x}_1, \bar{x}_2)$ is the interior steady state of (5). In the following we show that steady state $(\bar{x}_1, \bar{x}_2)$ is locally asymptotically stable for (5) when $n > 1$ but sufficiently close to 1. We shall use the Liapunov-Schmidt expansion method as given in [2]. Let $\epsilon = n - 1$ be positive and every small. Since $(\bar{x}_1, \bar{x}_2) \to (0, 0)$ as $\epsilon \to 0^+$, we have the following expansion

$$\begin{aligned} x_1(\epsilon) &= x_{11}\epsilon + x_{12}\epsilon^2 + O(\epsilon^3) \\ x_2(\epsilon) &= x_{21}\epsilon + x_{22}\epsilon^2 + O(\epsilon^3) \\ n(\epsilon) &= 1 + n_1\epsilon + n_2\epsilon^2 + O(\epsilon^3) \end{aligned}$$

for $\epsilon > 0$ sufficiently small. Substituting these expressions into the equilibrium equations and equating similar terms with respect to ϵ we arrive at

$$x_{11} = \frac{1}{\alpha}x_{21} \text{ and } x_{21} = \alpha x_{11}$$

and hence $x_{11} = 1$ and $x_{21} = \alpha$, up to scalar multiplications. Similarly, equating like terms of ϵ^2, we have

$$x_{12} = (-bx_{21}^2 + x_{22} + x_{21}n_1)/\alpha \text{ and } x_{22} = \alpha x_{12} - mx_{11}x_{21}.$$

Therefore $n_1 = m + c\alpha$ and thus

$$\begin{aligned} x_1(\epsilon) &= \epsilon + O(\epsilon^2) \\ x_2(\epsilon) &= \alpha\epsilon + O(\epsilon^2) \\ n(\epsilon) &= 1 + (m + c\alpha)\epsilon + O(\epsilon^2) \end{aligned} \tag{6}$$

for $\epsilon > 0$ small. Notice (6) corresponds to the x_1, x_2-components of an interior steady state for system (5) when $\epsilon > 0$ is sufficiently small, i.e., when $n > 1$ is sufficiently close to 1. The Jacobian matrix of system (5) evaluated at $(\bar{x}_1, \bar{x}_2)$ can be rewritten in terms of n and can be expressed as $\hat{J}(\bar{x}_1, \bar{x}_2) = \hat{J}_1 + \hat{J}_2\epsilon + O(\epsilon^2)$. Since

$$\frac{n}{\alpha}e^{-c\bar{x}_2} - \frac{bn}{\alpha}e^{-c\bar{x}_2} = \frac{1}{\alpha} + \frac{m - \alpha c}{\alpha}\epsilon - c(2m + \alpha c)\epsilon^2 + O(\epsilon^3)$$

and

$$\frac{\alpha}{(1 + m\bar{x}_1)^2} = \alpha - 2m\alpha\epsilon + 3m^2\alpha\epsilon^2 + O(\epsilon^3)$$

for $\epsilon > 0$ sufficiently small, we have

$$\hat{J}_1 = \begin{pmatrix} 0 & \frac{1}{\alpha} \\ \alpha & 0 \end{pmatrix} \text{ and } \hat{J}_2 = \begin{pmatrix} 0 & \frac{m - \alpha c}{\alpha} \\ -2m\alpha & 0 \end{pmatrix}. \tag{7}$$

Let $\lambda_i = \lambda_{i1} + \lambda_{i2}\epsilon + O(\epsilon^2)$ and $v_i = v_{i1} + v_{i2}\epsilon + O(\epsilon^2)$ be the corresponding eigenvalue and eigenvector of $\hat{J}_1 + \hat{J}_2\epsilon + O(\epsilon^2)$ respectively, $i = 1, 2$. Then

$$\begin{aligned} &\hat{J}_1 v_{i1} = \lambda_{i1} v_{i1} \\ &\hat{J}_1 v_{i2} + \hat{J}_2 v_{i1} = \lambda_{i1} v_{i2} + \lambda_{i2} v_{i1} \end{aligned} \tag{8}$$

for $i = 1, 2$. It follows that λ_{i1} is an eigenvalue of $\hat{J}_1$ with corresponding eigenvector v_{i1}, $i = 1, 2$. Consequently, we have $\lambda_{11} = 1$ with eigenvector $v_{11} = (1 \ \alpha)^T$ and $\lambda_{21} = -1$ with eigenvector $v_{21} = (-1 \ \alpha)^T$, where T denotes transpose.

Furthermore, the second equation of (8) can be rewritten as

$$(\hat{J}_1 - \lambda_{i1} I)v_{i2} = (\lambda_{i2} I - \hat{J}_2)v_{i1},$$

$i = 1, 2$. It follows from the Fredholm alternative that the right hand side of the above equation must be orthogonal to the solutions y of the homogeneous equation $(\hat{J}_1^T - \lambda_{i1} I)y = 0$ in order for the above equation to have a solution. This implies $\lambda_{12} = -\dfrac{m+\alpha c}{2}$, and $\lambda_{22} = \dfrac{m+\alpha c}{2}$. Consequently for $\epsilon > 0$ sufficiently small, we have

$$\lambda_1 = 1 - \frac{m+\alpha c}{2}\epsilon + O(\epsilon^2) \text{ and } \lambda_2 = -1 + \frac{m+\alpha c}{2}\epsilon + O(\epsilon^2).$$

Therefore if $\epsilon > 0$ is sufficiently small, i.e., if $n > 1$ is sufficiently close to 1, then the interior steady state $(\bar{x}_1, \bar{x}_2)$ is locally asymptotically stable for system (5). Furthermore, since the left hand side of (2) as a function of n is increasing, $\bar{x}_2$ is an increasing function of n. On the other hand, since $\bar{x}_1 = \frac{n}{\alpha}\bar{x}_2 e^{-c\bar{x}_2}$ and $\lim_{n\to 1^+} \bar{x}_2(n) = 0$, we see that $\bar{x}_1$ is an increasing function of n when $n > 1$ is sufficiently close to 1. Therefore, we can conclude that when $n > 1$ is small enough so that $\beta a \bar{x}_1 < 1$, then $E_1 = (\bar{x}_1, \bar{x}_2, 0)$ is locally asymptotically stable for the original system (1).

Lemma 1.2. *System (5) has a unique interior steady state $(\bar{x}_1, \bar{x}_2)$ when $n > 1$. The interior steady state is locally asymptotically stable for system (5) when $n > 1$ is sufficiently close to 1. Similarly, system (1) has a nontrivial boundary steady state $E_1 = (\bar{x}_1, \bar{x}_2, 0)$ if $n > 1$. The steady state E_1 is locally asymptotically stable for (1) if $n > 1$ is sufficiently close to 1.*

We next prove that both populations can survive when $n > 1$ and $\beta a \bar{x}_1 > 1$, where $\beta a \bar{x}_1 > 1$ can be interpreted as the reproductive number of the predator when the prey population is stabilized at the steady state. The predator population can survive if this reproductive number is larger than 1.

Theorem 1.3. *System (1) is uniformly persistent if $n > 1$ and $\beta a \bar{x}_1 > 1$, i.e., there exists $M > 0$ such that $\liminf_{t\to\infty} x_1(t) \geq M$, $\liminf_{t\to\infty} x_2(t) \geq M$, and $\liminf_{t\to\infty} p(t) \geq M$ for all solutions $(x_1(t), x_2(t), p(t))$ of (1) with $x_1(0) > 0, x_2(0) > 0$ and $p(0) > 0$.*

Proof. The proof of uniform persistence is standard. Indeed, it is clear that system (1) has a global attractor X. Let Y be the union of nonnegative coordinate planes. Then $R^3_+ \backslash Y$ is forward invariant for (1). The only invariant set in Y is $\{E_0, E_1\}$, where $\{E_0\}$ and $\{E_1\}$ are isolated in X. It remains to prove that the stable sets of E_0 and E_1 lie in Y.

Suppose there exists a solution $(x_1(t), x_2(t), p(t))$ of (1) with $x_1(0) > 0, x_2(0) > 0, p(0) > 0$ such that $\lim_{t\to\infty} (x_1(t), x_2(t), p(t)) = E_0$. Then for any $\epsilon > 0$ there exists $t_0 > 0$ such that $x_1(t) < \epsilon, x_2(t) < \epsilon$, and $p(t) < \epsilon$ for $t \geq t_0$. Since $n > 1$, we can choose $\epsilon > 0$ such that $ne^{-c\epsilon}e^{-a\epsilon} > 1$. Note

$$\begin{aligned} x_1(t+1) &> bx_2(t)e^{-c\epsilon} \\ x_2(t+1) &> \frac{\alpha e^{-a\epsilon}x_1(t)}{1+mx_1(t)} \end{aligned} \tag{9}$$

for $t \geq t_0$. Consider the following system of difference equations

$$\begin{aligned} y_1(t+1) &= by_2(t)e^{-c\epsilon} \\ y_2(t+1) &= \frac{\alpha e^{-a\epsilon}y_1(t)}{1+my_1(t)} \\ y_1(0) &= x_1(t_0), y_2(0) = x_2(t_0). \end{aligned} \tag{10}$$

Since $bxe^{-c\epsilon}$ and $\dfrac{\alpha e^{-a\epsilon}x}{1+mx}$ are increasing functions of x, we see that $x_1(t+t_0) \geq y_1(t)$ and $x_2(t+t_0) \geq y_2(t)$ for $t \geq 0$. Furthermore, since $ne^{-c\epsilon}e^{-a\epsilon} > 1$, system (10) has a unique interior steady state $(\bar{y}_1, \bar{y}_2)$ where $\bar{y}_1 > 0$ satisfies

$$1 = \frac{ne^{-a\epsilon}e^{-c\epsilon}}{1+my}. \tag{11}$$

On the other hand, (10) is equivalent to the following second order difference equation

$$y_1(t+2) = \frac{ne^{-c\epsilon}e^{-a\epsilon}y_1(t)}{1+my_1(t)}. \tag{12}$$

Since the map induced by (12) is increasing, using a result of [6, Theorem 1.10], one can show that solutions of (12) with $y_1(0), y_1(1) > 0$ converge to $\bar{y}_1$. As a result, we have $\liminf_{t\to\infty} x_1(t) \geq \bar{y}_1 > 0$. We obtain a contradiction and conclude that the stable set of E_0 lies on Y. Using $\beta a\bar{x}_1 > 1$, a similar argument can be applied to show that the stable set of E_1 lies on Y. Therefore, system (1) is uniformly persistent with respect to Y by [7, Theorem 4.1]. Q.E.D.

Using the same argument as in Theorem 1.3, one can show that system (5), the stage-structured prey population model, is uniformly persistent if $n > 1$. Notice the inherent net reproductive number of the prey, n, does not depend on the cannibalism rate c. However, it follows from equation (2) that $\bar{x}_2$ is a decreasing function of c. Therefore increasing cannibalism of the adult prey population can diminish its own population density at the steady state. In particular, $\lim_{c\to\infty} \bar{x}_2(c) = 0$ and cannibalism can have severe deleterious effect to the prey population at the population level. On the other hand, if $n > 1$ and c is large so that $\beta b\bar{x}_1 < 1$, then the full system may not be uniformly persistent. Consequently, cannibalism of the prey may drive the predator population to extinction.

Finally we use an example to study system (5). We choose $\alpha = 0.7$, $m = 0.5$, $a = 0.6$, $\beta = 20$ and $b = 100$. The bifurcation parameter is the cannibalism rate c, where c ranges from 0.01 to 13. Figure 1 although not shown here plots the bifurcation diagram for the juvenile prey population. From this numerical simulation it seems that the interior steady state is globally asymptotically stable when c is small and period-doubling bifurcations occur when c is large.

We acknowledge that our model is derived under very simple biological assumptions. Different models may reach different biological conclusions than stated here. We shall continue to study systems (1) and (5) in more details and to investigate a parallel model when the predator is stage-structured and the adult predator population can consume its own juveniles.

References

[1] B. Bogstad, G. R. Lily and S. Mehl, Cannibalism and year-class strength in Atlantic cod in Arcto-boreal ecosystems, ICES Mar. Sci. Symp., **198** (1994), 576–585.

[2] J. M. Cushing, An Introduction to Structured Population Dynamics, CBMS-NSF Regional Conf. Ser. in Appl. Math., **71**, SIAM, 1998.

[3] J. M. Cushing, A simple model of cannibalism, Math. Biosci., **107** (1991), 47–54.

[4] S. Elaydi, An Introduction to Difference Equations, Second ed., Springer, New York, 1999.

[5] P. Getto, O. Dickmann and A. de Roos, On the (dis)advantages of canibalism, J. Math. Biol., **51** (2005), 695–712.

[6] A. Grove and G. Ladas, Periodicities in Nonlincar Difference Equations, CRC Press, 2005.

[7] J. Hofbauer and J. So, Uniform persistence and repellors for maps, Proc. Amr. Math. Soc., **107** (1989), 1137–1142.

[8] S. Jang and J. Baglama, A nutrient-prey-predator model with intratrophic predation, Appl. Math. Comput., **129** (2002), 517–536.

[9] S. Jang, J. Baglama and P. Seshaiyer, Intratrophic predation in a simple food chain with fluctuating nutrient, Discrete Contin. Dyn. Syst. Ser. B, **5** (2005), 335–352.

[10] C. Kohlmeier and W. Ebenhoh, The stabilizing role of cannibalism in a predator-prey system, Bull. Math. Biol., **57** (1995), 401–423.
[11] K. Magnusson, Destabilizing effect of cannibalism on a structured predator-prey system, Math. Biosci., **155** (1999), 61–75.
[12] F. van der Bosch, A. M. de Roos and W. Gabriel, Cannibalism as a life boat mechanism, J. Math. Biol., **26** (1995), 619–621.

Department of Mathematics
University of Louisiana at Lafayette
Lafayette, LA 70504-1010
USA

Current address
Department of Mathematics and Statistics
Texas Tech University
Lubbock, TX 79409-1042
USA

E-mail address: Sophia.Jang@ttu.edu

Advanced Studies in Pure Mathematics 53, 2009
Advances in Discrete Dynamical Systems
pp. 95–112

3-dimensional i.i.d. binary random vectors governed by Jacobian elliptic space curve dynamics

Tohru Kohda

Abstract.

Sufficient conditions have been recently given for a classs of ergodic maps of an interval onto itself: $I = [0,1] \subset R \to I$ and its associated binary function to generate a sequence of independent and identically distributed (i.i.d.) binary random variables. Jacobian elliptic Chebyshev map, its derivative and second derivative induce Jacobian elliptic space curve. A mapping of the space curve with its coordinates, e.g., X, Y and Z, onto itself is introduced which defines 3 projective onto mappings, represented in the form of rational functions of $\{x_n, y_n, z_n\}_{n=0}^{\infty}$. Such mappings with their absolutely continuous invariant measures as functions of elliptic integrals and their associated binary function can generate a 3-dimensional sequence of i.i.d. binary random vectors.

§1. Introduction

Bernoulli shift and its associated binary function can produce a sequence of independent and identically distributed (i.i.d.) binary random variables (BRVs) [1], [2]. Tent map [3], closely related to the Bernoulli map, and its associated binary function can also generate a sequence of i.i.d. BRVs. Ulam and von Neumann[4] showed that the logistic map is topologically conjugate to the tent map via the homeomorphism $h^{-1}(\omega) = \dfrac{2}{\pi}\sin^{-1}\sqrt{\omega}$. They also pointed out that the logistic map is a strong candidate for pseudo-random number generation (PRNG) even though it has a non-uniform absolutely continuous invariant (ACI) measure. A number of analog chaos techniques, which use a chaotic real-valued trajectory itself, have also been proposed [5],[6]. Binary sequences

Received October 12, 2007.
Key words and phrases. chaos, duplication formula, pseudo-radom number, coin tossing, independence, Bernoulli shift, independent and identically distributed binary random variable, Jacobian elliptic function, ellpitic space curve, elliptic integral.

using chaos, however, play an important role in several applications such as spreading spectrum codes [7], [8], [9] , [10], pseudo-random number generators [11] and cryptosystems [12], [13].

Motivated by this situation, we have shown that a class of ergodic maps with its unique ACI measure satisfying equi-distributivity property (EDP) can generate a sequence of i.i.d. binary random variables if its associated binary function satisfies constant summation property (CSP) [14]. Fortunately, many well-known 1-dimensional maps, which are topologically conjugate to the tent map via homeomorphism [3], satisfy EDP. The Bernoulli map, logistic map and Chebyshev polynomial are good examples [15]. These maps are governed by duplication formulae. In other words, *a duplication formula gives chaotic dynamics.* It is well known that elliptic functions satisfy an addition theorem [16]. We introduced a Jacobian elliptic Chebyshev rational map as a rational function version of Chebyshev polynomial [17]. This map as well as the other well known maps mentioned above are mappings from an interval onto itself.

Modern cryptosystems, however, need more and more pseudo-random numbers. In fact, to break DES of 64 bits, it takes 2^{43} steps and the success rate is 85% if 2^{43} pairs (plaintext, ciphertext) are known [18]. Furthermore, the size of new block ciphers such as AES becomes large, e.g., 512 bits.

This situation motivated us to discuss a closed smooth space curve defined by an algebraic relation between the Jacobian elliptic function, its derivative and second derivative. These duplication formulae give a real-valued sequence $\{x_n, y_n, z_n\}_{n=0}^{\infty}$ generated by 3-dimensional dynamics with Cartesian coordinates, e.g., X, Y and Z. Such 3-dimensional dynamics forces us to define a mapping from such a space curve onto itself and three projection mappings from an interval onto itself associated with coordinates, i.e., $x_{n+1} = \tau_x(x_n)$, $y_{n+1} = \tau_y(y_n)$ and $z_{n+1} = \tau_z(z_n)$, respectively. The former $\tau_x(\cdot)$, being single-valued, is the same as the Jacobian elliptic Chebyshev rational map. On the contrary, the latter two $\tau_y(\cdot)$ and $\tau_z(\cdot)$, being multi-valued, consist of single-valued mappings, each of which is a rational function of x_n, y_n, z_n and have their ACI measures with EDP. Hence, every bit of binary expansion of real-valued vector (x_n, y_n, z_n) satisfies CSP. This implies that the mapping from the space curve onto itself governed by duplication formulae gives a sequence of i.i.d. 3-dimensional binary random vectors.

§2. Related theories

We will begin by describing some of the related theories which play an important role in evaluating statistical properties of a sequence of binary random variables generated by a real-valued sequence.

2.1. EDP and CSP

Perhaps the simplest mathematical object that can display chaotic behavior is a class of one-dimensional maps $\omega_{n+1} = \tau(\omega_n)$, where $\omega_n = \tau^n(\omega_0) \in I = [d, e]$, $n = 0, 1, 2, \ldots$ and $\tau(\cdot) : I \to I$.
Consider a piecewise monotonic (PM) onto ergodic map $\tau(\cdot)$ that satisfies the following properties:

- **i):** there is a (trvial) partitition $d = d_0 < \cdots < d_{N_\tau} = e$ of I such that for each integer $i = 1, \cdots, N_\tau$, $(N_\tau > 2)$ the restriction of $\tau(\cdot)$ to the interval $I_i = [d_{i-1}, d_i)$, denoted by $\tau_i(\omega)$, is a C^2 function; as well as
- **ii):** $\tau(I_i) = (d, e)$, that is, τ has N_τ monotonic onto maps τ_i;
- **iii):** τ has a unique ACI measure, denoted by $f^*(\omega)d\omega$.

The following four definitions are important to evaluate statistical properties of $\{\omega_n\}_{n=0}^{\infty}$.

Definition 1 (Perron-Frobenius operator [19]). *The Perron-Frobenius operator P_τ acting on function of bounded variation $F(\omega) \in L^\infty$ for $\tau(\omega)$ is defined as*

$$P_\tau F(\omega) = \frac{d}{d\omega} \int_{\tau^{-1}([d,\omega])} F(y)dy = \sum_{i=0}^{N_\tau - 1} |g_i'(\omega)| F(g_i(\omega)),$$

where $g_i(\omega)$ is the i-th preimage of ω and N_τ denotes the number of preimages.

The ACI measure $f^*(\omega)d\omega$ satisfies

$$P_\tau f^*(\omega) = f^*(\omega). \tag{1}$$

Birchoff Individual Ergodic Theorem [19] tells us that for a stationary real-valued sequence $\{F(\omega_n)\}_{n=0}^{\infty}$, the time average of $\{F(\omega_n)\}_{n=0}^{\infty}$, defined by

$$\langle F \rangle = \lim_{T\to\infty} (1/T) \sum_{n=0}^{T-1} F(\omega_n) \tag{2}$$

is equal almost everywhere to the expectation of $F(\omega)$, defined by

$$\mathbf{E}_\omega[F(\tau^n)] = \int_I F(\tau^n(\omega))f^*(\omega)d\omega. \tag{3}$$

From the stationarity of process, we denote $\mathbf{E}_\omega[F(\tau^n)]$ by $\mathbf{E}_\omega[F]$. Consider two sequences $\{G(\tau^n(\omega))\}_{n=0}^{\infty}$ and $\{H(\tau^n(\omega))\}_{n=0}^{\infty}$, where $G(\omega)$, $H(\omega) \in L^\infty$. The second-order cross-covariance function between these sequences from a seed $\omega = \omega_0$ is defined by

$$\rho(\ell, G, H) = \int_I (G(\omega) - \mathbf{E}_\omega[G]) \cdot (H(\tau^\ell(\omega)) - \mathbf{E}_\omega[H]) f^*(\omega)d\omega, \tag{4}$$

where $\ell = 0, 1, 2, \cdots$. The operator P_τ is useful in evaluating correlation functions because it has the following important property:

$$\int_I G(\omega)P_\tau\{H(\omega)\}d\omega = \int_I G(\tau(\omega))H(\omega)d\omega. \tag{5}$$

Using this property, we get

$$\rho(\ell, G, H) = \int_I P_\tau^\ell\{(G(\omega) - \mathbf{E}_\omega[G])f^*(\omega)\}(H(\omega) - \mathbf{E}_\omega[H])d\omega. \tag{6}$$

Bernoulli map with its uniform ACI measure $f^*(\omega)d\omega = d\omega$ is defined as

$$\tau_B(\omega) = 2\omega (\text{mod } 1) = \begin{cases} 2\omega, & 0 < \omega < \frac{1}{2}, \\ 2\omega - 1, & \frac{1}{2} \leqslant \omega < 1. \end{cases} \tag{7}$$

If ω is represented by its binary expansion as $\omega = 0.d_1(\omega)d_2(\omega)\cdots$, then the binary expansion of $\tau_B(\omega)$ is given by $\tau_B(\omega) = 0.d_2(\omega)d_3(\omega)\cdots$. This implies that $\tau_B(\cdot)$ shifts the digits one place to the left. The functions $d_k(\cdot)$, called Rademacher functions, furnish us with a model of independent tosses of a fair coin [2]. A sequence $\{d_k(\omega)\}_{k=0}^{\infty}$ can be regarded as a sequence of i.i.d. BRVs in the sense that for almost every ω, $d_k(\omega)$ can imitate coin tossing.

Another map and its associated binary function are as follows. Consider piecewise linear map of p branches with $f^*(\omega)d\omega = d\omega$, given by [3] $(N_\tau = p)$,

$$N_p(\omega) = (-1)^{\lfloor p\omega \rfloor} p\omega (\text{mod } p), \ \omega \in [0, 1]. \tag{8}$$

In particular, $N_2(\omega)$ is referred to as the tent map. Introduce its associated BRV defined as

$$a_k = \begin{cases} 0, & \text{for } N_2^k(\omega) \leqslant \frac{1}{2}, \\ 1, & \text{for } N_2^k(\omega) > \frac{1}{2}. \end{cases} \tag{9}$$

Then for $\omega = 0.d_1(\omega)d_2(\omega)\cdots$, we get $a_0(\omega) = d_1(\omega)$, $a_k(\omega) = d_k(\omega) \oplus d_{k+1}(\omega)$, $k \geqslant 1$, where $\oplus$ denotes a modulo 2 addition (or exclusiveor) operation. Hence $N_2(\omega)$ and its associated binary functions $a_k(\cdot)$ can generate a sequence of i.i.d. BRVs.

Naturally, the important question arises, that can any other map and its associated binary function generate a sequence of i.i.d. BRVs? We have got an affirmative answer to this question [14], [15], which is firstly, the map should satisfy EDP and secondly, the binary function should satisfy CSP.

Definition 2 (EDP [14]). *If a piecewise-monotonic onto map* $\tau(\omega)$ *satisfies*

$$(10) \qquad |g_i'(\omega)|f^*(g_i(\omega)) = \frac{1}{N_\tau} f^*(\omega), \quad 0 \leq i \leq N_\tau - 1,$$

then the map is said to satisfy equi-distributivity property (EDP).

Definition 3 (CSP [14],[15]). *For a class of maps with EDP, if its associated function* $G(\cdot)$ *satisfies*

$$(11) \quad \frac{1}{N_\tau} \sum_{i=0}^{N_\tau - 1} G(g_i(\omega)) = \mathbf{E}_\omega[G] \ \textit{or} \ P_\tau\{G(\omega)f^*(\omega)\} = \mathbf{E}_\omega[G]f^*(\omega)$$

then $G(\cdot)$ *is said to satisfy* constant summation property (CSP).

CSP guarantees no-correlation between two functions $G(\cdot)$ and ${}^\forall H(\cdot)$, i.e., $\rho(\ell, G, H) = 0$, $\ell > 0$ [15]. Fortunately, EDP is satisfied by many well-known maps and is invariant under topological conjugation.

Definition 4 (topological conjugation [19]). *Two transformations* $\bar{\tau} : \bar{I} \to \bar{I}$ *and* $\tau : I \to I$ *on intervals* $\bar{I}$ *and* I *are called* topological conjugate *if there is a homeomorphism* $h : \bar{I} \overset{onto}{\to} I$ *as* $\tau(\omega) = h \circ \bar{\tau} \circ h^{-1}(\omega)$.

Suppose $\tau(\cdot)$ and $\bar{\tau}(\cdot)$ have their ACI measures $f^*(\omega)d\omega$ and $\bar{f}^*(\bar{\omega})d\bar{\omega}$ respectively. Then, under the topological conjugation, these ACI measures have the relation

$$(12) \qquad f^*(\omega) = \left| \frac{dh^{-1}(\omega)}{d\omega} \right| \bar{f}^*(h^{-1}(\omega)).$$

The relation between $\tau(\cdot)$ and $\bar{\tau}(\cdot)$ via h is represented diagrammatically as follows :

$$(13) \qquad \begin{array}{ccc} I & \xrightarrow{\tau} & I \\ {\scriptstyle h^{-1}}\downarrow & & \uparrow{\scriptstyle h} \\ \bar{I} & \xrightarrow{\bar{\tau}} & \bar{I} \end{array}$$

Remark 1. If we take $N_2(\overline{\omega})$ as $\overline{\tau}(\overline{\omega})$, then $f^*(\omega)$ is simply represented by the derivative of $h^{-1}(\omega)$. Hence, if $h(\overline{\omega})$ can be given in an inverse function form, then its integrand gives an ACI measure within normalization factor. Most famous example of inverse functions is sin function, *i.e.*, $\omega = \int_0^{\sin\omega} \frac{du}{\sqrt{1-u^2}}$.
This remark provides a starting point for discussion. In fact, Ulam and von Neumann [4] gave the logistic map

$$L_2(\omega) = 4\omega(1-\omega),\ \omega \in [0,1] \tag{14}$$

with $f^*(\omega)d\omega = \frac{d\omega}{\pi\sqrt{\omega(1-\omega)}}$ which is topologically conjugate to $N_2(\overline{\omega})$ using $h^{-1}(\omega) = \dfrac{2}{\pi}\sin^{-1}\sqrt{\omega}$.

2.2. Binary function

In our previous study [14], we proposed methods to obtain binary sequences from chaotic real-valued sequences $\{\tau^n(\omega)\}_{n=0}^{\infty}$. We define a (non-trivial) partition $d = t_0 < t_1 < \cdots < t_{2M} = e$ of $[d,e]$ and T denotes the set of thresholds $\{t_r\}_{r=0}^{2M}$. Then the following binary function is obtained

$$C_T(\omega) = \sum_{r=0}^{2M} (-1)^r \Theta_{t_r}(\omega), \tag{15}$$

where $\Theta_t(\omega)$ is the threshold function such that

$$\Theta_t(\omega) = \begin{cases} 0, & \text{for } \omega < t \\ 1, & \text{for } \omega \geq t. \end{cases} \tag{16}$$

§3. Duplication formula gives chaos

The example mentioned above shows that duplication formula gives chaos. To observe it, several examples are listed as follows.
(1) logistic map: Transformation $x = \sin^2\theta$ gives $\left(\frac{dx}{d\theta}\right)^2 = 4x(1-x)$. Let $x_n = \sin^2\theta_n, \theta_{n+1} = 2\theta_n$. Then we get 2-dimensional sequences $\{(x_n, y_n)\}_{n=0}^{\infty}$, given by

$$\begin{aligned} x_{n+1} &= L_2(x_n) = 4x_n(1-x_n), \\ y_{n+1}^2 &= \left(\frac{1}{2}\cdot\frac{d\,L_2(x_n)}{d\theta_n}\right)^2 = 4L_2(x_n)(1-L_2(x_n)). \end{aligned} \tag{17}$$

(2) Chebyshev map of degree 2: Grossmann and Thomae [3] observed that Chebyshev polynomial maps of degree p ($p = 2, 3, \cdots$) [20] with its

ACI measure $f^*(\omega)d\omega = \dfrac{d\omega}{\pi\sqrt{1-\omega^2}}$, defined by

$$T_p(\omega) = \cos(p\cos^{-1}\omega),\ \omega \in [-1,1] \tag{18}$$

is topologically conjugate to $N_p(\omega)$ via $h(\overline{\omega}) = \cos\pi\overline{\omega}$. Transformation $x = \cos\theta$ gives $\left(\frac{dx}{d\theta}\right)^2 = 1 - x^2$. Let $x_n = \cos\theta_n,\ \theta_{n+1} = 2\theta_n$. Then we get 2-dimensional sequences $\{(x_n, y_n)\}_{n=0}^{\infty}$, given by

(19)

$$x_{n+1} = T_2(x_n) = 2x_n^2 - 1,\ y_{n+1}^2 = \left(\frac{1}{2}\cdot\frac{dT_2(x_n)}{d\theta_n}\right)^2 = 1 - (T_2(x_n))^2.$$

(3) Schröder and Böttcher map: [1] Schröder [22] and Böttcher [23] gave a rational function version of $L_2(\cdot)$ with parameter k, defined as

$$R_2^{\mathrm{sn}^2}(\omega,k) = \frac{4\omega(1-\omega)(1-k^2\omega)}{(1-k^2\omega^2)^2},\ \omega \in [0,1] \tag{20}$$

with its ACI measure

$$f^*(\omega,k)d\omega = \frac{d\omega}{2K(k)\sqrt{\omega(1-\omega)(1-k^2\omega)}} \tag{21}$$

via $h^{-1}(\omega) = \dfrac{1}{K(k)}\mathrm{sn}^{-1}(\sqrt{\omega},k)$, where $\mathrm{sn}(\omega,k)$ is the inverse function of the elliptic integral with modulus k ($|k| < 1$) and $K(k)$ is the complete elliptic integral, each of which is given respectively as

$$u = \int_0^{\mathrm{sn}(u,k)} \frac{dv}{\sqrt{(1-v^2)(1-k^2v^2)}},\ K(k) = \int_0^{\frac{\pi}{2}} \frac{d\theta}{\sqrt{1-k^2\sin^2\theta}}. \tag{22}$$

Transformation $x = \mathrm{sn}^2 u$ gives $\left(\frac{dx}{du}\right)^2 = 4x(1-x)(1-k^2x)$. Let $x_n = \mathrm{sn}^2 u_n, u_{n+1} = 2u_n$. Then we get 2-dimensional sequences $\{(x_n, y_n)\}_{n=0}^{\infty}$, given by

$$x_{n+1} = R_2^{\mathrm{sn}^2}(x_n,k) = \frac{4x_n(1-x_n)(1-k^2x_n)}{(1-k^2x_n^2)^2} \tag{23}$$

$$\begin{aligned} y_{n+1}^2 &= \left(\frac{1}{2}\cdot\frac{dR_2^{\mathrm{sn}^2}(x_n,k)}{du_n}\right)^2 \\ &= 4R_2^{\mathrm{sn}^2}(x_n,k)(1-R_2^{\mathrm{sn}^2}(x_n,k))(1-k^2R_2^{\mathrm{sn}^2}(x_n,k)). \end{aligned} \tag{24}$$

[1] see [21] for a historical review of rational maps.

§4. Jacobian elliptic space curve and 3-dimensional dynamics

We know that the Jacobian elliptic function $\mathrm{cn}(u,k)$ [2] is an inverse function of an elliptic integral of the first kind in the Legendre-Jacobi normal form [16]

$$u = \int_{\mathrm{cn}(u,k)}^{1} \frac{dt}{\sqrt{(1-t^2)(1-k^2+k^2t^2)}}. \tag{25}$$

Kohda and Fujisaki [17] introduced the Jacobian elliptic Chebyshev rational map with positive integer p

$$R_p^{\mathrm{cn}}(\omega,k) = \mathrm{cn}(p\,\mathrm{cn}^{-1}(\omega,k),k), \quad \omega \in [-1,1] \tag{26}$$

which is topologically conjugate to the tent map $N_p(u)$ via homeomorphism $h^{-1}(\omega,k) = \frac{\mathrm{cn}^{-1}(\omega,k)}{2K(k)}$ and has its ACI measure

$$f^*(\omega,k)d\omega = \frac{d\omega}{2K(k)\sqrt{(1-\omega^2)(1-k^2+k^2\omega^2)}}. \tag{27}$$

This map is a rational function version of the Chebyshev polynomial

$$T_p(\omega) = \cos(p\cos^{-1}\omega), \quad \omega \in [-1,1]. \tag{28}$$

We know that $R_p^{\mathrm{cn}}(\omega,k)$ satisfies the semi-group property

$$R_r^{\mathrm{cn}}(R_s^{\mathrm{cn}}(\omega,k),k) = R_{rs}^{\mathrm{cn}}(\omega,k) \tag{29}$$

for integers r, s and when $p = 2$,

$$R_2^{\mathrm{cn}}(\omega,k) = \frac{1-2(1-\omega^2)+k^2(1-\omega^2)^2}{1-k^2(1-\omega^2)^2}. \tag{30}$$

Let us concentrate on the Jacobian real elliptic function with $p = 2$ [16]. As shown in Fig. 1, the Jacobian elliptic function $X = \mathrm{cn}(u,k)$, its derivative $Y = \frac{d}{du}\mathrm{cn}\,u = -\mathrm{sn}\,u\,\mathrm{dn}\,u$ and the second derivative $Z = \frac{d^2}{du^2}\mathrm{cn}\,u$ give the Jacobian elliptic space curve, given by

$$Y^2 = (1-X^2)(1-k^2+k^2X^2), \; Z = X(-1+2k^2(1-X^2)). \tag{31}$$

[2] $\mathrm{cn}(u,0)$ simply reduces to $\cos u$.

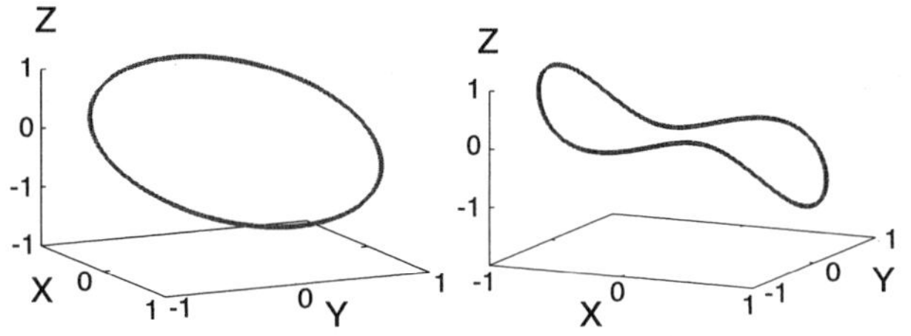

(a) $k = 0.1$ (b) $k = 0.9$

Fig. 1. Two Jacobian elliptic space curves (X, Y, Z).

Let $u_{n+1} = 2u_n$, $x_n = \operatorname{cn} u_n, y_n = \frac{dx_n}{du_n}$ and $z_n = \frac{d^2x_n}{du_n^2}$. Then we get a 3-dimensional dynamics, given by

(32)

$$\begin{cases} x_{n+1} & = R_2^{\mathrm{cn}}(x_n, k) = \tau_x(x_n, k), \\ y_{n+1}^2 & = (\frac{1}{2}\frac{dx_{n+1}}{du_n})^2 = (1 - x_{n+1}^2)(1 - k^2 + k^2x_{n+1}^2) = \tau_y^2(y_n, k), \\ z_{n+1} & = \frac{1}{4}\frac{d^2x_{n+1}}{du_n^2} = \tau_z(z_n(x_n), k) = \tau_z(x_n, k) \\ & = \frac{k^2-1+2(1-k^2)x_n^2+k^2x_n^4}{1-k^2(1-x_n^2)^2}\{1 - 2(\frac{1-k^2+k^2x_n^4}{1-k^2(1-x_n^2)^2})^2\}. \end{cases}$$

This gives a mapping from such a space curve onto itself which induces three projective onto mappings associated with coordinates,e.g.,X, Y, Z, denoted by $\tau_x(\cdot), \tau_y(\cdot), \tau_z(\cdot)$. The first one is shown in Fig.2(a), which has a symmetric ACI measure, defined by

$$f_X^*(x, k)dx = \frac{dx}{2K(k)\sqrt{(1 - x^2)(1 - k^2 + k^2x^2)}}$$

in Fig.3(a).

In addition, it has been shown [24] that the projective onto map τ_y is symmetric and has a symmetric ACI measure as shown in Figs.2(b) and 3(b), respectively. (see Appendix A for theoretical expression of τ_y) Its associated symmetric binary function, e.g., binary expansion of real-valued orbit $\{x_n\}_{n=0}^{\infty}$ or $\{y_n\}_{n=0}^{\infty}$ can generate a sequence of i.i.d. binary random variables [24].

Here we consider the map τ_z and examine whether it has its symmetric ACI measure [25]. Squaring the second expression of Eq.(31) with $k \neq 0$ gives the relation

$$X^6 - \frac{1}{k^2}(-1 + 2k^2)X^4 + \frac{1}{4k^4}(-1 + 2k^2)^2X^2 - \frac{Z^2}{4k^4} = 0 \tag{33}$$

which implies that for a given Z, X^2 has the following three real-valued solutions at most.

$$(34)\quad X^2(Z) = \begin{cases} \xi_1^2(Z), & \text{for } k \le \sqrt{1/2}\ (R(Z,k) > 0) \\ \xi_1^2(Z), & \text{for } k > \sqrt{1/2} \text{ and } R(Z,k) > 0 \\ \xi_i^2(Z),\ 2 \le i \le 4, \text{for } k > \sqrt{1/2} \text{ and } R(Z,k) < 0, \end{cases}$$

where $R(Z,k) = \frac{b^2(Z,k)}{4} + \frac{a^3(k)}{27}$, $a(k) = -\frac{1}{12k^4}(-1+2k^2)^2, b(Z,k) = \frac{1}{4\cdot 27}\{\frac{(-1+2k^2)^3}{k^6} - \frac{27}{k^4}Z^2\}$.

On the space curve, 3-dimensional dynamics has a unique ACI measure with respect to each coordinate. Fig. 3(c) shows comparison between the marginal distribution taken from experiments and theoretical calculations, where the theoretical distributions of τ_z is given as follows

(35)

$$f_Z^*(z,k)dz = \begin{cases} \frac{1}{2K(k)} f_Z(\xi_1(Z),k)dz, & \text{for } 0 < k \le \sqrt{1/2} \\ \frac{1}{2K(k)} f_Z(\xi_1(Z),k)dz, & \text{for } k > \sqrt{1/2},\ r(k) \le |z| < 1 \\ \frac{1}{2K(k)} \sum_{\ell=2}^{4} f_Z(\xi_\ell(Z),k)dz & \text{for } k > \sqrt{1/2}, |z| \le r(k) \end{cases}$$

where

$$(36)\quad \begin{aligned} r(k) &= \sqrt{\frac{2}{27}(-1+2k^2)^3}, \\ f_Z(\xi_\ell(Z),k)dz & \\ &= \frac{dz}{\sqrt{(1-\xi_\ell^2(Z))(1-k^2+k^2\xi_\ell^2(Z))}\,|-6k^2\xi_\ell^2(Z)+2k^2-1|}. \end{aligned}$$

Finally, we notice that theoretical distribution $f_X^* dx$ is also given by integrand of elliptic integral for inverse function $\mathrm{cn}^{-1}(u,k)$ (see Eq.(25)). The same is true for $f_Y^* dy$. In fact, inverse function $\left(\frac{d\,\mathrm{cn}(u,k)}{du}\right)^{-1} = (-\mathrm{sn}(u,k)\,\mathrm{dn}(u,k))^{-1}$ is defined by Eq.(45) and Eq.(46) (see Appendix B). Similarly $f_Z^* dz$ is expressed in the inverse function form, as given by Eq.(49) and Eq.(50) (see Appendix C).

§5. I.I.D. binary random vectors

We shall now look into the relation between (z_n, z_{n+1}). Eqs.(33) and (34) tell us that the relation $z_{n+1} = \tau_z(\xi_1(z_n))$ is one-to-one when $k < \sqrt{1/2}$ but the graph of z_n versus z_{n+1} is *one-to-many* when $k >$

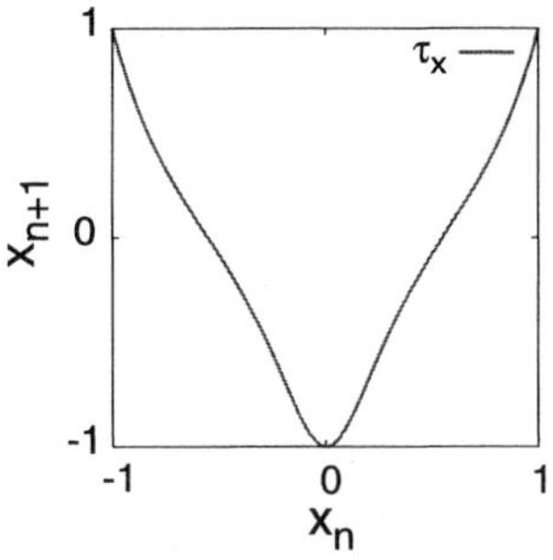

(a) $x_{n+1} = \tau_x(x_n)$

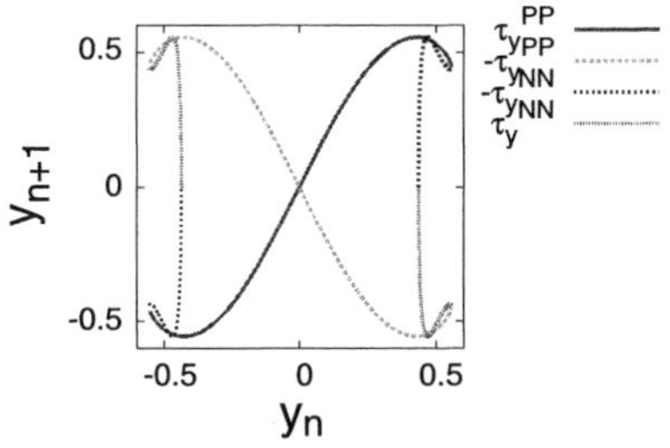

(b) $y_{n+1} = \tau_y(y_n)$

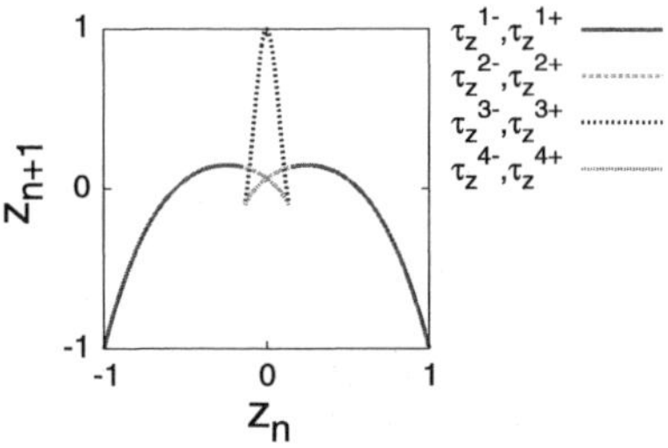

(c) $z_{n+1} = \tau_z(z_n)$

Fig. 2. Three projection mappings when $k = 0.9$.

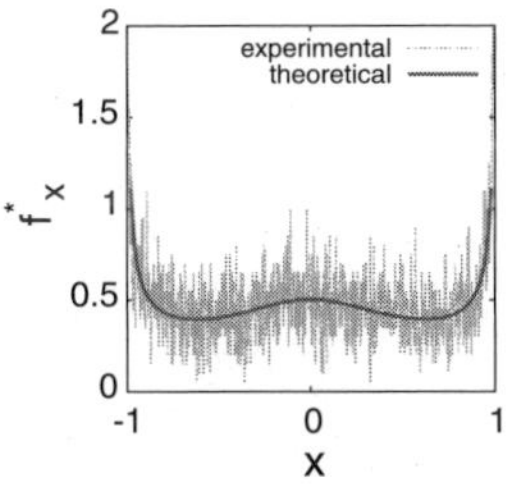

(a) $f_X^*(x)dx$

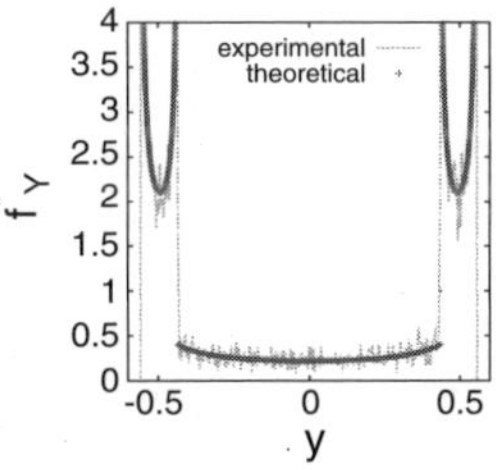

(b) $f_Y^*(y)dy$

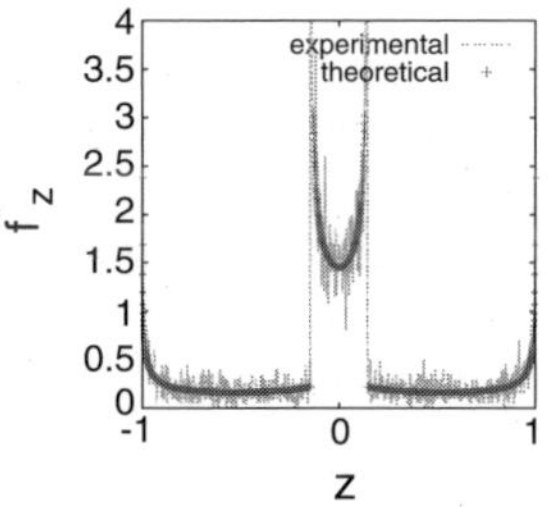

(c) $f_Z^*(z)dz$

Fig. 3. Three marginal distributions when $k = 0.9$

$\sqrt{1/2}$. Namely, the latter case gives a closed curve as shown in Fig. 2(c). Suppose that $k > \sqrt{1/2}$ and that $X_1(x)$ is the first bit of normalized x in binary representation, such as

$$\frac{x+1}{2} = 0.X_1(x)X_2(x)\cdots X_i(x)\cdots, X_i(x) \in \{0,1\}.$$

We denote $X_1(x)$ by X_1 and $1 - X_1(x)$ by $\overline{X_1}$. Similarly $Z_1(z)$ and $1 - Z_1(z)$ are denoted by Z_1 and $\overline{Z_1}$ respectively. In addition, $D(\frac{dz}{dx})$ and $1 - D(\frac{dz}{dx})$ are represented by D_z and $\overline{D_z}$ respectively, where $D(\frac{dz}{dx}) = 0(\text{or } 1)$ when $\frac{dz}{dx} < 0$ (or when $\frac{dz}{dx} \geq 0$).

Then, we can obtain a piecewise-monotonic onto map τ_z defined by

$$\begin{aligned}(37)\quad \tau_z &= X\,\overline{Z}\,\overline{D}\tau_z^{1-} + \overline{X}Z\overline{D}\tau_z^{1+} + X\overline{Z}\,\overline{D}\tau_z^{2-} + \overline{X}\,Z\overline{D}\tau_z^{2+}\\ &+ \overline{X}\,\overline{Z}\,D\tau_z^{3-} + XZD\tau_z^{3+} + \overline{X}\,\overline{Z}\,\overline{D}\tau_z^{4-} + XZ\overline{D}\tau_z^{4+}\end{aligned}$$

where $\tau_z^{i-} = \tau_z(-\xi_i(z)), \tau_z^{i+} = \tau_z(\xi_i(z))$, $1 \leq i \leq 4$ and where $\xi_i^2(z)$ is defined by Eq.(34).

It can be shown that for uniform ACI measure $f_U^*(u)du = du$,

$$(38)\quad \left.\begin{array}{lll} P_{\tau_x}\{C_{T_x}(x)f_X^*(x)\} & = \mathbf{E}_u[C_{T_x}]f_X^*(x), & x = \operatorname{cn} u \\ P_{\tau_y}\{C_{T_y}(y)f_Y^*(y)\} & = \mathbf{E}_u[C_{T_y}]f_Y^*(y), & y = -\operatorname{sn} u \operatorname{dn} u \\ P_{\tau_z}\{C_{T_z}(z)f_Z^*(z)\} & = \mathbf{E}_u[C_{T_z}]f_Z^*(z), & z = \dfrac{d(-\operatorname{sn} u \operatorname{dn} u)}{du} \end{array}\right\}$$

holds, where $\{C_{T_x}(x_n)\}_{n=0}^{\infty}, \{C_{T_y}(y_n)\}_{n=0}^{\infty}$ and $\{C_{T_z}(z_n)\}_{n=0}^{\infty}$ are symmetric binary sequences with their sets of symmetric thresholds T_x, T_y and T_z associated with real-valued sequences $\{x_n\}_{n=0}^{\infty}, \{y_n\}_{n=0}^{\infty}$ and $\{z_n\}_{n=0}^{\infty}$.

This implies that $\rho(\ell, C_{T_x}, C_{T_x}) = \rho(\ell, C_{T_y}, C_{T_y}) = \rho(\ell, C_{T_z}, C_{T_z}) = 0$, for $\ell \geq 0$. [14]

It should be noted that $C_{T_x}(x)$, $C_{T_y}(\tau_y^{\ell}(y))$, $C_{T_z}(\tau_z^{m}(z))$ are not always independent each other for $\ell = m = 0$, that is, e.g., $\mathbf{E}_u[C_{T_x}C_{T_y}] \neq \mathbf{E}_u[C_{T_x}]\mathbf{E}_u[C_{T_y}]$ even if each of them is a sequence of i.i.d. BRVs. This is inevitable as long as these sequences are generated from a single seed $u = u_0$. However, we can design appropriate sets of thresholds T_x, T_y, T_z satisfying $\mathbf{E}_u[C_{T_x}C_{T_y}] = \mathbf{E}_u[C_{T_x}]\mathbf{E}_u[C_{T_y}]$ (see [14] for details).

§6. Conclusion

We discussed a real-valued dynamics on the Jacobian elliptic space curve between Jacobian elliptic function, its derivative and second derivative, governed by their duplication formulae. Furthermore, we showed that a mapping of the space curve onto itself: $R^3 \to R^3$ which defines 3 projective onto mappings with their ACI measures satisfying EDP and can generate sequences of 3-dimensional i.i.d. binary random vectors when using their associated symmetric binary functions, e.g., bits of binary expansions of these real-valued x_n, y_n, z_n as shown in Fig. 4.

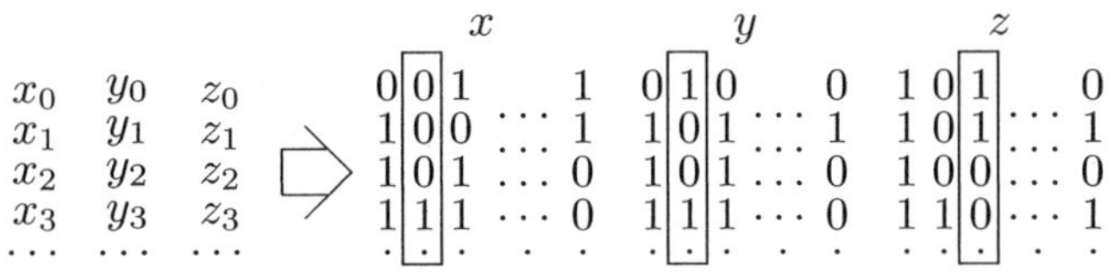

Fig. 4. Method of generating multidimensional i.i.d. binary vectors

References

[1] P. Billingsley, Probability and Measure, 3rd ed., Wiley-Interscience, 1995.
[2] M. Kac, Statistical Independence in Probability, Analysis and Number Theory, Mathematical Association of America, 1959.
[3] S. Grossmann and S. Thomae, Invariant distributions and stationary correlation functions of one-dimensional discrete processes, Z. Naturforsch., **32a** (1977), 1353–1363.
[4] S. M. Ulam and J. Von Neumann, On combination of stochastic and deterministic processes, Bull. Amer. Math. Soc., **53** (1947), 1120.
[5] J. A. Gonzalez, L. I. Reyes, L. E. Guerrero and G. Gutierrez, From exactly solvable chaotic maps to stochastic dynamics, Phys. D, **178** (2003), 26–50.
[6] L. Pecora and T. Caroll, Synchronization in Chaotic Systems, Phys. Rev. Lett., **64** (1990), 821–824.
[7] G. Heidari-Bateni and C. D. McGillem, A chaotic direct-sequence spread-spectrum communication system, IEEE Trans. Comm., **42** (1994), 1524–1527.
[8] T. Kohda and A. Tsuneda, Pseudonoise sequences by chaotic nonlinear maps and their correlation properties, IEICE Trans. Comm., **E76-B** (1993), 855–862.
[9] G. Mazzini, G. Setti and R. Rovatti, Chaotic complex spreading sequences for asynchronous DS-CDMA part I: system modeling and results, IEEE Trans. Circuit Syst., **CAS-44** (1997), 937–947.
[10] D. S. Broomhead, J. P. Huke and M. R. Muldoon, Codes for spread spectrum applications generated using chaotic dynamical systems, Dynam. Stability Systems, **14** (1999), 95–105.
[11] L. Kocarev and G. Jakimoski, Pseudorandom bits generated by chaotic maps, IEEE Trans. Circuits Systems I Fund. Theory Appl., **50** (2003), 123–126.

[12] G. Jakimoski and L. Kocarev, Chaos and Cryptography: Block Encryption Ciphers Based on Chaotic Maps, IEEE Trans. Circuits Systems I Fund. Theory Appl., **48** (2001), 163–169.

[13] N. Masuda and K. Aihara, Cryptosystems with discretized chaotic maps, IEEE Trans. Circuits Systems I Fund. Theory Appl., **49** (2002), 28–40.

[14] T. Kohda and A. Tsuneda, Statistics of chaotic binary sequences, IEEE Trans. Inform. Theory, **43** (1997), 104–112.

[15] T. Kohda, Information Sources using chaotic dynamics, Proceedings of the IEEE, **90** (2002), 641–661.

[16] M. Abramowitz and I. A. Stegun, Handbook of Mathematical Functions with Formulas, Graphs, and Mathematical Tables, Dover Publ., New-York, 1972.

[17] T. Kohda and H. Fujisaki, Jacobian elliptic Chebyshev rational maps, Phys. D, **148** (2001), 242–254.

[18] J. Pieprzyk, T. Hardjono and J. Seberry, Fundamentals of Computer Security, Springer, 2003.

[19] A. Lasota, M. C. Mackey, Chaos, Fractals and Noise, Springer-Verlag, 1994.

[20] R. L. Adler and T. J. Rivlin, Ergodic and mixing properties of Chebyshev polynomials, Proc. Amer. Math. Soc., **15** (1964), 794–796.

[21] J. Milnor, On Lattés Maps, Stony Brook IMS Reprint, pp.1–29, #2004/01, Feb. 2004, revised Sept. 2004.

[22] E. Schröder, Ueber iterirte Functionen, Math. Ann., **3** (1871), 296–322.

[23] L. E. Böttcher, The principal laws of convergence of iterates and their application to analysis (Russian), Izv. Kazan. Fiz.-Mat. Obsch., **14** (1904), 155–234.

[24] A. Kato and T. Kohda, 2-D i.i.d. Binary Random Vectors Generated by Jacobian Elliptic Rational Map, Proceedings of 2004 International Symposium on Nonlinear Theory and its Applications, 2004, pp. 613–616.

[25] A. Ono and T. Kohda, Solvable Three-dimensional Rational Chaotic Map Defined by Jacobian Elliptic Function, Internat. J. Bifur. Chaos Appl. Sci. Engrg., **17** (2007), 3645–3650.

§**Appendix A.** Derivation of the theoretical expression of τ_y

The first expression in Eq.(31) gives

$$y_n^2 = (1 - x_n^2)(1 - k^2 + k^2 x_n^2). \tag{39}$$

Solving Eq.(39), we get for $k \neq 0$

$$x_n^2 = \frac{2k^2 - 1 \pm \sqrt{1 - 4k^2 y_n^2}}{2k^2}. \tag{40}$$

Eq.(30) and Eq.(32) give

$$R_2^{\mathrm{cn}}(x_n, k) = \frac{1 - 2(1 - x_n^2) + k^2(1 - x_n^2)^2}{1 - k^2(1 - x_n^2)^2} \tag{41}$$

and

$$y_{n+1} = \sqrt{(1 - (R_2^{\mathrm{cn}}(x_n, k))^2)\,(1 - k^2 + k^2(R_2^{\mathrm{cn}}(x_n, k))^2)}. \tag{42}$$

Substituting Eq.(40) and Eq.(41) into Eq.(42), we have

$$\begin{aligned} y_{n+1} &= \frac{2\sqrt{2}ky\sqrt{2k^2 - 1 \pm \sqrt{1 - 4k^2y^2}}}{(2k^2 - 1 + 2k^2y^2 \pm \sqrt{1 - 4k^2y^2})^2} \\ &\times \left\{1 - 2k^2y^2 \pm (2k^2 - 1)\sqrt{1 - 4k^2y^2}\right\}. \end{aligned} \tag{43}$$

where three $\pm$ signs on R.H.S. are either $+$ or $-$. Denote two maps by $\tau_y^{PP}(y)$ and $\tau_y^{NN}(y)$ when $+$ and $-$ are chosen on the R.H.S. of Eq.(43), respectively. Then

$$\begin{aligned} \tau_y(y) &= X_1(D \oplus Y_1)\tau_y^{PP}(y) + \overline{X_1}\overline{(D \oplus Y_1)}(-\tau_y^{PP}(y)) \\ &+ X_1\overline{(D \oplus Y_1)}\tau_y^{NN}(y) + \overline{X_1}(D \oplus Y_1)(-\tau_y^{NN}(y)) \end{aligned} \tag{44}$$

§Appendix B. Inverse function Y [24]

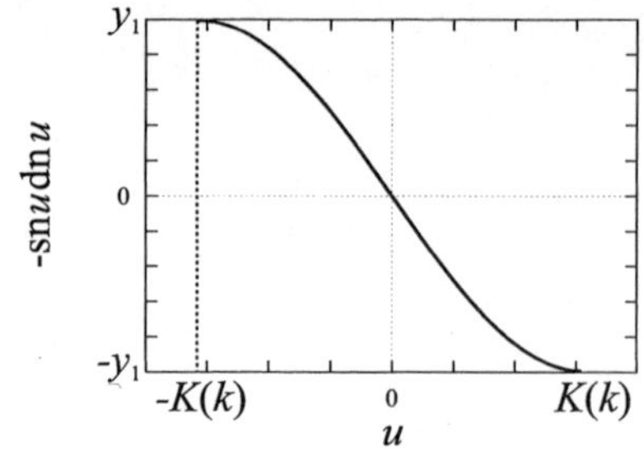

(a) when $k \leq \sqrt{1/2}$

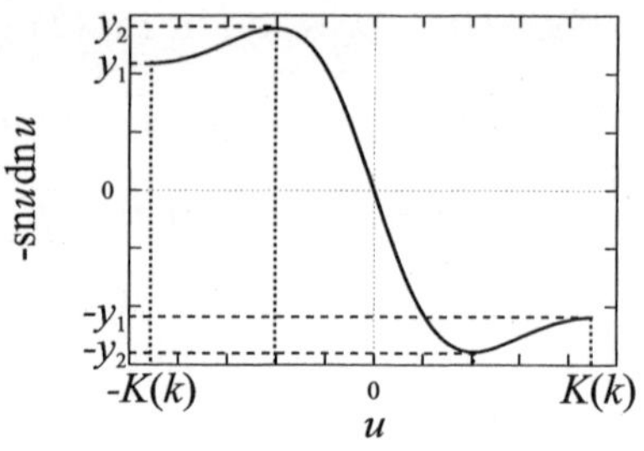

(b) when $k > \sqrt{1/2}$

Fig. 5. $y = -\operatorname{sn} u \operatorname{dn} u$ ($y_1 = \sqrt{1-k^2}$ and $y_2 = 1/2k$, $k \neq 0$).

When $0 < k \leq \sqrt{1/2}$,

$$u = \int_{-\operatorname{sn} u \operatorname{dn} u}^{0} f_Y^+(y)dy. \tag{45}$$

When $k > \sqrt{1/2}$,

$$u = \begin{cases} \displaystyle\int_{-\operatorname{sn} u \operatorname{dn} u}^{0} f_Y^+(y)dy, \quad \text{for } |u| \leq \operatorname{cn}^{-1}\sqrt{\frac{2k^2-1}{2k^2}} \\ \displaystyle\int_{\frac{1}{2k}}^{0} f_Y^+(y)dy - \int_{-\operatorname{sn} u \operatorname{dn} u}^{\frac{1}{2k}} f_Y^-(y)dy, \\ \qquad \text{for } -K(k) \leq u < -\operatorname{cn}^{-1}\sqrt{\frac{2k^2-1}{2k^2}} \\ \displaystyle\int_{-\frac{1}{2k}}^{0} f_Y^+(y)dy - \int_{-\operatorname{sn} u \operatorname{dn} u}^{-\frac{1}{2k}} f_Y^-(y)dy, \\ \qquad \text{for } \operatorname{cn}^{-1}\sqrt{\frac{2k^2-1}{2k^2}} < u \leq K(k) \end{cases} \tag{46}$$

where[24]

$$f_Y^{\pm}(y)dy = \frac{\sqrt{2}k}{\sqrt{(2k^2-1\pm\sqrt{1-4k^2y^2})(1-4k^2y^2)}}dy$$

where the $\pm$ sign on R.H.S is either $+$ or $-$ and is to be decided on the basis whether there is f_Y^+ or f_Y^- on the L.H.S.

§**Appendix C.** Inverse function Z [25]

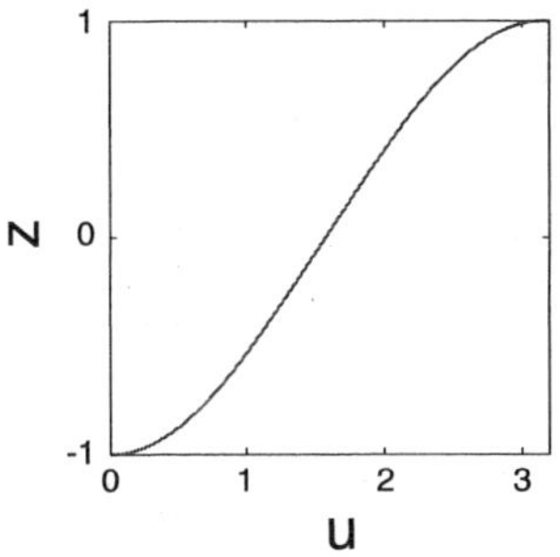

(a) $k \leq \sqrt{1/2}$

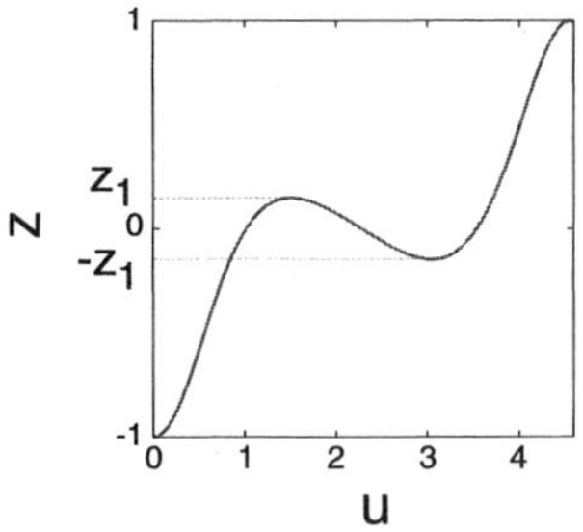

(b) $k > \sqrt{1/2}$

Fig. 6. $z = \operatorname{cn} u(-1 + 2k^2 - 2k^2 \operatorname{cn}^2 u)$, $(z_1 = r(k))$

When $k \leq \sqrt{1/2}$ (see Fig.6(a)), simple differential calculation gives

$$\begin{aligned}(47)\qquad &\frac{d(\operatorname{cn} u(-1 + 2k^2 - 2k^2\operatorname{cn}^2 u))}{du} \\ &= \sqrt{(1 - \operatorname{cn}^2 u)(1 - k^2 + k^2\operatorname{cn}^2 u)} \times \{6k^2\operatorname{cn}^2 u - 2k^2 + 1\}.\end{aligned}$$

Integrating each side of Eq.(47) over u, we have

$$(48)\qquad u = \int_{-1}^{\operatorname{cn} u(-1+2k^2-2k^2\operatorname{cn}^2 u)} \frac{dZ}{\sqrt{(1-X^2(Z))(1-k^2+k^2X^2(Z))}\{6k^2X^2(Z)-2k^2+1\}},$$

where $X^2(Z)$ is given by Eq.(34). ACI measure of the map τ_z is defined in the form of inverse of elliptic functions, i.e., elliptic integral.

$$(49)\qquad u(z) = \int_{-1}^{z} f_Z(\xi_1(Z))dZ, \qquad \text{for } -1 \leq z \leq 1, k \leq \sqrt{1/2}.$$

The same discussion applies to $k > \sqrt{1/2}$ case with care to constants of integration (see Fig.6(b)).

(50)

$$\begin{cases} u_1(z) = \int_{-1}^{z} f_Z(\xi_1(Z))dZ, & \text{for} \quad -1 \le z < -r(k) \\ u_2(z) = u_1(-r(k)) + \int_{-r(k)}^{z} f_Z(\xi_2(Z))dZ, & \text{for} - r(k) \le z < 0 \\ u_3(z) = u_2(0) + \int_{0}^{z} f_Z(\xi_4(Z))dZ, & \text{for} \quad 0 \le z < r(k) \\ u_4(z) = u_3(r(k)) - \int_{r(k)}^{z} f_Z(\xi_3(Z))dZ, & \text{for} \quad r(k) \ge z > -r(k) \\ u_5(z) = u_4(-r(k)) + \int_{-r(k)}^{z} f_Z(\xi_4(Z))dZ, & \text{for} \quad -r(k) \le z < 0 \\ u_6(z) = u_5(0) + \int_{0}^{z} f_Z(\xi_2(Z))dZ, & \text{for} \quad 0 \le z < r(k) \\ u_7(z) = u_6(r(k)) + \int_{r(k)}^{z} f_Z(\xi_1(Z))dZ, & r(k) \le z \le 1 \end{cases}$$

where $f_Z(X_i(Z))$ is given by Eq.(35) and

$$r(k) = \sqrt{\frac{2}{27}(-1 + 2k^2)^3}.$$

Department of Computer Science and Communication Engineering
Kyushu University
Fukuoka
Japan

E-mail address: kohda@csce.kyushu-u.ac.jp

Advanced Studies in Pure Mathematics 53, 2009
Advances in Discrete Dynamical Systems
pp. 113–120

Inefficacy of temporary policy in Neumeyer=Yano's monetary model

Atsumasa Kondo

Abstract.

In this paper, we explore the effects of temporary policy in a dynamic general equilibrium by using Neumeyer and Yano's money-in-utility model. Even under dynamic interaction between consumers through markets, the impact of temporary policy both on the present and future economy is very small if the long-run interest rate level is low, which coincides with the intuitive explanation by the permanent income hypothesis.

§1. Introduction

The permanent income hypothesis proposed by Friedman [1958] implies that a temporary income change has a relatively small impact on consumers' consumption behavior compared with a permanent income change if the long-run interest rate is low. Yano [1984, 91, 98] formulates a temporary policy and proves a similar result using a dynamic general equilibrium theory, in which social interactions generated through markets are explicitly counted in. Although his frameworks hinge on a no restrictive assumption, they are real models without money.

The main purpose of this paper is to demonstrate the inefficacy of temporary policy by using a dynamic model with many consumers

Received October 30, 2006.
Revised October 12, 2007.
Key words and phrases. dynamic general equilibrium, many consumers, temporary policy.
I would like to thank my advisor professor Makoto Yano for his helpful suggestions and advice, and the seminar participants of International Conference on Difference Equations and Applications. In the conference this paper was entitled "On the Inefficacy of Temporary Policy in a Monetary Economy". Financial aid from the 21st century COE program at Kyoto University is gratefully acknowledged.

and with money in their utility functions. To that end, we use a closed-economy version of Neumeyer and Yano [1995]'s money-in-utility model. Neumeyer and Yano [1995] develop a two-country monetary model and explore how unanticipated temporary policy works across national borders. In this paper, we derive a general equilibrium path and conduct a comparative dynamic analysis highlighting the limit case, in which the long-run interest rate vanishes. We find that, in the near limit, a temporary policy that redistributes wealth among consumers has almost no impact on consumption and saving allocations of every consumer. Even with a complex dynamic interaction among any finite number of consumers through markets, the real effects of temporary fiscal and monetary policies can be ignored throughout the whole time horizon. Thus, this study is on the same line as Friedman and Yano, extending to a monetary dynamic general equilibrium theory.

The rest of this paper is organized as follows: First, we set up the model and define an equilibrium in Section 2. In Section 3, we derive the closed-form solution of an equilibrium path. In Section 4, we conduct comparative dynamics and establish the inefficacy theorem.

§2. Model

Think of an economy with an infinite time horizon and discrete time points $t = -1, 0, 1, ...$ We call the period between time $t-1$ and t the period t. There are H consumers and the government, and they transact a consumption good, money and bond at each period. The government imposes a lump-sum tax, τ_{ht}, that is in the consumption-good form on every consumer, and spends it, g_t, $t = 0, 1, ...$ The government provides money M_t^g and bond B_t^g to the consumers. Money brings liquidity service to the consumers, and a unit of bond generates an interest rate $1 + i_t$ from the period t to $t+1$.

Consumer h initially holds money $M_{h,-1}$ and bond $B_{h,-1}$ at $t = -1$, and gets a consumption good endowment y_{ht} for the period t. The stream $\{y_{ht}\}$ is supposed to be bounded over time. Every consumer has his own periodwise preference over his consumption $c_{ht} \in \mathbb{R}_+$ and real money balances $m_{ht} (= M_{ht}/p_t) \in \mathbb{R}_+$ that is represented by a utility function $u_h : \mathbb{R}_+^2 \to \mathbb{R}$. Every consumer at time t, $t = 0, 1, ...$, must be subject to the flow budget constraints:

$$(1) \quad \begin{aligned} &M_{ht+s} + B_{ht+s} \\ &\le p_{t+s}(y_{ht+s} - \tau_{ht+s} - c_{ht+s}) + M_{ht+s-1} + (1 + i_{t+s-1})B_{ht+s-1}, \end{aligned}$$

$s = 0, 1, ...,$ and the no Ponzi-Game condition:

$$\lim_{T\to\infty} \inf \left(\prod_{j=1}^{T} \frac{1}{1 + r_{t+j-1}}\right) \frac{A_{ht+T}}{p_{t+T}} \geq 0, \tag{2}$$

where $1 + r_t = (1 + i_t)\, p_t/p_{t+1}$ is the real interest rate from period t to $t+1$ and $A_{ht} = M_{ht-1} + (1 + i_{t-1})\, B_{ht-1}$ is the consumer h's financial asset at the beginning of the period t, $t = 0, 1, ...$ Following Neumeyer and Yano [1995], we integrate the flow budget constraints and the no Ponzi-Game condition at period t into the intertemporal budget constraints. (See (2).)

Consumer h's behavior is summarized as the following maximizing problem:

$$\begin{aligned} &\left\{c^*_{ht+s}, M^*_{ht+s}, B^*_{ht+s}\right\}_{s=0}^{\infty} \\ &\in \arg \max_{\{c_{ht+s}, M_{ht+s}, B_{ht+s}\}} \sum_{s=0}^{\infty} \beta^s u_h\left(c_{ht+s}, m_{ht+s}\right) \\ \text{s.t. } &\sum_{s=0}^{\infty} \left(\prod_{j=1}^{s} \frac{1}{1 + r_{t+j-1}}\right) \left(c_{ht+s} + \delta_{t+s} m_{ht+s}\right) \leq w_{ht}, \\ \text{where } w_{ht} &= A_{ht}/p_t + \sum_{s=0}^{\infty} \left(\prod_{j=1}^{s} \frac{1}{1 + r_{t+j-1}}\right) \left(y_{ht+s} - \tau_{ht+s}\right), \end{aligned} \tag{3}$$

$t = 0, 1, ...,$ where $\beta \in (0, 1)$ is a discount factor of the future utility and $\delta_t = i_t/\left(1 + i_t\right)$ is the opportunity cost of real money holding at the period t. To shorten notation we define an operator taking the discounted value $\widetilde{x}_t = \sum_{s=0}^{\infty} \left(\prod_{j=1}^{s} \frac{1}{1+r_{t+j-1}}\right) x_{t+s}$ for given t and $\{x_{t+s}\}_{s=0}^{\infty}$. Using this operator, we can rewrite $w_{ht} = A_{ht}/p_t + \widetilde{y}_{ht} - \widetilde{\tau}_{ht}$.

The government sets up a stream of its policy variables $\{M^g_{t+s}, B^g_{t+s}, g_{t+s}, \tau_{ht+s}; h \in \mathcal{H}\}$, $\mathcal{H} = \{1, ..., H\}$. The stream must be subject to the flow budget constraints

$$\begin{aligned} &M^g_{t+s} + B^g_{t+s} \\ &\geq p_{t+s}\left(g_{t+s} - \sum_{h=1}^{H} \tau_{ht+s}\right) + M^g_{t+s-1} + \left(1 + i_{t+s-1}\right) B^g_{t+s-1}, \end{aligned} \tag{4}$$

$s = 0, 1, ...$ and the no Ponzi-Game condition

$$\lim_{T\to\infty} \sup \left(\prod_{j=1}^{T} \frac{1}{1 + r_{t+j-1}}\right) \frac{D^g_{t+T}}{p_{t+T}} \leq 0 \tag{5}$$

for $t = 0, 1, ...$ in much the same way as for consumers. In this paper, the government is supposed to be subject to (4)-(5) with equality.

The stream of price and allocation of the economy is determined so that all markets are cleared:

$$(6) \qquad c_t^* + g_t = y_t; \quad M_t^* = M_t^g; \quad B_t^* = B_t^g; \quad \text{for every } t = 0, 1, ...$$

We denote by $x_t = \sum_{h=1}^{H} x_{ht}$ an aggregate variables. The prices in an equilibrium are distinguished with an asterisk. Since, by Walras's law, the bond market clearing condition is redundant for every $t = 0, 1, ...$, this is ignored in the following analyses.

§3. Equilibrium path

This section derives a closed-form solution, which will be presented in Proposition 1. To this end, following Neumeyer and Yano [1995], we will specify the utility function of the consumer h as $u_h(c, m) = \log c + \gamma_h \log m$. Then, the optimal path is obtained as follows

$$(7) \qquad c_{ht}^* = \frac{1}{1+\gamma_h}(1-\beta)\, w_{ht}; \qquad \delta_t m_{ht}^* = \frac{\gamma_h}{1+\gamma_h}(1-\beta)\, w_{ht};$$

for $h \in \mathcal{H}$ and $t = 0, 1, ...$

As was seen in (7), to get the consumption path we need to obtain the dynamics of the real wealth $\{w_{ht}\}$, which are determined as follows:

Lemma 1. *It holds that* $w_{ht+1} = \beta(1+r_t)\, w_{ht}$ *for every* $h \in \mathcal{H}$ *and* $t = 0, 1, ...$

Proof. By the flow budget constraint, it holds that

$$(8) \qquad c_{ht} + \delta_t m_{ht} \leq y_{ht} - \tau_{ht} + \frac{A_{ht}}{p_t} - \frac{1}{1+r_t}\frac{A_{ht+1}}{p_{t+1}}.$$

Since (7) holds, the left-hand-side of (8) is equal to $LHS = (1-\beta)\, w_{ht}$. By the definition of the real wealth $w_{ht} = \frac{A_{ht}}{p_t} + \widetilde{y}_{ht} - \widetilde{\tau}_{ht}$, the right-hand-side of (8) can be rewritten as $RHS = w_{ht} - \frac{1}{1+r_t} w_{ht+1}$. Hence, $w_{ht+1} = \beta(1+r_t)\, w_{ht}$. Q.E.D.

Thus, by (7) and Lemma 1, $c_{ht+1}^* = \beta(1+r_t)\, c_{ht}^*$ and $\delta_{t+1} m_{ht+1}^* = \beta(1+r_t)\, \delta_t m_{ht}^*$ for every $h \in \mathcal{H}$ and $t = 0, 1, ...$ The real interest rate, which is crucial to the equilibrium path as well as the real wealth, is determined in an equilibrium by the disposable products and the discount factor β; it holds that $1 + r_t^* = \frac{1}{\beta}\frac{y_{t+1}-g_{t+1}}{y_t - g_t}$ for every $t = 0, 1, ...$ since $\frac{y_{t+1}-g_{t+1}}{y_t-g_t} = \frac{c_{t+1}^*}{c_t^*}$ in an equilibrium.

The nominal variables — the price level, the opportunity cost of the real money balances and the nominal interest rate — must satisfy the following relationships in an equilibrium; $\frac{p^*_{t+1}}{p^*_t} = \frac{M^g_{t+1}}{M^g_t}\frac{\delta^*_{t+1}}{\delta^*_t}\frac{y_t - g_t}{y_{t+1} - g_{t+1}}$ and $1 + i^*_t = \frac{1}{\beta}\frac{M^g_{t+1}}{M^g_t}\frac{\delta^*_{t+1}}{\delta^*_t}$ for every $t = 0, 1, ...$, which can be easily ascertained.

Now, we denote the growth rate of money supply by $\mu_t = \left(M^g_{t+1}/M^g_t\right) - 1$ and make the following assumptions:

- **Assumption.**
  A. $y_{ht} = y_h,\ g_t = g,\ \tau_{ht} = \tau_h,\ \mu_t = \mu,\ \gamma_h = \gamma,$ for every $h \in \mathcal{H}$ and $t = 0, 1, ...$
  B. $\beta < 1 + \mu$.
  C. $\lim_{T\to\infty}\left(\prod_{j=1}^{T}\frac{1}{1+i^*_{t+j-1}}\right) = 0$.

By Assumption A, $1 + r_t = \beta^{-1}$ for any t. Without Assumption B, no equilibrium may exist in money-in-utility models. For this point, see Brock [1974]. Assumption C excludes hyperdeflationary paths[1].

Further, we can obtain the equilibrium level of the real money balance m^*_{ht} (Lemma 2).

Lemma 2. *It holds that* $m^*_t = \frac{\gamma(1-\beta)}{1+\gamma}w^*_t\frac{1+\mu}{1+\mu-\beta}$ *for every* $t = 0, 1, ...$

Proof. First, note that $\frac{1}{1+\mu_t}\frac{1}{1+r_t} = \frac{1}{1+i_t}\frac{m^g_t}{m^g_{t+1}}$. Let $t = 0, 1, ...$ By using (7), we can obtain, by induction, that

$$m^*_t - \left(\prod_{j=1}^{T}\frac{1}{1+\mu_{t+j-1}}\frac{1}{1+r_{t+j-1}}\right)m^*_{t+T} = \gamma\sum_{s=0}^{T-1}\left(\prod_{j=1}^{s}\frac{1}{1+\mu_{t+j-1}}\frac{1}{1+r_{t+j-1}}\right)c^*_{t+s}$$

for every T=1, 2, ... In an equilibrium, the limit of both sides of the above expression as $T \to \infty$ can be obtained as follows. Since $\mu_t = \mu$ and $\beta < 1 + \mu$,

$$RHS = \gamma\sum_{s=0}^{T-1}\left(\prod_{j=1}^{s}\frac{1}{1+\mu_{t+j-1}}\frac{1}{1+r^*_{t+j-1}}\right)\beta^s\left(\prod_{j=1}^{s}\left(1+r^*_{t+j-1}\right)\right)c^*_t$$

$$= \gamma c^*_t\prod_{s=0}^{T-1}\left(\frac{\beta}{1+\mu}\right)^s \longrightarrow \gamma c^*_t\frac{1+\mu}{1+\mu-\beta} \text{ as } T \to \infty.$$

---

[1]Brock [1974] proved that a hyperdeflationary path may be an equilibrium that is self-fulfilling in a money-in-the-utility-function model.

Since $\lim_{T\to\infty}\left(\prod_{j=1}^{T}\frac{1}{1+i_{t+j-1}^*}\right)=0$, it holds that

$$LHS = m_t^* - \left(\prod_{j=1}^{T}\frac{1}{1+i_{t+j-1}^*}\frac{m_{t+j-1}^*}{m_{t+j}^*}\right)m_{t+T}^*$$
$$= m_t^* - m_t^*\left(\prod_{j=1}^{T}\frac{1}{1+i_{t+j-1}^*}\right) \longrightarrow m_t^* \text{ as } T\to\infty.$$

Thus, $m_t^* = \gamma c_t^* \frac{1+\mu}{1+\mu-\beta} = \frac{\gamma(1-\beta)}{1+\gamma} w_{ht}^* \frac{1+\mu}{1+\mu-\beta}$. Q.E.D.

As a direct consequence, the equilibrium price of the real money $\delta_t^* = \frac{1+\mu-\beta}{1+\mu}$ is obtained. Now, we can derive an equilibrium path. Before that though, we normalize the equilibrium initial price to be constant $p_0^* = p_0$.[2]

Proposition 1. *An equilibrium path is given as below*

$$(i)\ p_t^* = (1+\mu)^t p_0; \quad (ii)\ 1+r_t^* = \beta^{-1}; \quad (iii)\ 1+i_t^* = \frac{1}{\beta}(1+\mu);$$
$$(iv)\ w_{ht}^* = \frac{A_{h0}}{p_0} + \frac{y_h-\tau_h}{1-\beta}; \quad (v)\ c_{ht}^* = \frac{1-\beta}{1+\gamma}\frac{A_{h0}}{p_0} + \frac{y_h-\tau_h}{1+\gamma};$$
$$(vi)\ m_{ht}^* = \frac{\gamma}{1+\gamma}\frac{1+\mu}{1+\mu-\beta}\left[(1-\beta)\frac{A_{h0}}{p_0} + y_h - \tau_h\right];$$
$$(vii)\ M_{ht}^* = \frac{\gamma}{1+\gamma}\frac{(1+\mu)^{t+1}}{1+\mu-\beta}\left[(1-\beta)A_{h0} + p_0(y_h-\tau_h)\right];$$

for every $t = 0, 1, \ldots$

The proof of the Proposition 1 is so easy that we omit it here. Proposition 1 implies that, as was recognized by Bewley [1982] and Yano [1984, 91, 98], even a stationary path depends on the initial conditions A_{h0} of the model in dynamic general equilibrium models with many consumers. A similar result robustly holds for the present monetary model. Correspondingly, Proposition 1 indicates that both temporary fiscal and monetary policies that redistribute wealth among consumers have prolonged periods of impact on the economy. (The precise definition for the terminology "temporary policy" in this paper is given in the next section.)

[2] Previous studies have often adopted the normalization $\sum_{h=1}^{H}(1/\lambda_h)=1$ (See, e.g. Bewley [1982] or Yano [1998]), where λ_h is the Lagrange multiplier of consumer h's maximizing problem.

§4. Inefficacy of temporary policy

This study highlights the near limit as $\beta \to 1$, which directly implies that the economies with low long-run interest rate level are considered. To conduct meaningful comparative dynamics, we need to suppose $\mu \geq 0$ since Assumption B: $\beta < 1 + \mu$ is essential for the analyses given in the previous section. In addition, it is necessary that national bond issuance level and the fiscal deficit $g - \tau$ be adequately low in the low interest rate case.

Temporary policy in this paper is thought to be conducted in a lump-sum fashion at the initial time point $t = -1$. That consists of fiscal policy $(\tau_{h,-1};\ h \in \mathcal{H})$ and monetary policy $(M_{h,-1};\ h \in \mathcal{H})$ satisfying $\sum_{h=1}^{H} \tau_{h,-1} = 0$ and $\sum_{h=1}^{H} M_{h,-1} = 0$, i.e. they purely redistribute the physical and financial assets among consumers.

Now, we are in an appropriate position to proceed to the main analysis in this paper. By Proposition 1, it holds in the limit as $\beta \to 1$ that

$$1 + r_t^* \to 1; \qquad 1 + i_t^* \to 1 + \mu; \qquad c_{ht}^* \to \frac{y_h - \tau_h}{1 + \gamma};$$

$$m_{ht}^* \to \frac{\gamma}{1+\gamma} \frac{1+\mu}{\mu} (y_h - \tau_h); \qquad M_{ht}^* \to \frac{\gamma}{1+\gamma} \frac{(1+\mu)^{t+1}}{\mu} p_0 (y_h - \tau_h);$$

for every $t = 0, 1, \ldots$ This, in comparison with Proposition 1, demonstrates that the dependence of the allocation path on the initial conditions A_{h0} will disappear for the vanishing discount rate $\beta \to 1$. In other words, temporary fiscal policy $(\tau_{h,-1};\ h \in \mathcal{H})$ and monetary policy $(M_{h,-1};\ h \in \mathcal{H})$ have almost no impact on current and future allocations if the long-run interest rate is very low.

We formally summarize the result of the comparative dynamic analysis as follows:

Theorem 1. *(Inefficacy of Temporary Policy; Friedman=Yano) Temporary policies have almost no impact both on current and future consumption allocations in a dynamic general equilibrium if the interest rate level is low.*

As already discussed in the introduction, the result is on the same line as Friedman and Yano. However, my analyses here take the two following aspects into account. First, price effect generated through markets, which is ignored in the arguments concerning the standard permanent income hypothesis, is considered. Second, money that helps consumers to transact, which is not counted in by the existing works of Yano [1984, 91, 98], is also considered.

References

[1] T. Bewley, An integration of equilibrium theory and turnpike theory, J. Math. Econom., **10** (1982), 233–267.

[2] W. Brock, Money and growth: the case of long run perfect foresight, Internat. Econom. Rev., **15** (1974), 750–777.

[3] M. Friedman, A Theory of Consumption Function, Princeton Univ. Press, 1958.

[4] P. Neumeyer and M. Yano, Cross Border Nominal Asset and International Monetary Independence, mimeo, 1995.

[5] M. Yano, The turnpike of dynamic general equilibrium paths and its insensitivity to initial conditions, J. Math. Econom., **13** (1984), 235–254.

[6] ———, Temporary transfers in a simple dynamic general equilibrium model, J. Econom. Theory, **54** (1991), 372–388.

[7] ———, On the dual stability of a von Neumann facet and the inefficacy of temporary fiscal policy, Econometrica, **66** (1998), 427–451.

Shiga University
1-1-1 Banba, Hikone
Shiga, 522-8522
Japan

E-mail address: a-kondo@biwako.shiga-u.ac.jp

Advanced Studies in Pure Mathematics 53, 2009
Advances in Discrete Dynamical Systems
pp. 121–128

Discrete potential theory for iterated maps of the interval

C. Correia Ramos, Nuno Martins, J. Sousa Ramos and Ricardo Severino

Abstract.

Using Markov partitions and algebraic graph theory we introduce, in the context of discrete dynamical systems, some laws which characterize the nonlinear dynamics of iterated maps of the interval. In the Markov digraphs we assume that each directed edge has a weight associated to it, given by the Markov invariant measure. This system of weights produces a diffusion process determined by a transition matrix. In this setting, we define a current and a potential which are dynamical invariants.

§1. Introduction and preliminaries

Nonlinear dynamical systems can be effectively studied and modelled by difference equations. An important problem is to describe the behavior of the discrete dynamical system determined by the iteration of a map f,

$$x_{k+1} = f(x_k).$$

Our approach uses symbolic dynamics to obtain a digraph associated with the original discrete dynamical system, as in [3]. Several invariants for the discrete dynamical system can be obtained and interpreted in the digraph setting, see for example [7] and also [4], [5]. The notions of current, potential, conductance, have its origins in the electric circuits and have been generalized to graph theory. We extend some of this notions to discrete dynamical systems.

Given an interval map f, and using the orbits of the critical points, we obtain a graph, the *Markov digraph*, directly from the transition matrix A_f and Markov partition for f. Using the Parry measure, we

Received November 15, 2006.
Revised October 12, 2007.

obtain a weighted digraph, and by analogy with the electric circuits we introduce a current and a potential which satisfies the general Kirchoff laws, and are important dynamical invariants. These are very general laws that can be applied in many different contexts, see [2] (and also [1]).

We define the (ergodic) potential as the Parry invariant measure p. The reason is that if we think the time evolution as an iteration, this is a stationary potential. It can be seen as the density of points in the vertices (which corresponds to the partition intervals). We define also the (ergodic) current q in a way that do not change under iteration. In a certain sense, it corresponds to the conditional probability of a vertex to have a transition defined by an edge.

1.1. Interval maps

Let $f : I \to I$ be a piecewise monotone map of the interval into itself. The *singular points* of f are the points of I in which f' is zero, not defined or discontinuous and constitutes a finite set $\{c_1, ..., c_m\}$. Therefore, there is a finite partition

$$I = I_1 \cup C_1 \cup ... \cup C_m \cup I_{m+1}$$

so that I_j is a maximal interval of monotonicity for f (*a lap*) and $C_j = \{c_j\}$. Every point in I has a unique symbolic expansion in the alphabet

$$\mathfrak{A} = \{I_1, ..., I_{m+1}, C_1, ..., C_m\},$$

denoted by $it(x)$ (*itinerary of* x), given as follows: the *address of* x, $ad(x) \in \mathfrak{A}$, is determined by $x \in ad(x)$. The itinerary is then defined by

$$it(x) = (ad(x), ad(f(x)), ad(f^2(x), ...) \in \mathfrak{A}^{\mathbb{N}}.$$

Consider the orbits of the singular points. The itineraries of the images of the singular points are called the *kneading sequences* of f, $K_{c_j} = it(f(c_j))$, (see [6]), and define the *kneading invariant* for f,

$$\mathcal{K}_f = (K_{c_1}, ..., K_{c_m}).$$

This is a very important invariant for the discrete dynamical system defined by f.

We assume that f is such that the orbits of the singular points are finite, i.e., such that all the kneading sequences are eventually periodic. Let $\{x_1, x_2, ..., x_{n'}\}$ be the union of the singular points orbits, ordered according to

$$x_1 < x_2 < ... < x_{n'}.$$

Define the intervals $J_i = [x_i, x_{i+1}]$ with $i = 1, ..., n' - 1$. The set $\{J_1, ..., J_n\}$, with $n = n' - 1$, is a Markov partition for f. Then we define the transition matrix $A_f = (a_{ij})$ by

$$a_{ij} = \begin{cases} 1 \text{ if } f(\mathrm{Int}(J_i)) \supseteq f(\mathrm{Int}(J_j)) \\ 0 \text{ otherwise.} \end{cases}$$

1.2. Digraphs

Consider a graph $\mathcal{G} = (G_0, G_1, s, r)$ where G_0 is the set of vertices and G_1 the set of edges. The maps $s, r : G_1 \to G_0$, (source and range) assign to every edge its head and its tail, respectively.

Let $C^0 = C^0(\mathcal{G}; \mathbb{R})$ be the linear space of real functions on the set G_0, and $C^1 = C^1(\mathcal{G}; \mathbb{R})$ be the linear space of real functions on the set G_1. The canonical basis for C^0 is given by $\{\delta_v : v \in G_0\}$. The same for C^1, with $\{\delta_e : e \in G_1\}$.

Let us assume $G_0 = \{1, ..., n\}$. The adjacency matrix of $\mathcal{G}$ is the $n \times n$ matrix, $A = (a_{ij})$, with

$$a_{ij} = \begin{cases} 1 \text{ if there is } e \in G_1 : s(e) = i, r(e) = j \\ 0 \text{ otherwise.} \end{cases}$$

The incidence operator $D : C^1 \to C^0$ is defined by $D\delta_e = \delta_{r(e)} - \delta_{s(e)}$. It is represented by the incidence matrix D. The adjoint $D^t : C^0 \to C^1$ is defined as follows: Given $\alpha \in C^0$ we have $(D^t\alpha)(e) = \alpha(r(e)) - \alpha(s(e))$. The linear space $\ker D^t$ is the space of constant functions on the connected components of $\mathcal{G}$, see [2]. Therefore, if G is connected $\ker D^t$ is the one-dimensional space generated by the function $\sum_{i=1}^n \delta_i$. The operator D also has an explicit formula

$$(Dg)(v) = \sum_{e:r(e)=v} g(e) - \sum_{e:s(e)=v} g(e).$$

The kernel of D is the set of functions $g \in C^1$ so that

$$\sum_{e:r(e)=v} g(e) = \sum_{e:s(e)=v} g(e)$$

for every $v \in G_0$. An element in $\ker D$ is usually called a *flow*, or a *current.* Let $Z = \ker D$ and let $Z^\perp$ be the orthogonal complement of Z, that is,

$$C^1 = Z \oplus Z^\perp = \ker D \oplus (\ker D)^\perp .$$

A *path* is a sequence of vertices (or edges) in the graph $\mathcal{G}$. A path $(v_0, ..., v_k)$ is called a closed path if $v_0 = v_k$ and is called a *cycle* if all other vertices are distinct. To each cycle γ we associate the function $z_\gamma \in C^1$

$$z_\gamma(e) = \begin{cases} 1, & \text{if the sequence } s(e), r(e) \text{ occurs in } \gamma \\ -1, & \text{if the sequence } r(e), s(e) \text{ occurs in } \gamma \\ 0, & \text{otherwise.} \end{cases}$$

Let S be a non-empty subset of G_0. Define the function $b_S \in C^1$ by

$$b_S(e) = \begin{cases} 1, & \text{if } \{s(e), r(e)\} \cap S = \{r(e)\} \\ -1, & \text{if } \{s(e), r(e)\} \cap S = \{s(e)\} \\ 0, & \text{otherwise.} \end{cases}$$

Therefore, given $S \subset G_0$, the function b_S determines which edges "go outside" or "go inside" the region determined by the vertices in S. The set of edges which have exactly one vertex in S is called a *cut* or a *cocycle* of G.

We have that $z_\gamma \in Z$, for every cycle γ and $b_S \in Z^\perp$, for every proper subset $S \subset G_0$, see [2]. Therefore, Z is usually called the *cycle space* and $Z^\perp$ the *cocycle space.*

Let $\mathcal{T}$ be a spanning tree for $\mathcal{G}$, that is, a connected subgraph containing every vertex of $\mathcal{G}$ and with no cycle. If we remove an edge from $\mathcal{T}$ we obtain two components, $\mathcal{T}^+$ and $\mathcal{T}^-$, one containing $r(e)$ and the other containing $s(e)$. The cocycle associated with the set of vertices in $\mathcal{T}^+$, denoted by $S(e) = S(\mathcal{T}, e)$, is called a *fundamental cocycle* determined by $\mathcal{T}$ and e. The importance of the fundamental cocycles is that for any spanning tree $\mathcal{T}$ the functions $b_{S(e)}$, with e being an edge in $\mathcal{T}$, constitute a basis for the cocycle space $Z^\perp$, see [2]. Now, given a spanning tree $\mathcal{T}$ for each edge e not in $\mathcal{T}$ there is a unique path in $\mathcal{T}$ with initial vertex $s(e)$ and final vertex $r(e)$. This path together with the edge e determines a cycle $\gamma(e) = \gamma(\mathcal{T}, e)$ called a *fundamental cycle* determined by $\mathcal{T}$ and e. Given a spanning tree $\mathcal{T}$, the functions $z_{\gamma(e)}$, with e an edge not in $\mathcal{T}$, determines a base for the cycle space Z. Then, we denote $C = C(\mathcal{T})$ as the matrix of the fundamental cycles, formed by the vectors in the base of Z determined by $\mathcal{T}$. In the same manner, we define $B = B(\mathcal{T})$ as the matrix of the fundamental cocycles, formed by the vectors in the base of $Z^\perp$ determined by $\mathcal{T}$.

§2. Current and potential for interval maps

Let f be a Markov map of the interval I into itself, with Markov partition $\{I_1, ..., I_n\}$, and transition matrix A_f. There is a special measure associated with f, the *Parry measure*, which is an invariant measure and the maximal entropy measure, see [8]. It can be given as follows: Let λ_f be the Perron eigenvalue of A_f. Let u be the Perron left eigenvalue of A_f and v the Perron left eigenvalue of A_f. Let $P_f = (p_{ij})$, where

$$p_{ij} = \frac{a_{ij}v_j}{\lambda v_i},$$

and $p = (p_i)$ where $p_i = \frac{u_i v_i}{\sum u_j v_j},$. We have $p = pP_f$, that is, p is the left Perron eigenvalue of P_f, and $\sum_{i=1}^{n} p_i = 1$. Therefore, p represents an invariant probability measure, which is precisely the Parry measure.

Let $\mathcal{G}_f = (G_0, G_1, r, s)$ be the digraph associated with the Markov partition of f so that the transition matrix A_f is the adjacency matrix of $\mathcal{G}_f$. Following the notation introduced in Section 1, let D_f be the incidence matrix of $\mathcal{G}_f$. Note that G_1 coincides with the set of admissible words of size 2. More precisely, there is a bijection

$$ij \text{ admissible word} \longleftrightarrow (ij) \in G_1.$$

Definition 1. *The (ergodic) current is the function $q_f \in C^1_f$ defined by*

$$q_f(e) = p_{s(e)} P_{s(e)r(e)}.$$

The (ergodic) potential is the function $p_f \in C^0_f$.

We will see that q is in fact a function of Z_f, therefore, is completely justified the name of current or flow. Next, we present our main result:

Theorem 2. *Let f be a Markov interval map. Then we have the following:*

$$D_f \; q_f = 0.$$

For every $b \in Z_f^{\perp}$, $\langle b, q_f \rangle = 0$ (current Kirchoff law).

For every $z \in Z_f$, $\left\langle z, D_f^t \; p_f \right\rangle = 0$ (voltage Kirchoff law).

Proof. We have

$$\begin{aligned}
(D_f q_f)(j) &= \sum_{e:r(e)=j} q_f(e) - \sum_{e:s(e)=j} q_f(e) \\
&= \sum_{e:r(e)=j} p_{s(e)} P_{s(e)r(e)} - \sum_{e:s(e)=j} p_{s(e)} P_{s(e)r(e)} \\
&= \sum_i p_i P_{ij} - \sum_i p_j P_{ji} = \sum_i p_i P_{ij} - p_j \sum_i P_{ji}.
\end{aligned}$$

Since $\sum_i P_{ji} = 1$ (the vector $(1, ..., 1)$ is a right eigenvector of P_f) and $\sum_i p_i P_{ij} = p_j$ (p_f is a left eigenvector of P_f) we have $D_f q_f = 0$. Since

$(D_f q_f)(j) = 0$ corresponds to the equation $\langle b, q_f \rangle = 0$ for the cocycle $b = (0, ..., 1, .., 0)$ (1 in j position) we have by linearity that $\langle b, q_f \rangle = 0$, for every $b \in Z_f^{\perp}$. Now, since $\mathrm{Im}(D_f^t) \cong \ker(D_f)^{\perp}$ is orthogonal to Z, therefore the result follows. Q.E.D.

Example 3. *Let us consider the unimodal map $f_b(x) = 4bx(1-x)$, with $b = 0.9764266...$, i.e., such that the orbit of the critical point is periodic of period 5 with kneading sequence $RLLRC$. Its topological entropy is equal to*

$$h_{top}(f_b) = \log(1.72208...).$$

The transition matrix associated with the Markov partition (using the partition generated by the critical point orbit) is

$$A_f = \begin{pmatrix} 0 & 1 & 1 & 1 & 0 \\ 0 & 0 & 0 & 0 & 1 \\ 0 & 0 & 0 & 1 & 1 \\ 0 & 0 & 1 & 0 & 0 \\ 1 & 1 & 0 & 0 & 0 \end{pmatrix}.$$

The spectral radius is $\lambda_{\max}(A_f) = 1.72208...$ and the normalized Perron right eigenvector is (approximately)

$$v = (0.277916, 0.141392, 0.213326, 0.123877, 0.243489).$$

The normalized Perron left eigenvector is (approximately)

$$u = (0.141392, 0.223497, 0.195811, 0.195811, 0.243489).$$

The matrix P_f is (approximately) given by

$$P_f = \begin{pmatrix} 0 & 0.295431 & 0.445734 & 0.258834 & 0 \\ 0 & 0 & 0 & 0 & 1 \\ 0 & 0 & 0 & 0.337203 & 0.662797 \\ 0 & 0 & 1 & 0 & 0 \\ 0.662797 & 0.337203 & 0 & 0 & 0 \end{pmatrix},$$

and the vector p_f, which corresponds to the Parry measure, is (approximately)

$$p_f = (0.20027, 0.161055, 0.212892, 0.123624, 0.302159).$$

The incidence matrix of the digraph is

$$D_f^t = \begin{pmatrix} -1 & -1 & 0 & 0 & 0 & 0 & 0 & -1 & 1 \\ 1 & 0 & 0 & 0 & 1 & -1 & 0 & 0 & 0 \\ 0 & 1 & -1 & -1 & 0 & 0 & 1 & 0 & 0 \\ 0 & 0 & 1 & 0 & 0 & 0 & -1 & 1 & 0 \\ 0 & 0 & 0 & 1 & -1 & 1 & 0 & 0 & -1 \end{pmatrix}.$$

The matrix of a basis of the fundamental cycles is given by

$$C_f = \begin{pmatrix} -1 & 1 & 0 & 1 & 1 & 0 & 0 & 0 & 0 \\ 1 & -1 & 0 & -1 & 0 & 1 & 0 & 0 & 0 \\ 0 & 0 & 1 & 0 & 0 & 0 & 1 & 0 & 0 \\ 0 & -1 & -1 & 0 & 0 & 0 & 0 & 1 & 0 \\ 0 & 1 & 0 & 1 & 0 & 0 & 0 & 0 & 1 \end{pmatrix}.$$

The matrix of a basis of the fundamental cocycles (cuts) is given by

$$B_f = \begin{pmatrix} 1 & 0 & 0 & 0 & 1 & -1 & 0 & 0 & 0 \\ 0 & 1 & 0 & 0 & -1 & 1 & 0 & 1 & -1 \\ 0 & 0 & 1 & 0 & 0 & 0 & -1 & 1 & 0 \\ 0 & 0 & 0 & 1 & -1 & 1 & 0 & 0 & -1 \end{pmatrix}.$$

The current vector, $q_f = p_f P_f$, is (approximately)

$$\begin{aligned} q_f \quad = \quad & (0.059166, 0.0892672, 0.0717877, 0.141104, \\ & 0.101889, 0.161055, 0.123624, 0.0518367, 0.20027). \end{aligned}$$

The divergence $D_f\ q = 0$ and $B_f\ q_f = 0$, the total current that crosses each cocycle is zero. Moreover the difference of potential along each cycle is zero, $C_f D_f^t\ p_f = 0$.

References

[1] R. Bacher, P. de la Harpe and T. Nagnibeda, The lattice of integral flows and the lattice of integral cuts on a finite graph, Bull. Soc. Math. France, **125** (1997), 167–198.

[2] N. Biggs, Algebraic potential theory on graphs, Bull. London Math. Soc., **29** (1997), 641–682.

[3] C. C. Ramos, N. Martins, R. Severino and J. S. Ramos, Critical groups for iterated maps, Tenth International Conference on Difference Equations, Special Functions and Applications, ICDEA2005, Munich, Germany, 25-30 July 2005.

[4] C. C. Ramos, N. Martins and J. S. Ramos, Conductance and noncommutative dynamical systems, Nonlinear Dynam., **44** (2006), 127–134.

[5] S. Fernandes and J. S. Ramos, Conductance, laplacian and mixing rate in discrete dynamical systems, Nonlinear Dynam., **44** (2006), 117–126.

[6] J. Milnor and W. Thurston, On iterated maps of the interval, In: Dynamical Systems, Proceedings Univ. Maryland, 1986-1987, (ed. J. C. Alexander), Lect. Notes in Math., **1342**, Springer-Verlag, 1988, pp. 465–563.

[7] J. S. Ramos, Introduction to nonlinear dynamics of electronic systems: tutorial, Nonlinear Dynam., **44** (2006), 3–14.

[8] P. Walters, An Introduction to Ergodic Theory, Grad. Texts in Math., **79**, Springer-Verlag, 1982.

C. Correia Ramos
CIMA
Centro de Investigação em Matemática e Aplicações
Department of Mathematics
Universidade de Évora, Rua Romão Ramalho, 59
7000-671, Évora
Portugal

Nuno Martins
Centro de Análise Matemática
Geometria e Sistemas Dinâmicos
Department of Mathematics
Instituto Superior Técnico
Av. Rovisco Pais 1
1049-001 Lisboa
Portugal

J. Sousa Ramos
Centro de Análise Matemática
Geometria e Sistemas Dinâmicos
Department of Mathematics
Instituto Superior Técnico
Av. Rovisco Pais 1
1049-001 Lisboa
Portugal

Ricardo Severino
CIMA
Centro de Investigação em Matemática e Aplicações
Department of Mathematics
Universidade de Évora, Rua Romão Ramalho, 59
7000-671, Évora
Portugal

E-mail address: `ccr@uevora.pt`
`nmartins@math.ist.utl.pt`
`sramos@math.ist.utl.pt`
`ricardo@math.uminho.pt`

Advanced Studies in Pure Mathematics 53, 2009
Advances in Discrete Dynamical Systems
pp. 129–139

Global behavior of a two-dimensional monotone difference system

Momoe Inoue and Hideaki Matsunaga

Abstract.

We investigate global behavior of solutions of a nonlinear difference system

$$x_{n+1} = px_n(1 + y_n), \quad y_{n+1} = qy_n(1 + x_n), \quad n = 0, 1, 2, \ldots,$$

where parameters p, q and initial values x_0, y_0 are positive. We give sufficient conditions for every solution of the system to be unbounded and sufficient conditions for the global stable manifold of the positive equilibrium to exist, which is a unbounded separatrix for the system. Some related conjectures are also given.

§1. Introduction and preliminaries

Consider a nonlinear difference system

$$u_{n+1} = \frac{au_nv_n}{b + cv_n}, \quad v_{n+1} = \frac{du_nv_n}{e + fu_n}, \quad n = 0, 1, 2, \ldots, \tag{1}$$

where all parameters a, b, c, d, e, f are positive and initial values u_0, v_0 are nonnegative. The system (1) may be regarded as a cooperative system (see [4, 5]). By the change of variables $x_n = e/(fu_n)$ and $y_n = b/(cv_n)$, the system (1) can be transformed into a *monotone* difference system

$$x_{n+1} = px_n(1 + y_n), \quad y_{n+1} = qy_n(1 + x_n), \quad n = 0, 1, 2, \ldots, \tag{2}$$

where $p = c/a > 0$ and $q = f/d > 0$. For general study of monotone dynamical systems, one can refer to [3, 6, 7] and the references contained therein.

The purpose of this paper is to classify global behavior of solutions of the system (2) with positive initial values x_0, y_0. Our results are

Received October 31, 2006.

closely related to results in [5] due to Kulenović and Nurkanović for the system (1) with $b = c = e = f = 1$.

The equilibria of (2) are $(0, 0)$ for any p and q, and $(1/q - 1, 1/p - 1)$ for $0 < p < 1$ and $0 < q < 1$. In addition, if $p = 1$, then every point on the x-axis is an equilibrium point, and if $q = 1$, then every point on the y-axis is an equilibrium point. The map $T : \mathbf{R}^2_+ \to \mathbf{R}^2_+$ of (2) is given by

$$T(x, y) = (px(1 + y), qy(1 + x)), \tag{3}$$

where $\mathbf{R}^2_+ = \{(x, y) \mid x \geq 0, y \geq 0\}$. Then, for any $(x, y) \in \mathbf{R}^2_+$, the Jacobian matrix for T is given by

$$J_T(x, y) = \begin{pmatrix} p(1 + y) & px \\ qy & q(1 + x) \end{pmatrix}$$

and the characteristic equation of the Jacobian evaluted at $E_0 = (0, 0)$ is

$$(\lambda - p)(\lambda - q) = 0.$$

As is well known (see, e.g. [1, 2]), the equilibrium E_0 of (2) is locally asymptotically stable if $0 < p < 1$ and $0 < q < 1$ and it is unstable if $p > 1$ or $q > 1$. On the other hand, the characteristic equation of the Jacobian evaluted at $E_{p,q} = (1/q - 1, 1/p - 1)$ is

$$\lambda^2 - 2\lambda + 1 - (1 - p)(1 - q) = 0$$

with roots

$$\lambda_\pm = 1 \pm \sqrt{(1 - p)(1 - q)}.$$

Obviously, $|\lambda_+| > 1$ and $|\lambda_-| < 1$ if $0 < p < 1$ and $0 < q < 1$, which implies that the equilibrium $E_{p,q}$ of (2) is a saddle point. Therefore, we summarize local stability properties of the equilibria of (2) as follows:

Cases	Equilibria of (2)
$p > 1$ or $q > 1$	E_0 (unstable)
$0 < p < 1$ and $0 < q < 1$	E_0 (locally AS) and $E_{p,q}$ (unstable)
$p = q = 1$	x-axis and y-axis
$p = 1$ and $0 < q < 1$	x-axis
$0 < p < 1$ and $q = 1$	y-axis

Table 1. Linearized stability analysis of (2)

In the following, we will give some notations to state a basic property on the map T given by (3). For $v = (v_1, v_2)$, $w = (w_1, w_2) \in \mathbf{R}^2_+$, we say that $v \leq w$ if $v_1 \leq w_1$ and $v_2 \leq w_2$. Two points v, $w \in \mathbf{R}^2_+$ are said to be *related* if $v \leq w$ or $w \leq v$. Also, a strict inequality between points is defined as $v < w$ if $v \leq w$ and $v \neq w$. A strong inequality is defined as $v \ll w$ if $v_1 < w_1$ and $v_2 < w_2$.

Proposition 1. *Let v, $w \in \operatorname{Int}\mathbf{R}^2_+$. If $v < w$, then $T^n(v) \ll T^n(w)$ for $n = 1, 2, \ldots$.*

Proof. Let $v = (v_1, v_2)$, $w = (w_1, w_2) \in \operatorname{Int}\mathbf{R}^2_+$. If $v < w$, then

$$v_1(1 + v_2) < w_1(1 + w_2), \quad v_2(1 + v_1) < w_2(1 + w_1),$$

which implies $T(v) \ll T(w)$. By induction, we have $T^n(v) \ll T^n(w)$ for $n = 1, 2, \ldots$. Q.E.D.

Proposition 1 shows that if two points in $\operatorname{Int}\mathbf{R}^2_+$ are related, then all iterates of these points are related.

§2. Global results

2.1. The case $p > 1$ or $q > 1$

In this case, there exists a unique equilibrium E_0 which is unstable. Then we have the following result on global behavior of solutions of (2).

Theorem 1. *Assume that $p > 1$ or $q > 1$. Then every solution (x_n, y_n) of (2) satisfies $\lim_{n\to\infty} x_n = \lim_{n\to\infty} y_n = \infty$.*

Proof. We consider the case $p > 1$. In case $q > 1$, the proof is similar and will be omitted. By (2), we have

$$x_{n+1} > px_n, \quad n = 0, 1, 2, \ldots,$$

which implies that $x_n > p^n x_0 \to \infty$ as $n \to \infty$, and so $\lim_{n\to\infty} x_n = \infty$. There are two possible cases to consider.

Case (i): $p > 1$ and $q > 1$. An argument similar to that above yields $\lim_{n\to\infty} y_n = \infty$.

Case (ii): $p > 1$ and $q \leq 1$. Since $\lim_{n\to\infty} x_n = \infty$, there exists a positive integer N such that

$$x_n > \frac{1}{q} - 1 \quad \text{for } n \geq N.$$

Then we have

$$\frac{y_{n+1}}{y_n} = q(1 + x_n) > 1 \quad \text{for } n \geq N,$$

which implies that y_n is an increasing sequence for $n \geq N$. Suppose that y_n is bounded above. Then there exists a positive number β such that $\lim_{n\to\infty} y_n = \beta$, and hence

$$q(1 + x_n) = \frac{y_{n+1}}{y_n} \to 1 \quad \text{as} \quad n \to \infty.$$

This is a contradiction to the fact that the left-hand side tends to ∞ as $n \to \infty$, and so $\lim_{n\to\infty} y_n = \infty$. This completes the proof. Q.E.D.

2.2. The case $0 < p < 1$ and $0 < q < 1$

In this case, there exist two equilibria E_0 which is locally asymptotically stable and $E_{p,q}$ which is a saddle point. For each $v = (v_1, v_2) \in \operatorname{Int} \mathbf{R}_+^2$, we define $Q_i(v)$ for $i = 1, \ldots, 4$ to be the usual four quadrants centered at v and numbered in a counterclockwise direction, e.g., $Q_1(v) = \{(x, y) \in \operatorname{Int} \mathbf{R}_+^2 \mid v_1 \leq x, v_2 \leq y\}$.

The first lemma deals with behavior of solutions of (2) in $Q_1(E_{p,q})$ or $Q_3(E_{p,q})$.

Lemma 1. *Assume that* $0 < p < 1$ *and* $0 < q < 1$. *Let* (x_n, y_n) *be a solution of* (2). *Then the following statements hold:*

(a) *If* $(x_0, y_0) \in Q_1(E_{p,q}) \setminus E_{p,q}$, *then* $\lim_{n\to\infty} x_n = \lim_{n\to\infty} y_n = \infty$.

(b) *If* $(x_0, y_0) \in Q_3(E_{p,q}) \setminus E_{p,q}$, *then* $\lim_{n\to\infty} x_n = \lim_{n\to\infty} y_n = 0$.

Proof. We will prove the statement (a). The proof of the statement (b) is similar and will be omitted. Let $(x_0, y_0) \in Q_1(E_{p,q}) \setminus E_{p,q}$, that is, $x_0 > 1/q - 1$ and $y_0 > 1/p - 1$. Then

$$\frac{x_1}{x_0} = p(1 + y_0) > 1, \quad \frac{y_1}{y_0} = q(1 + x_0) > 1,$$

which, by induction, implies

$$x_{n+1} > x_n > \frac{1}{q} - 1, \quad y_{n+1} > y_n > \frac{1}{p} - 1, \quad n = 0, 1, 2, \ldots.$$

Therefore x_n and y_n are increasing sequences. Suppose that x_n and y_n are bounded above. Then there exist positive numbers α, β such that $\lim_{n\to\infty} x_n = \alpha > 1/q - 1$, $\lim_{n\to\infty} y_n = \beta > 1/p - 1$, and thus, the point (α, β) is another equilibrium in $Q_1(E_{p,q})$. This is a contradiction to the fact that (2) has only two equilibria E_0 and $E_{p,q}$. Suppose that y_n (resp. x_n) is bounded above and x_n (resp. y_n) tends to ∞ as $n \to \infty$. Then a contradiction also arises by a similar fashion in the proof of Theorem 1. Consequently, $\lim_{n\to\infty} x_n = \lim_{n\to\infty} y_n = \infty$. The proof is complete. Q.E.D.

Lemma 1 shows that a stable eigenvector of $E_{p,q}$ can be chosen to have components of opposite signs. Thus there exists a monotonically decreasing function $h(x)$ such that the local stable manifold W^s_{loc} of $E_{p,q}$ is the graph of h. In fact, the following lemma holds.

Lemma 2. *Assume that* $0 < p < 1$ *and* $0 < q < 1$. *Then the global stable manifold* W^s *of* $E_{p,q}$ *is a graph of a decreasing function of* x.

Proof. It follows from Lemma 1 that $W^s \subset Q_2(E_{p,q}) \cup Q_4(E_{p,q})$. Hence we will show that W^s contains no related points, then the proof will be complete. To this end, we have only to verify that W^s_{loc} contains no related points by Proposition 1. Suppose that there exist related points $v = (v_1, v_2)$, $w = (w_1, w_2) \in W^s_{\text{loc}}$ such that $v < w$. Then $h(v_1) = v_2 \le w_2 = h(w_1)$, which contradicts the fact that h is monotonically decreasing, and thus, W^s_{loc} contains no related points. This completes the proof. Q.E.D.

Next we will observe behavior of solutions of (2) in $Q_2(E_{p,q})$ or $Q_4(E_{p,q})$.

Lemma 3. *Assume that* $0 < p < 1$ *and* $0 < q < 1$. *Let* (x_n, y_n) *be a solution of* (2). *Then the following statements hold:*

(a) *If* $(x_0, y_0) \in Q_2(E_{p,q}) \setminus E_{p,q}$ *and* $x_1 > 1/q - 1$, *then* $(x_1, y_1) \in Q_1(E_{p,q})$ *and* $\lim_{n\to\infty} x_n = \lim_{n\to\infty} y_n = \infty$.
(b) *If* $(x_0, y_0) \in Q_4(E_{p,q}) \setminus E_{p,q}$ *and* $y_1 > 1/p - 1$, *then* $(x_1, y_1) \in Q_1(E_{p,q})$ *and* $\lim_{n\to\infty} x_n = \lim_{n\to\infty} y_n = \infty$.
(c) *If* $(x_0, y_0) \in Q_2(E_{p,q}) \setminus E_{p,q}$ *and* $y_1 < 1/p - 1$, *then* $(x_1, y_1) \in Q_3(E_{p,q})$ *and* $\lim_{n\to\infty} x_n = \lim_{n\to\infty} y_n = 0$.
(d) *If* $(x_0, y_0) \in Q_4(E_{p,q}) \setminus E_{p,q}$ *and* $x_1 < 1/q - 1$, *then* $(x_1, y_1) \in Q_3(E_{p,q})$ *and* $\lim_{n\to\infty} x_n = \lim_{n\to\infty} y_n = 0$.

Proof. We will prove the statement (a). The proof of the statements (b), (c) and (d) are similar and will be omitted. Let $(x_0, y_0) \in Q_2(E_{p,q}) \setminus E_{p,q}$ and $x_1 = px_0(1 + y_0) > 1/q - 1$. Then we have

$$1 + x_0 > 1 + \frac{1}{p(1 + y_0)} \left(\frac{1}{q} - 1 \right),$$

which implies

$$y_1 = qy_0(1+x_0) > qy_0 + \frac{(1-q)y_0}{p(1+y_0)}. \tag{4}$$

Then it follows from $y_0 > 1/p - 1$ that

$$\begin{aligned} & qy_0 + \frac{(1-q)y_0}{p(1+y_0)} - \left(\frac{1}{p} - 1\right) \\ &= \frac{1}{1+y_0}\left\{qy_0(1+y_0) + \frac{1-q}{p}y_0 - \left(\frac{1}{p}-1\right)(1+y_0)\right\} \\ &= \frac{1}{1+y_0}\left\{qy_0^2 + \left(q+1-\frac{q}{p}\right)y_0 - \frac{1}{p} + 1\right\} \\ &= \frac{(qy_0+1)(y_0-1/p+1)}{1+y_0} > 0. \end{aligned}$$

This, together with (4), yields $y_1 = qy_0(1+x_0) > 1/p - 1$, that is, $(x_1, y_1) \in Q_1(E_{p,q})$. By virtue of Lemma 1, we therefore conclude that $\lim_{n\to\infty} x_n = \lim_{n\to\infty} y_n = \infty$ and the statement (a) holds. The proof is complete. Q.E.D.

Lemma 4. *Assume that* $0 < p < 1$ *and* $0 < q < 1$*. Then the global stable manifold* W^s *of* $E_{p,q}$ *is unbounded and separates* $\mathrm{Int}\,\mathbf{R}^2_+$ *into two invariant regions.*

Proof. Let W_1^s denote the connected component of W^s which contains $E_{p,q}$. Lemma 2 shows that $W_1^s \subset Q_2(E_{p,q}) \cup Q_4(E_{p,q})$ and W_1^s is a graph of a decreasing function of x. Each of the components $Q_2(E_{p,q})$ and $Q_4(E_{p,q})$ may contain only one limiting point of W_1^s which does not belong to W_1^s. Suppose that $v_2 \in Q_2(E_{p,q})$ and $v_4 \in Q_4(E_{p,q})$ denote these limiting points. Since W_1^s is invariant under T, it follows that the set $\{v_2, v_4\}$ is invariant. Hence $\{v_2, v_4\}$ is a period-two solution or each point is fixed. Lemma 3, however, implies that if $v_4 \in Q_4(E_{p,q})$ and $T(v_4)$ belongs to the region $\{(x,y) \in \mathrm{Int}\,\mathbf{R}^2_+ \mid y > 1/p - 1\}$, then $T(v_4) \in Q_1(E_{p,q})$; if $v_2 \in Q_2(E_{p,q})$ and $T(v_2)$ belongs to the region $\{(x,y) \in \mathrm{Int}\,\mathbf{R}^2_+ \mid y < 1/p - 1\}$, then $T(v_2) \in Q_3(E_{p,q})$. Thus $\{v_2, v_4\}$ is not a period-two solution. Also v_2 and v_4 are not fixed points because the map T has only two fixed points E_0 and $E_{p,q}$. Therefore, each of the components $Q_2(E_{p,q})$ and $Q_4(E_{p,q})$ contain no limiting points of W_1^s except for $E_{p,q}$. This means that W^s is unbounded and separates $\mathrm{Int}\,\mathbf{R}^2_+$ into two invariant regions and the proof is complete. Q.E.D.

Now the following theorem is the main result of this subsection.

Theorem 2. *Assume that* $0 < p < 1$ *and* $0 < q < 1$. *Then the following statements hold:*

(a) *Every solution* (x_n, y_n) *of* (2) *starting above* W^s *remains above* W^s *and satisfies* $\lim_{n\to\infty} x_n = \lim_{n\to\infty} y_n = \infty$.

(b) *Every solution* (x_n, y_n) *of* (2) *starting below* W^s *remains below* W^s *and satisfies* $\lim_{n\to\infty} x_n = \lim_{n\to\infty} y_n = 0$.

Proof. We will prove the statement (a). The proof of the statement (b) is similar and will be omitted. Let v denote the positive initial point (x_0, y_0) above W^s. By virtue of Lemma 1, we have only to consider the case where $v \in Q_2(E_{p,q}) \cup Q_4(E_{p,q})$. Let w be the point where the vertical line through v intersects W^s. Then $w < v$, and so, Proposition 1 shows that $T^n(w) \ll T^n(v)$ for $n = 0, 1, 2, \ldots$. This, together with Lemma 2, implies that $T^n(v)$ remains above W^s. Suppose that $T^n(v)$ remains in $Q_2(E_{p,q})$ or $Q_4(E_{p,q})$. Then $T^n(v)$ tends to $E_{p,q}$ as $n \to \infty$ because $T^n(w) \ll T^n(v)$ and $T^n(w)$ tends to $E_{p,q}$ as $n \to \infty$. This means that $v \in W^s$, which contradicts the definition of v. Therefore, there exists a positive integer N such that $T^N(v) \in Q_3(E_{p,q})$. By virtue of Lemma 3, we thus conclude that $T^n(v)$ tends to (∞, ∞) as $n \to \infty$. This completes the proof. Q.E.D.

2.3. The case $p = q = 1$

In this case, every point on each coordinate axis is an equilibrium point.

Theorem 3. *Assume that* $p = q = 1$. *Then every solution* (x_n, y_n) *of* (2) *satisfies* $\lim_{n\to\infty} x_n = \lim_{n\to\infty} y_n = \infty$.

Proof. By (2), we have $x_{n+1} > x_n$, $y_{n+1} > y_n$ for $n = 0, 1, 2, \ldots$, which implies that x_n and y_n are increasing sequences. Suppose that x_n and y_n are bounded above. Then there exist positive numbers α, β such that $\lim_{n\to\infty} x_n = \alpha > x_0$, $\lim_{n\to\infty} y_n = \beta > y_0$, and hence, the point (α, β) is another equilibrium in $\operatorname{Int} \mathbf{R}^2_+$. This contradicts the fact that the equilibria of (2) only exist on each coordinate axis. Suppose that y_n (resp. x_n) is bounded above and x_n (resp. y_n) tends to ∞ as $n \to \infty$. Then a contradiction also arises by a similar fashion in the proof of Theorem 1. Consequently, $\lim_{n\to\infty} x_n = \lim_{n\to\infty} y_n = \infty$ and the proof is complete. Q.E.D.

2.4. The case $p = 1$ and $0 < q < 1$

In this case, every point on the x-axis is an equilibrium point. For simplicity, we divide $\mathrm{Int}\,\mathbf{R}^2_+$ into the following two regions:

$$D_1 = \left\{(x, y) \mid x \geq \frac{1}{q}-1,\ y > 0\right\}, \quad D_2 = \left\{(x, y) \mid 0 < x < \frac{1}{q}-1,\ y > 0\right\}.$$

Let (x_n, y_n) be a solution of (2) with $(x_0, y_0) \in D_1$. Then we have $x_{n+1} > x_n > x_0 \geq 1/q - 1$ for $n = 1, 2, \ldots$, which implies that $y_{n+1} > y_n > 0$ for $n = 1, 2, \ldots$. Hence it is easily seen that $\lim_{n\to\infty} x_n = \lim_{n\to\infty} y_n = \infty$ by a similar manner in the proof of Lemma 1 (a), and therefore, the following theorem holds.

Theorem 4. *Assume that $p = 1$ and $0 < q < 1$. Then every solution (x_n, y_n) of (2) starting from D_1 satisfies $\lim_{n\to\infty} x_n = \lim_{n\to\infty} y_n = \infty$.*

A computer simulation suggests that the following conjecture is true.

Conjecture 1. *Assume that $p = 1$ and $0 < q < 1$. Then there exists a decreasing function $\varphi(x)$ such that $\varphi(1/q - 1) = 0$ and the graph of φ lies in D_2 and has the y-axis as the asymptote, and the following statements hold:*

(a) *Every solution (x_n, y_n) of (2) starting above the graph of φ enters D_1 and satisfies $\lim_{n\to\infty} x_n = \lim_{n\to\infty} y_n = \infty$.*

(b) *Every solution (x_n, y_n) of (2) starting below the graph of φ remains below this curve and satisfies $\lim_{n\to\infty} x_n = x^* < 1/q - 1$ and $\lim_{n\to\infty} y_n = 0$ where x^* is some positive number depending on (x_0, y_0).*

2.5. The case $0 < p < 1$ and $q = 1$

In this case, every point on the y-axis is an equilibrium point. For simplicity, we divide $\mathrm{Int}\,\mathbf{R}^2_+$ into the following two regions:

$$\Delta_1 = \left\{(x, y) \mid x > 0,\ y \geq \frac{1}{p}-1\right\}, \quad \Delta_2 = \left\{(x, y) \mid x > 0,\ 0 < y < \frac{1}{p}-1\right\}.$$

Theorem 5. *Assume that $0 < p < 1$ and $q = 1$. Then every solution (x_n, y_n) of (2) starting from Δ_1 satisfies $\lim_{n\to\infty} x_n = \lim_{n\to\infty} y_n = \infty$.*

Conjecture 2. *Assume that $0 < p < 1$ and $q = 1$. Then there exists a decreasing function $\psi(x)$ such that $\psi(0) = 1/p - 1$ and the graph* *of ψ lies in Δ_2 and has the x-axis as the asymptote, and the following statements hold:*

(a) *Every solution (x_n, y_n) of (2) starting above the graph of ψ enters Δ_1 and satisfies $\lim_{n\to\infty} x_n = \lim_{n\to\infty} y_n = \infty$.*

(b) *Every solution (x_n, y_n) of (2) starting below the graph of ψ remains below this curve and satisfies $\lim_{n\to\infty} x_n = 0$ and $\lim_{n\to\infty} y_n = y^* < 1/p - 1$ where y^* is some positive number depending on (x_0, y_0).*

Finally, to illustrate our global results for (2), we consider the following cases:

(i) $p = 2$, $q = 0.5$, $(x_0, y_0) = (0.005, 1)$, $(0.02, 1)$, $(0.06, 1)$, $(0.12, 1)$, $(0.2, 1)$, $(0.4, 1)$, $(0.7, 1)$;

(ii) $p = q = 0.5$, $(x_0, y_0) = (0.03, 5)$, $(0.07, 5)$, $(0.13, 5)$, $(0.3, 5)$, $(0.6, 5)$, $(5, 0.03)$, $(5, 0.07)$, $(5, 0.13)$, $(5, 0.3)$, $(5, 0.6)$, $(1, 1)$;

(iii) $p = q = 1$, $(x_0, y_0) = (0.2, 0.2)$, $(0.2, 1)$, $(0.2, 2)$, $(0.2, 3)$, $(0.2, 4)$, $(1, 0.2)$, $(2, 0.2)$, $(3, 0.2)$, $(4, 0.2)$;

(iv) $p = 1$, $q = 0.5$, $(x_0, y_0) = (0.06, 1)$, $(0.1, 1)$, $(0.18, 1)$, $(0.35, 1)$, $(0.6, 1)$, $(2, 0.05)$, $(3, 0.05)$, $(4, 0.05)$.

We give some portraits of solution orbits of (2) drawn by a computer. Figures 1–4 correspond to the cases (i)–(iv), respectively. The dotted curve in Figure 2 is the global stable manifold of the positive equilibrium.

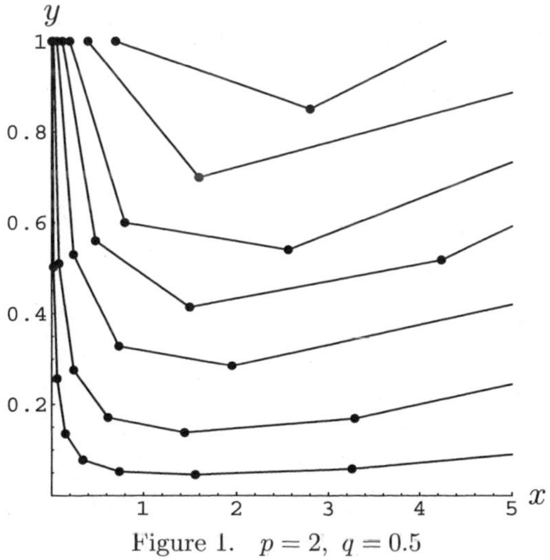

Figure 1. $p = 2$, $q = 0.5$

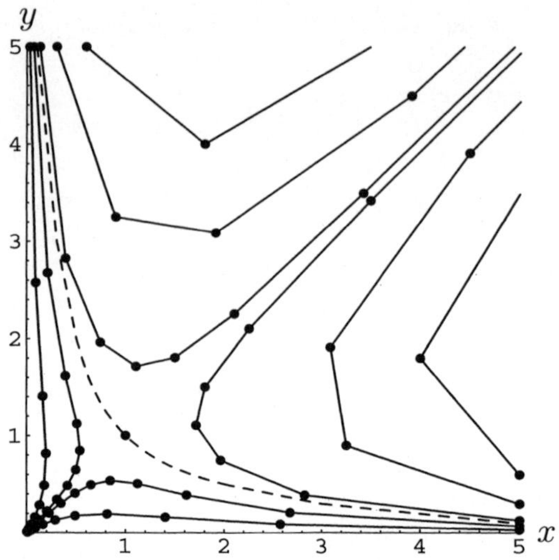

Figure 2. $p = q = 0.5$

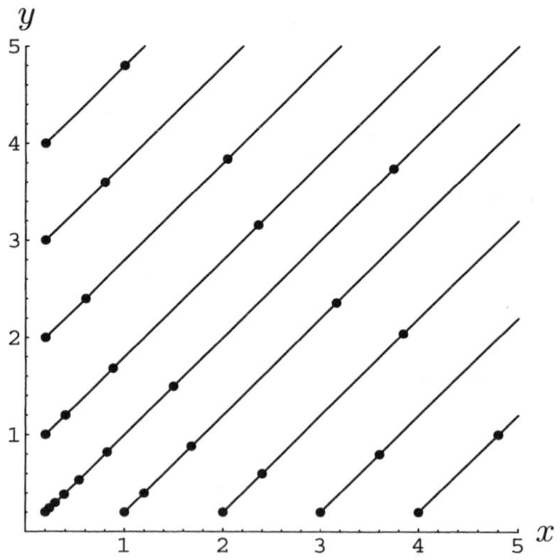

Figure 3. $p = q = 1$

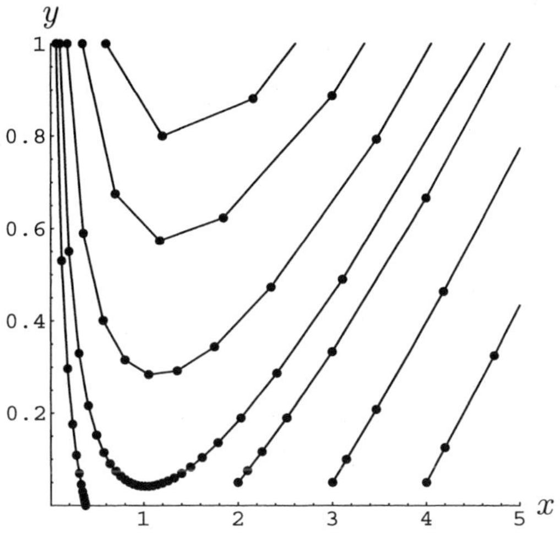

Figure 4. $p = 1,\ q = 0.5$

References

[1] S. Elaydi, Discrete Chaos, Chapman & Hall/CRC, Florida, 2000.

[2] J. Hale and H. Kocak, Dynamics and Bifurcations, Springer-Verlag, New York, 1991.

[3] M. W. Hirsch and H. L. Smith, Monotone maps: a review, J. Difference Equ. Appl., **11** (2005), 379–398.

[4] M. R. S. Kulenović and O. Merino, Discrete Dynamical Systems and Difference Equations with Mathematica, Chapman & Hall/CRC, Florida, 2002.

[5] M. R. S. Kulenović and M. Nurkanović, Global asymptotic behavior of a two-dimensional system of difference equations modeling cooperation, J. Difference Equ. Appl., **9** (2003), 149–159.

[6] H. L. Smith, Monotone Dynamical Systems: An Introduction to the Theory of Competitive and Cooperative Systems, Amer. Math. Soc., Providence, RI, 1995.

[7] H. L. Smith, Planar competitive and cooperative difference equations, J. Difference Equ. Appl., **3** (1998), 335–357.

Momoe Inoue and Hideaki Matsunaga
Department of Mathematical Sciences
Osaka Prefecture University
Sakai, Osaka 599-8531
Japan

E-mail address: `momoe@ms.osakafu-u.ac.jp`
`hideaki@ms.osakafu-u.ac.jp`

Advanced Studies in Pure Mathematics 53, 2009
Advances in Discrete Dynamical Systems
pp. 141–149

Asymptotic stability conditions for a delay difference system

Fumiyuki Kawahigashi and Hideaki Matsunaga

Abstract.

In this paper we obtain some necessary and sufficient conditions for the asymptotic stability of the zero solution of a delay difference system

$$x_n = A(x_{n-k} + x_{n-k-l}), \quad n = 0, 1, 2, \ldots,$$

where A is a 2×2 real constant matrix and k and l are positive integers.

§1. Introduction

The purpose of this paper is to establish some necessary and sufficient conditions for the zero solution of a delay difference system

$$x_n = A(x_{n-k} + x_{n-k-l}), \quad n = 0, 1, 2, \ldots, \tag{1}$$

to be asymptotically stable. Here, A is a 2×2 real constant matrix, and k and l are positive integers. In the scalar case, Kaewong et al. [3] showed the following result.

Theorem A. *Let $A = a$ be a real number. Then the zero solution of* (1) *is asymptotically stable if and only if*

$$a^-_{\max} < a < \frac{1}{2}, \tag{2}$$

where $a^-_{\max}$ is the negative maximum value of $1/(2\cos\frac{2kp\pi}{2k+l})$ for $p = 0, 1, \ldots, 2k+l-1$.

By the transformation $x_n = Py_n$ with a nonsingular matrix P, the system (1) can be written as

$$y_n = P^{-1}AP(y_{n-k} + y_{n-k-l}), \quad n = 0, 1, 2, \ldots.$$

Received November 14, 2006.

Consequently, we have only to consider (1) where the matrix A is either of the following two matrices in Jordan form:

$$\text{(I)}\quad A = a\begin{pmatrix}\cos\theta & -\sin\theta\\ \sin\theta & \cos\theta\end{pmatrix}, \qquad \text{(II)}\quad A = \begin{pmatrix}d_1 & b\\ 0 & d_2\end{pmatrix},$$

where a, b, d_1, d_2 and θ are real numbers with $|\theta| \le \pi/2$.

§2. Main results

The following theorems are our main results. For simplicity, we put

$$a_p = \frac{1}{2\cos\frac{2kp\pi - l|\theta|}{2k+l}} \tag{3}$$

for $p = 0, 1, \ldots, 2k+l-1$.

Theorem 1. *Suppose that the matrix A is given by* (I). *Then the zero solution of* (1) *is asymptotically stable if and only if*

$$a^-_{\max} < a < a^+_{\min}, \tag{4}$$

where $a^+_{\min}$ and $a^-_{\max}$ are the positive minimum value of a_p and the negative maximum value of a_p for $p = 0, 1, \ldots, 2k+l-1$, respectively.

Theorem 2. *Suppose that the matrix A is given by* (II). *Then the zero solution of* (1) *is asymptotically stable if and only if*

$$a^-_{\max} < d_j < \frac{1}{2} \quad \textit{for } j = 1, 2, \tag{5}$$

where $a^-_{\max}$ is the negative maximum value of a_p with $\theta = 0$ for $p = 0, 1, \ldots, 2k+l-1$.

Our results are proved by using the fact that the zero solution of (1) is asymptotically stable if and only if all the roots of its associated characteristic equation

$$F(\lambda) \equiv \det(\lambda^{k+l} I - \lambda^l A - A) = 0 \tag{6}$$

are inside the unit disk. Here, I is the 2×2 identity matrix. In this paper we only give the proof of Theorem 1 because Theorem 2 is proved in a similar way. Thus we consider the matrix A given by (I). An easy calculation yields that

$$F(\lambda) = \begin{vmatrix} \lambda^{k+l} - a(\lambda^l + 1)\cos\theta & a(\lambda^l + 1)\sin\theta \\ -a(\lambda^l + 1)\sin\theta & \lambda^{k+l} - a(\lambda^l + 1)\cos\theta \end{vmatrix}$$
$$= \{\lambda^{k+l} - a(\lambda^l + 1)\cos\theta\}^2 - \{ia(\lambda^l + 1)\sin\theta\}^2$$
$$= \{\lambda^{k+l} - ae^{i\theta}(\lambda^l + 1)\}\{\lambda^{k+l} - ae^{-i\theta}(\lambda^l + 1)\}.$$

Hence, all the roots of the characteristic equation (6) are inside the unit disk if and only if all the roots of the equation

$$f(a, \lambda) \equiv \lambda^{k+l} - ae^{i\theta}(\lambda^l + 1) = 0 \tag{7}$$

are inside the unit disk, since $F(\lambda) = f(a, \lambda)\overline{f(a, \bar{\lambda})}$ where $\bar{\lambda}$ denotes the complex conjugate of any complex λ. Consequently, the following proposition holds.

Proposition 1. *The zero solution of* (1) *is asymptotically stable if and only if all the roots of* (7) *are inside the unit disk.*

Note that, in case $a = 0$, all the roots of (7) are 0 (multiplicity $k+l$). Now, we will discuss the location of the roots of (7) as a varies. The first two lemmas deal with the value of a and the roots of (7) on the unit circle. Hereafter, we may assume that $\theta \geq 0$.

Lemma 1. *Let λ be a root of* (7) *on the unit circle. Then the root λ and the real number a are expressed as*

$$\lambda = e^{i\omega_p}, \quad \text{where } \omega_p = \frac{2p\pi + 2\theta}{2k + l}, \tag{8}$$

and

$$a = \frac{1}{2\cos\frac{2kp\pi - l\theta}{2k+l}} \equiv a_p \tag{9}$$

for some $p = 0, 1, \ldots, 2k + l - 1$. Conversely, if $a = a_p$, then $\lambda = e^{i\omega_p}$ is a root of (7).

Proof. Let λ be a root of (7) on the unit circle. Then $\lambda^l + 1 \neq 0$ and (7) becomes

$$a = \frac{\lambda^{k+l}}{e^{i\theta}(\lambda^l + 1)}. \tag{10}$$

Since a is real and $\bar{\lambda} = 1/\lambda$, we have

$$a = \frac{\bar{\lambda}^{k+l}}{e^{-i\theta}(\bar{\lambda}^l + 1)} = \frac{e^{i\theta}/\lambda^{k+l}}{1/\lambda^l + 1} = \frac{e^{i\theta}}{\lambda^k(1 + \lambda^l)}. \tag{11}$$

From (10) and (11), we obtain $\lambda^{2k+l} = e^{i2\theta}$, which implies that (8) is valid for some integer p. Then it follows from (10) that

$$a = \frac{e^{i2\theta}\lambda^{-k}}{e^{i\theta}(e^{i2\theta}\lambda^{-2k}+1)} = \frac{1}{e^{i\theta}\lambda^{-k}+e^{-i\theta}\lambda^{k}} = \frac{1}{2\cos(k\omega_p-\theta)} \equiv a_p.$$

Conversely, if $a = a_p$, it is clear that $\lambda = e^{i\omega_p}$ is a root of (7). The proof is complete. Q.E.D.

Lemma 2. *The roots of* (7) *on the unit circle are simple.*

Proof. Let λ be given by (8). Note that $\lambda^l + 1 \neq 0$. By (10), we have

$$\frac{\partial f(a,\lambda)}{\partial\lambda} = (k+l)\lambda^{k+l-1} - lae^{i\theta}\lambda^{l-1} = \frac{\lambda^{k+l-1}(k\lambda^l+k+l)}{\lambda^l+1}.$$

Suppose that $\partial f(a,\lambda)/\partial\lambda = 0$. Then

$$k\lambda^l + k + l = 0. \tag{12}$$

In view of $\bar{\lambda} = 1/\lambda$, (12) yields $k/\lambda^l + k + l = 0$, that is,

$$(k+l)\lambda^l + k = 0. \tag{13}$$

By adding (12) and (13), we obtain $(2k+l)(\lambda^l+1) = 0$, which contradicts that $\lambda^l + 1 \neq 0$. Hence $\partial f(a_p, e^{i\omega_p})/\partial\lambda \neq 0$, and therefore, the assertion of this lemma holds. Q.E.D.

Next, we will observe how the roots of (7) cross the unit circle when the real number a varies.

Lemma 3. *The absolute values of the roots of* (7) *at* $\lambda = e^{i\omega_p}$ *increase as* $|a|$ *increses.*

Proof. By virtue of Lemma 2, we may regard the root λ of (7) as a holomorphic function of a in a neighborhood of $a = a_p$, and thus, the implicit function theorem shows that

$$\frac{d\lambda}{da} = -\frac{\frac{\partial f(a,\lambda)}{\partial a}}{\frac{\partial f(a,\lambda)}{\partial\lambda}}. \tag{14}$$

Let λ be expressed in the polar form, that is, $\lambda = re^{i\omega}$. Then we have

$$\frac{d\lambda}{da} = \frac{\lambda}{r}\left(\frac{dr}{da} + ir\frac{d\omega}{da}\right). \tag{15}$$

Hence it follows from (14) and (15) that

$$\frac{dr}{da} = \operatorname{Re}\left\{\frac{r}{\lambda}\cdot\frac{d\lambda}{da}\right\} = \frac{r}{\left|\frac{\partial f(a,\lambda)}{\partial\lambda}\right|^2}\operatorname{Re}\left\{-\frac{1}{\lambda}\cdot\frac{\partial f(a,\lambda)}{\partial a}\cdot\overline{\frac{\partial f(a,\lambda)}{\partial\lambda}}\right\}. \tag{16}$$

By using the fact that

$$\frac{\partial f(a,\lambda)}{\partial a} = -e^{i\theta}(\lambda^l+1) = -\frac{\lambda^{k+l}}{a},$$

we find

$$\begin{aligned}
&\operatorname{Re}\left\{-\frac{1}{\lambda}\cdot\frac{\partial f(a,\lambda)}{\partial a}\cdot\overline{\frac{\partial f(a,\lambda)}{\partial\lambda}}\right\} \\
&= \frac{1}{2}\left\{-\frac{1}{\lambda}\cdot\frac{\partial f(a,\lambda)}{\partial a}\cdot\overline{\frac{\partial f(a,\lambda)}{\partial\lambda}} - \lambda\cdot\overline{\frac{\partial f(a,\lambda)}{\partial a}}\cdot\frac{\partial f(a,\lambda)}{\partial\lambda}\right\} \\
&= \frac{1}{2}\left\{\frac{(k+l)\lambda^l+k}{a(\lambda^l+1)} + \frac{k\lambda^l+k+l}{a(\lambda^l+1)}\right\} = \frac{2k+l}{2a}.
\end{aligned} \tag{17}$$

From (16) and (17), we therefore obtain

$$\left.\frac{dr}{da}\right|_{a=a_p} = \frac{(2k+l)r}{2a_p\left|\frac{\partial f(a,\lambda)}{\partial\lambda}\right|^2},$$

which, together with Lemma 1, implies that the absolute values of the roots of (7) at $\lambda = e^{i\omega_p}$ increase as $|a|$ increases. This completes the proof. Q.E.D.

Now we are in a position to prove Theorem 1.

Proof of Theorem 1. By virtue of Proposition 1, we verify that the condition (4) is the necessary and sufficient condition for all the roots of (7) to be inside the unit disk. Recall that in case $a = 0$, all the roots of (7) lie on the origin. This, together with the continuity of the roots with respect to a, implies that if $|a|$ is sufficiently small, then all the roots of (7) lie inside the unit disk. In addition, Lemma 1 asserts that $a^+_{\min}$ (resp. $a^-_{\max}$) is the positive minimum (resp. negative maximum) value of a such that a root of (7) intersects the unit circle as a increases (resp. decreases) from 0. Lemma 3 also shows that if $a \geq a^+_{\min}$ (resp. $a \leq a^-_{\max}$), then there exists a root λ^* of (7) such that $|\lambda^*| \geq 1$. Thus, we conclude that all the roots of (7) are inside the unit disk if and only if the condition (4) holds. The proof of Theorem 1 is complete. Q.E.D.

§3. The case of $l = mk$

Our asymptotic stability conditions (4) and (5) in Theorems 1 and 2, unfortunately, involve constrained maximization and minimization problems concerning a_p. However, when l is a multiple of k, we can settle these problems. In the following, we consider the case of $l = mk$ where m is a positive integer. By virtue of Lemma 1, we notice that if a root of (7) lies on the unit circle, the real number a can be written as

$$(18) \qquad a = \frac{1}{2\cos\xi_p} \equiv \alpha_p, \quad \text{where } \xi_p = \frac{2p\pi - m\theta}{m+2},$$

for some $p = 0, 1, \ldots, m+1$. Thus, we have only to deal with α_p instead of a_p to obtain explicit asymptotic stability conditions. Throughout the remainder of this paper, we denote by $\alpha^+_{\min}$ and $\alpha^-_{\max}$ the positive minimum value of α_p and the negative maximum value of α_p, respectively.

Remark 1. In case $m = 1$, it follows from (18) that $\alpha_p = 1/(2\cos\frac{2p\pi-\theta}{3})$ for $p = 0, 1, 2$, and hence, we can immediately conclude that $\alpha^+_{\min} = \alpha_0$ and $\alpha^-_{\max} = \alpha_2$.

In case $m \geq 2$, the explicit values of $\alpha^+_{\min}$ and $\alpha^-_{\max}$ are given by the next lemma. For the sake of convenience, we divide the interval $[0, \pi/2]$ into intervals I_j defined by

$$I_0 = \left[0, \frac{\pi}{m}\right],$$

$$I_j = \left(\frac{j}{m}\pi, \frac{j+1}{m}\pi\right] \quad \text{for } j = 1, 2, \ldots, \left[\frac{m}{2}\right] - 1,$$

$$I_{[m/2]} = \begin{cases} \emptyset & \text{if } m \text{ is even,} \\ \left(\left[\frac{m}{2}\right]\frac{\pi}{m}, \frac{\pi}{2}\right] & \text{if } m \text{ is odd and } m \geq 3, \end{cases}$$

where $[\cdot]$ denotes the greatest integer function.

Lemma 4. *Let $m \geq 2$. Suppose that $\theta \in I_{2h}$ or $\theta \in I_{2h-1}$ for some integer h. Then*

$$\alpha^+_{\min} = \alpha_h \quad \textit{and} \quad \alpha^-_{\max} = \begin{cases} \alpha_{h+\frac{m+2}{2}} = -\alpha_h & \textit{if } m \textit{ is even}, \\ \alpha_{h+\frac{m+3}{2}} & \textit{if } m \textit{ is odd and } \theta \in I_{2h}, \\ \alpha_{h+\frac{m+1}{2}} & \textit{if } m \textit{ is odd and } \theta \in I_{2h-1}. \end{cases}$$

Proof. In view of $\theta \in I_{2h}$ or $\theta \in I_{2h-1}$, we notice that $(2h-1)\pi < m\theta \le (2h+1)\pi$. This, together with (18), yields that

$$\frac{(2p-2h-1)\pi}{m+2} \le \xi_p < \frac{(2p-2h+1)\pi}{m+2}. \tag{19}$$

First, we find the explicit value of $\alpha^+_{\min}$. By the definition of ξ_p and (19), we have

$$-\pi < \xi_0 < \cdots < \xi_{h-1} < -\tfrac{\pi}{m+2} \le \xi_h$$
$$< \tfrac{\pi}{m+2} < \xi_{h+1} < \cdots < \xi_{m+1} < 2\pi + \xi_0 < 2\pi,$$

which implies $\alpha^+_{\min} = \alpha_h$.

Next, we investigate the explicit value of $\alpha^-_{\max}$. There are two cases to consider.

Case 1: m is even. By the definition of ξ_p and (19), we have $-\pi < \xi_{m+1} - 2\pi < \xi_0$ and

$$-\pi < \xi_0 < \cdots < \xi_{h+\frac{m}{2}} < \tfrac{(m+1)\pi}{m+2} \le \xi_{h+\frac{m+2}{2}}$$
$$< \tfrac{(m+3)\pi}{m+2} \le \xi_{h+\frac{m+4}{2}} < \cdots < \xi_{m+1} < 2\pi,$$

which yield that

$$\alpha^-_{\max} = \alpha_{h+\frac{m+2}{2}} = \frac{1}{2\cos\xi_{h+\frac{m+2}{2}}} = \frac{1}{2\cos(\xi_h+\pi)} = -\frac{1}{2\cos\xi_h} = -\alpha_h.$$

Case 2: m is odd. By the definition of ξ_p and (19), we have

$$-\pi < \xi_{m+1} - 2\pi < \xi_0 < \cdots < \xi_{h+\frac{m+1}{2}} < \pi \le \xi_{h+\frac{m+3}{2}} < \cdots < \xi_{m+1} < 2\pi,$$

which implies $\alpha^-_{\max} = \alpha_{h+\frac{m+1}{2}}$ or $\alpha_{h+\frac{m+3}{2}}$. Note that

$$\gamma \equiv \left|\pi - \xi_{h+\frac{m+3}{2}}\right| - \left|\pi - \xi_{h+\frac{m+1}{2}}\right| = \xi_{h+\frac{m+3}{2}} + \xi_{h+\frac{m+1}{2}} - 2\pi = \frac{2(2h\pi - m\theta)}{m+2}. \tag{20}$$

If $\theta \in I_{2h}$, then $2h\pi < m\theta \le (2h+1)\pi$ and therefore $\gamma < 0$ by (20). This implies $\alpha^-_{\max} = \alpha_{h+\frac{m+3}{2}}$. On the other hand, if $\theta \in I_{2h-1}$, then $(2h-1)\pi < m\theta \le 2h\pi$ and thus $\gamma \ge 0$ by (20). This yields $\alpha^-_{\max} = \alpha_{h+\frac{m+1}{2}}$. The proof is complete. Q.E.D.

By virtue of Remark 1 and Lemma 4, we have the following result by Theorem 1.

Theorem 3. *Suppose that $l = mk$ and the matrix A is given by* (I) *where $|\theta| \in I_{2h}$ or $|\theta| \in I_{2h-1}$ for some integer h. Then the following statements hold:*

(i) *If* $m = 1$, *then the zero solution of* (1) *is asymptotically stable if and only if* $\alpha_2 < a < \alpha_0$.
(ii) *If* m *is even, then the zero solution of* (1) *is asymptotically stable if and only if* $-\alpha_h < a < \alpha_h$.
(iii) *If* m *is odd,* $m \geq 3$ *and* $|\theta| \in I_{2h}$, *then the zero solution of* (1) *is asymptotically stable if and only if* $\alpha_{h+\frac{m+3}{2}} < a < \alpha_h$.
(iv) *If* m *is odd,* $m \geq 3$ *and* $|\theta| \in I_{2h-1}$, *then the zero solution of* (1) *is asymptotically stable if and only if* $\alpha_{h+\frac{m+1}{2}} < a < \alpha_h$.

In case $\theta = 0$, it follows from (18) that $\alpha_p = 1/(2\cos\frac{2p\pi}{m+2})$ for $p = 0, 1, \ldots, m+1$. Then Lemma 4 and $\theta \in I_0$ assert that

$$\alpha_{\max}^{-} = \begin{cases} -\alpha_0 = -1/2 & \text{if } m \text{ is even,} \\ \alpha_{\frac{m+3}{2}} = -1/(2\cos\frac{\pi}{m+2}) & \text{if } m \text{ is odd,} \end{cases}$$

and hence, we have the following result by Theorem 2.

Theorem 4. *Suppose that* $l = mk$ *and the matrix* A *is given by* (II). *Then the following statements hold:*

(i) *If* m *is even, then the zero solution of* (1) *is asymptotically stable if and only if* $-1/2 < d_j < 1/2$ *for* $j = 1, 2$.
(ii) *If* m *is odd, then the zero solution of* (1) *is asymptotically stable if and only if* $-1/(2\cos\frac{\pi}{m+2}) < d_j < 1/2$ *for* $j = 1, 2$.

§4. Concluding remarks

We have established several results on the asymptotic stability of (1). In paticular, we have also obtained the explicit asymptotic stability conditions for (1) with $l = mk$.

Finally, we state some remarks on other related works for the asymptotic stability of delay difference systems. In [1], Dannan presented asymptotic stability conditions for a delay difference equation

$$x_n = ax_{n-k} + bx_{n-k-l}, \quad n = 0, 1, 2, \ldots, \tag{21}$$

where a and b are real numbers. In [4], the second author gave explicit asymptotic stability conditions for a delay difference system

$$x_n = ax_{n-1} + Bx_{n-k}, \quad n = 0, 1, 2, \ldots, \tag{22}$$

where a is a real number and B is a 2×2 real constant matrix. So it is interesting to extend these results. Indeed, the asymptotic stability problem for a delay difference system

$$x_n = Ax_{n-k} + Bx_{n-k-l}, \quad n = 0, 1, 2, \ldots, \tag{23}$$

where A and B are 2×2 real constant matrices, has not yet been solved.

References

[1] F. M. Dannan, The asymptotic stability of $x(n+k)+ax(n)+bx(n-l)=0$, J. Difference Equ. Appl., **10** (2004), 589–599.

[2] S. Elaydi, An Introduction to Difference Equations, 3rd ed., Springer-Verlag, New York, 2005.

[3] T. Kaewong, Y. Lenbury and P. Niamsup, A note on asymptotic stability conditions for delay difference equations, Int. J. Math. Math. Sci., **2005** (2005), 1007–1013.

[4] H. Matsunaga, Stability regions for a class of delay difference systems, Fields Inst. Commun., **42** (2004), 273–283.

Fumiyuki Kawahigashi and Hideaki Matsunaga
Department of Mathematical Sciences
Osaka Prefecture University
Sakai, Osaka 599-8531
Japan

E-mail address: fumiyuki@ms.osakafu-u.ac.jp
hideaki@ms.osakafu-u.ac.jp

Advanced Studies in Pure Mathematics 53, 2009
Advances in Discrete Dynamical Systems
pp. 151–158

Characterization of equilibrium paths in the two-sector model with sector specific externality

Miki Matsuo and Tomoya Sakagami

Abstract.

We study the two-sector economy with sector-specific externality following Benhabib, Nishimura and Venditti [2002]. We focus on the external effect of capital-labor ratio and provide the characterization of equilibrium paths in the case Benhabib, Nishimura and Venditti did not discuss explicitly.

§1. Introduction

The aim of this paper is to characterize the local behavior of the equilibrium paths around the steady state in the two-sector model with sector specific externality. It is well known that externalities may cause the indeterminacy of equilibrium paths in an infinite horizon model. Benhabib and Farmer [1994] has showed that indeterminacy could occur in the one-sector growth model with both externality and increasing returns. In their model, the production function is constant return to scale from the private perspective, while it is increasing return to scale from the social perspective. Since then, there have been many papers about the existence of indeterminate equilibria in dynamic general equilibrium models. However, most of the literature dealt with models in which the production function is increasing return to scale from the social perspective, until the publication of Benhabib and Nishimura [1998, 1999]. They proved that indeterminacy may arise in an economy in which the production function from the social perspective is constant return to scale in the continuous time framework. Benhabib, Nishimura

Received November 6, 2006.
Revised October 12, 2007.
We thank Professor Kazuo Nishimura for helpful comments and constructive suggestions on the earlier version of this paper.

and Venditti [2002] studied the two-sector model with sector specific external effects in discrete time. They assumed that each sector has Cobb-Douglas technology with positive sector specific externalities and there is an infinitely-lived representative agent with linear utility function. Under these assumptions, they provided conditions in which indeterminacy may occur even if the production function is decreasing return to scale from the social perspective.

In this paper, we study the same model as in Benhabib, Nishimura and Venditti [2002][1], focus on the external effect of capital-labor ratio in the pure capital goods sector and provide the characterization of equilibrium paths in the case allowing negative externality, as was not explicitly discussed. We will show how the degree of externality affects the local behavior of the equilibrium path around the steady state.

In Section 2 we describe the model. We discuss the existence of a steady state and give the local characterization of equilibrium path around the steady state in Section 3. Section 4 is the Appendix.

§2. The model

We consider the two-sector model introduced by Benhabib, Nishimura and Venditti [2002]. There exists an infinitely-lived representative agent with single period utility function given by

$$u(C_t) = C_t.$$

There are goods : consumption goods, C, and capital goods, K. Following Benhabib, Nishimura and Venditti, we assume that each good is produced with a Cobb-Douglas technology.

$$C_t = K_{1,t}^{\alpha_1} L_{1,t}^{\alpha_2}, \qquad \alpha_1 + \alpha_2 = 1 \tag{1}$$

$$Y_t = A_t K_{2,t}^{\beta_1} L_{2,t}^{\beta_2}, \qquad \beta_1 + \beta_2 = 1 \tag{2}$$

where A_t represents the externality and varies each period. We formulate the externality as follows:

$$A_t = \left(\frac{\bar{K}_{2,t}}{\bar{L}_{2,t}} \right)^b = \bar{K}_{2,t}^b \bar{L}_{2,t}^{-b2}. \tag{3}$$

[1]See also Nishimura and Venditti [2002] for derivation of the results.

[2]We focus on the external effect of capital-labor ratio in the capital goods sector.

A bar over a variable denotes the economy-wide average. We assume that the representative firms take as given these economy -wide average.

Definition 2.1. *We call* $Y_t = A_t K_{2,t}^{\beta_1} L_{2,t}^{\beta_2}$ *the production function from the private perspective, and* $Y_t = A_t K_{2,t}^{\beta_1+b} L_{2,t}^{\beta_2-b}$ *the production function from the social perspective.*

The aggregate capital is divided between sectors,

$$K_t = K_{1,t} + K_{2,t},$$

and the labor endowment is normalized to one and divided between sectors,

$$L_{1,t} + L_{2,t} = 1.$$

The capital accumulation equation is

$$K_{t+1} = Y_t,$$

that is, the capital depreciates completely in one period.

2.1. Profit maximization

We denote by p_2, w_1, and w_2 respectively the price of capital goods, the rental price of the capital goods and the wage rate of labor[3]. Then, each representative firm maximizes its profits:

$$\pi_1 = K_1^{\alpha_1} L_1^{\alpha_2} - w_1 K_1 - w_2 L_1, \tag{4}$$

$$\pi_2 = p_2 A K_2^{\beta_1} L_2^{\beta_2} - w_1 K_2 - w_2 L_2. \tag{5}$$

From the first order condition with respect to K_i, L_i $(i = 1,\ 2)$, we have the following conditions:

$$\frac{w_1}{w_2} = \frac{\alpha_1 L_1}{\alpha_2 K_1}, \tag{6}$$

$$\frac{w_1}{w_2} = \frac{\beta_1 L_2}{\beta_2 K_2}. \tag{7}$$

From equations (6) and (7), we derive the next equation.

---

[3]We normalize the price of consumption goods to one.

(8) $$\frac{\frac{\alpha_1}{\alpha_2}}{\frac{\beta_1}{\beta_2}} = \frac{\frac{K_1}{L_1}}{\frac{K_2}{L_2}}.$$

Factor intensities may be determined by the coefficients of the Cobb-Douglas functions. If $\frac{\alpha_1}{\alpha_2} > (<) \frac{\beta_1}{\beta_2}$, the consumption (capital) goods sector is capital intensive from the private perspective.

Using equation (8), $L_1 + L_2 = 1$, and $K = K_1 + K_2$. L_2 is expressed the follows :

(9) $$L_2 = \frac{\alpha_1 \beta_2 K_2}{\alpha_2 \beta_1 K + (\alpha_1 \beta_2 - \alpha_2 \beta_1) K_2}.$$

Substituting equation (9) into equation (2), and solving for K_2, K_2 is exhibited as a function of K, Y, and A.

(10) $$K_2 = \hat{K}_2 (K,\ Y;\ A).$$

Then, by equation (9), $L_1 + L_2 = 1$, and $K = K_1 + K_2$, K_1, L_1 and L_2 also are exhibited as functions of K, Y, and A, respectively. Therefore,

(11) $$K_1 = \hat{K}_1 (K,\ Y;\ A),\quad L_1 = \hat{L}_1 (K,\ Y;\ A),\quad L_2 - \hat{L}_2 (K,\ Y;\ A).$$

Denote by Δ^* the denominator of equation (9),

(12) $$\Delta^* (K,\ y;\ A) \equiv \alpha_2 \beta_1 K + (\alpha_1 \beta_2 - \alpha_2 \beta_1) \hat{K}_2 (K,\ Y;\ A).$$

We define the social production function as below:

(13)

$$\begin{aligned} T^* (Y,\ K;\ A) &= A \hat{K}_1 (K,\ Y;\ A)^{\alpha_1} \hat{L}_1 (K,\ Y;\ A)^{\alpha_2} \\ &= A \left(\frac{\alpha_2 \beta_1}{\Delta^* (K,\ Y;\ A)} \right)^{\alpha_2} \left(K - \hat{K}_2 (K,\ Y;\ A) \right)^{\alpha_1 + \alpha_2}. \end{aligned}$$

2.2. Utility maximization

The consumer optimization problem will be given by

(14) $$\max \sum_{t=0}^{\infty} \rho^t T^* (K_{t+1},\ K_t;\ A_t),$$
$$\textit{subject to} \quad k_0,\ \{A_t\}_{t=0}^{\infty} \ \textit{given}$$

where $\rho \in (0,\ 1)$ is the discount factor.

Then, the Euler equation in this model[4] is

$$T_1^*\left(K_{t+1},\ K_t;\ A_t\right)+\rho T_2^*\left(K_{t+2},\ K_{t+1};\ A_t\right)=0. \tag{15}$$

The solution of equation (14) satisfies the following transversality condition

$$\lim_{t\to+\infty}\rho^t K_{t+1}T_1^*\left(K_{t+1},\ K_t;\ A_t\right)=0. \tag{16}$$

We express the solution of this problem in $\{K_t\}_{t=0}^{\infty}$. This path depends on the choice of sequence $\{A_t\}_{t=0}^{\infty}$. If the sequence $\{A_t\}_{t=0}^{\infty}$ satisfies

$$A_t=\left[\hat{K}_2\left(K_{t+1},\ K_t;\ A_t\right)\right]^b\left[\hat{L}_2\left(K_{t+1},\ K_t;\ A_t\right)\right]^{-b}, \tag{17}$$

then the sequence $\{K_t\}_{t=0}^{\infty}$ is called an equilibrium path[5]. Solving equation (17) for A_t, A_t is given as a function of $(K_{t+1},\ K_t)$, namely $A_t=\hat{A}_t\left(K_{t+1},\ K_t\right)$. Substituting this expression into T^*,

$$T\left(K_{t+1},\ K_t\right)\equiv T^*\left(K_{t+1},\ K_t;\ \hat{A}_t\left(K_{t+1},\ K_t\right)\right). \tag{18}$$

This is the same as the function obtained by solving the first order conditions with respect to $K_1,\ K_2,\ L_1$, and L_2 of the Lagrangian below:

$$\begin{aligned}\mathcal{L}=&K_{1,t}^{\alpha_1}L_{1,t}^{\alpha_2}+p_{2t}\left(A_tK_{2,t}^{\beta_1}L_{2,t}^{\beta_2}-K_{t+1}\right)\\&+w_{1t}\left(K_t-K_{1,t}-K_{2,t}\right)+w_{2t}\left(1-L_{1,t}-L_{2,t}\right).\end{aligned} \tag{19}$$

Using the envelope theorem we derive the equilibrium prices[6].

$$T_1\left(K_{t+1},\ K_t\right)=-p_{2t}, \tag{20}$$

$$T_2\left(K_{t+2},\ K_{t+1}\right)=w_{1t+1}. \tag{21}$$

Then the Euler equation becomes the following:

$$T_1\left(K_{t+1},\ K_t\right)+\rho T_2\left(K_{t+2},\ K_{t+1}\right)=0. \tag{22}$$

---

[4]Where $T_1^*\left(K_{t+1},\ K_t;\ A_t\right)=\partial T^*\left(K_{t+1},\ K_t;\ A_t\right)/\partial K_{t+1}$ and $T_2^*\left(K_{t+1},\ K_t;\ A_t\right)=\partial T^*\left(K_{t+1},\ K_t;\ A_t\right)/\partial K_t$.

[5]On an equilibrium path, the representative firm's expectations correspond with the realized value.

[6]Using the envelope theorem, that is, $\frac{\partial T}{\partial K_{t+1}}=\frac{\partial \mathcal{L}}{\partial K_{t+1}}$ and $\frac{\partial T}{\partial K_t}=\frac{\partial \mathcal{L}}{\partial K_t}$, we get equations (20) and (21).

§3. Steady state

Definition 3.1. *A steady state is defined* $K_t = K_{t+1} = Y_t = K^*$ *and is given by the solution of* $T_1(K_{t+1}, K_t) + \rho T_2(K_{t+2}, K_{t+1}) = 0$.

The following lemmas are direct consequences of the results in Baierl, Nishimura and Yano [1998], Benhabib, Nishimura and Venditti [5] and Nishimura and Venditti [6].

Lemma 1. *In this model, there exists a unique stationary capital stock* K^* *such that:*

$$K^* = \frac{\alpha_1\beta_2(\rho\beta_1)^{\frac{1}{\beta_2-b}}}{\beta_1[\alpha_2 + (\alpha_1\beta_2 - \alpha_2\beta_1)]}. \tag{23}$$

To study local behavior of equilibrium path around a steady state K^*, we linearize the Euler equation (22) at the steady state K^* and obtain the following characteristic equation

$$\rho T_{21}\lambda^2 + [\rho T_{22} + T_{11}]\lambda + T_{12} = 0. \tag{24}$$

Lemma 2. *The characteristic equation (24) is equivalent to the expression*

$$\lambda^2 + \left[\frac{\alpha_2}{\rho(\alpha_2\beta_1 - \alpha_1\beta_2)} - \frac{\alpha_2 - (\beta_2 - b)}{\alpha_2}\right] - \frac{\alpha_2}{\rho(\alpha_2\beta_1 - \alpha_1\beta_2)} \cdot \frac{\alpha_2 - (\beta_2 - b)}{\alpha_2} = 0. \tag{25}$$

Lemma 3. *The characteristic roots of equation (26) are*

$$\lambda_1 = \frac{\alpha_2}{\rho(\alpha_2\beta_1 - \alpha_1\beta_2)}, \quad \lambda_2 = \frac{\alpha_2 - (\beta_2 - b)}{\alpha_2}. \tag{26}$$

The roots of the characteristic equation determine the local behavior of the equilibrium paths. The sign of λ_1 is determined by factor intensity differences from the private perspective[7], while the sign of λ_2 is determined by factor intensity differences from the social perspective.

In what follows, we provide the characterization of equilibrium paths in this model. In particular we can show that the local behavior of equilibrium path around the steady state changes with the degree of external effect in the capital goods sector.

---

[7]If $\alpha_2\beta_1 - \alpha_1\beta_2 > (<)\, 0$, the capital gooss sector (the consumption goods) is capital intensive from the private perspective.

We define local indeterminacy following Benhabib, Nishimura and Venditti [2002].

Definition 3.2. *A steady state k^* is called locally indeterminate if there exists ε such that for any $k_0 \in (k^* - \varepsilon,\ k^* + \varepsilon)$, there are infinitely many equilibrium paths converging to the steady state.*

Proposition 1. *Suppose that the capital goods sector is capital intensive from the private perspective, that is $\alpha_2\beta_1 - \alpha_1\beta_2 > 0$. Then, the steady state is a saddle for $b < \beta_2$, and it is totally unstable for $b > \beta_2$.*

Proof. Note that $\alpha_1 = 1-\alpha_2$ and $\beta_1 = 1-\beta_2$. By substituting into equation (26), we obtain $\lambda_1 = \frac{\alpha_2}{\rho(\alpha_2-\beta_2)} > 0$. Denoting $\rho_1 \equiv \frac{\alpha_2}{(\alpha_2-\beta_2)} > 1$, then $\lambda_1 = \frac{\rho_1}{\rho}$ is always greater than 1 as $\rho_1 > \rho > 0$.

Since $\lambda_1 > 1$ and λ_2 depends on the degree of externality b, we have the following cases.

(i) Let $b < \beta_2$. $0 < \frac{(\beta_2-b)}{\alpha_2} < 1$. Hence $0 < \lambda_2 < 1$. Therefore, the steady state is a saddle point.

(ii) Let $b > \beta_2$. Then $\lambda_2 > 1$. Hence the steady state is totally unstable. Q.E.D.

Remark 1. The production function from the social perspective is represented as follow:

$$Y = \left(\frac{K_2}{L_2}\right)^{\beta_1+b} L_2. \tag{27}$$

Divide both sides by L_2, $b + \beta_1 > 1$ and denote $\left(\frac{K_2}{L_2}\right)^{\beta_1+b}$ by $h\left(\frac{K_2}{L_2}\right)$. When b is larger than β_2, the function $h\left(\frac{K_2}{L_2}\right)$ exhibits increasing return.

$$h\left(\frac{K_2}{L_2}\right) = \left(\frac{K_2}{L_2}\right)^{\beta_1+b} > \frac{K_2}{L_2}.$$

Next we state results under the assumption that consumption goods are capital intensive from the private perspective with $\alpha_2\beta_1 - \alpha_1\beta_2 < 0$.

Proposition 2. *Let $\rho_2 \equiv \frac{\alpha_2}{(\beta_2-\alpha_2)}$. Suppose that $\rho_2 < \rho < 1$ and $2\alpha_2 < \beta_2$. Then, the steady state is a saddle for $b < \beta_2 - 2\alpha_2$, it is locally indeterminate for $0 < \beta_2 - 2\alpha_2 < b < \beta_2$, and it is a saddle for $b > \beta_2$.*

Proof. In this case, $\lambda_1 = \frac{\alpha_2}{\rho(\alpha_2-\beta_2)} < 0$. From $2\alpha_2 < \beta_2$, $\rho_2 = \frac{\alpha_2}{(\beta_2-\alpha_2)} < 1$. Hence λ_1 can be rewritten as $\lambda_1 = \frac{-\rho_2}{\rho}$. Then, $-1 < \lambda_1 < 0$. The size of $\lambda_2 = 1 - \frac{(\beta_2-b)}{\alpha_2}$ is determined in the following way.

(i) For $b < \beta_2 - 2\alpha_2$, $2 < \frac{(\beta_2-b)}{\alpha_2}$. Hence $\lambda_2 < -1$. Then the steady state is a saddle point.

(ii) For $0 < \beta_2 - 2\alpha_2 < b < \beta_2$, $0 < \frac{(\beta_2-b)}{\alpha_2} < 2$. Hence $-1 < \lambda_2 < 1$.Therefore the steady state is locally indeterminate.

(iii) For $b > \beta_2$, $\frac{(\beta_2-b)}{\alpha_2} < 0$. Hence $1 < \lambda_2$. Therefore the steady state is a saddle point. Q.E.D.

Remark 2. If $0 < \rho < \rho_2$, the steady state is unstable for $b < \beta_2 - 2\alpha_2$, it is a saddle for $0 < \beta_2 - 2\alpha_2 < b < \beta_2$, and it is totally unstable for $b > \beta_2$. Therefore $b > \beta_2$ or increasing returns of $h\left(\frac{K_2}{L_2}\right)$ implies the total instability of the steady state.

References

[1] G. Baierl, K. Nishimura and M. Yano, The role of capital depreciation in multi-sectoral models, J. Econom. Behavior Organization, **33** (1998), 467–479.

[2] J. Benhabib and R. E. Farmer, Indeterminacy and increasing returns, J. Econom. Theory, **20** (1994), 19–41.

[3] J. Benhabib and K. Nishimura, Indeterminacy and sunspots with constant returns, J. Econom. Theory, **81** (1998), 58–96.

[4] J. Benhabib and K. Nishimura, Indeterminacy arising in multi-sector economies, Jpn. Econ. Rev., **50** (1999), 485–506.

[5] J. Benhabib, K. Nishimura and A. Venditti, Indeterminacy and cycles in two-sector discrete-time model, Econom. Theory, **20** (2002), 217–235.

[6] K. Nishimura and A. Venditti, Dynamical systems arising from infinite time horizon optimization models, J. Differ. Equations Appl., **6** (2000), 753–773.

[7] K. Nishimura and A. Venditti, Intersectoral externalities and indeterminacy, J. Econom. Theory, **105** (2002), 140–157.

M. Matsuo
Institute of Economic Research
Kyoto University

T. Sakagami
Department of Economics
Kumamoto Gakuen University

E-mail address: `matsuo@kier.kyoto-u.ac.jp`
`tomoya@kumagaku.ac.jp`

Advanced Studies in Pure Mathematics 53, 2009
Advances in Discrete Dynamical Systems
pp. 159–168

Computing topological entropy in asymmetric Cournot duopoly games with homogeneous expectations

Diana A. Mendes, Vivaldo M. Mendes, J. Sousa Ramos and Orlando Gomes

Abstract.

The main aim of this paper is to analyse the dynamics of nonlinear discrete-time maps generated by duopoly games in which players have homogeneous expectations and heterogenous nonlinear cost functions. This framework leads to reaction functions that are non-monotonic and asymmetric and, in the particular case of naïve expectations, the model takes the form of an anti-triangular map, $T(x, y) = (f(y), g(x))$ characterized by a rich dynamical behavior, from stable to chaotic Nash equilibria. We also present the computation of topological entropy of this nonlinear Cournot model by using tools from symbolic dynamics and tensor products.

§1. Introduction

Oligopoly models are very simple economic structures (or games) that may lead to very complex dynamics, like multiple equilibria, indeterminacy, or chaotic dynamics, even in its most simple form which is usually denominated by duopoly (two firms producing and supplying a homogeneous good in a common market). This complexity may arise from the combination of various sources, such as the way in which each firm formulates expectations about the rival's decisions (concerning the levels of goods produced and supplied in the market or the price at which these goods are supplied by both firms), the consideration of heterogeneous cost structures, the introduction of dynamics into the firms's decision making process, or the adoption of sophisticated cooperative strategies by both players and their impact upon the equilibrium of the game, among others. The literature on these issues is extremely

Received December 31, 2006.
Revised October 12, 2007.

large by now and important references are, e.g., [11], [10], [5], [3], [4], [1], [9], and [8].

This paper deals with a duopoly game in a dynamic setting, assuming that firms have homogenous expectations, and production displays heterogenous cost functions which are nonlinear due to positive externalities. For simplicity, we also assume that expectations are formulated in a naive way by each firm, meaning that at every period of time each player expects his rival to offer the same quantity for sale in the current period as it did in the preceding one.[1] As we will show, a very simple economic structure may lead to extremely complex dynamics, to multiple equilibria and to chaotic dynamics. A major result obtained is that the Nash equilibrium (the noncooperative solution of the game) changes from stable and periodic to chaotic, through period-doubling bifurcations, when the system parameters are changed. The existence of chaotic motion is demonstrated by computing the topological entropy of the model through the application of tools from symbolic dynamics and tensor products [7], [6].

§2. Homogeneous naive expectations

We consider a dynamic version of a simple Cournot-type duopoly market where firms (players) produce homogeneous goods which are perfect substitutes and supply them at discrete-time periods $t = 0, 1, 2, ...$ in a common market. At each period t, every firm must formulate an expectation of the rival's output in the next time period in order to determine the corresponding profit-maximizing production quantities for period $t + 1$.

If we denote by $q_i(t)$, $i = 1, 2$ the output of firm i at time t, then its production $q_i(t+1)$, $i = 1, 2$ for the next period $t + 1$ is decided by solving the following optimization problem

$$(1) \qquad \begin{cases} q_1(t+1) = \arg\max_{q_1} \Pi_1\left(q_1(t), q_2^e(t+1)\right) \\ q_2(t+1) = \arg\max_{q_2} \Pi_2\left(q_1^e(t+1), q_2(t)\right), \end{cases}$$

where the function $\Pi(\cdot, \cdot)$ denotes the profit of the ith firm and $q_j^e(t+1)$ represents the expectations of firm i about the production decision of firm j ($j = 1, 2, j \neq i$). If the optimization problem has a unique

[1] Notice that with adaptive, rational or bounded rational expectations, the dynamics would become even more complicated than in the case of naive expectations. See, eg., Mendes et al. [8].

solution, then we denote the former by

$$\begin{cases} q_1(t+1) = f(q_2^e(t+1)) \\ q_2(t+1) = g(q_1^e(t+1)), \end{cases}$$

where f and g are the reaction functions.

The crucial point here is how each firm formulates expectations about the future production levels of its rival. Four different approaches have been found in the literature. Expectations can be *adaptive* if they are formulated in accordance with the rule $q_i^e(t+1) = q_i^e(t) + \xi(q_i(t) - q_i^e(t)), \xi \neq 1$ is the parameter that reflects the speed at which the expectation mistakes are corrected over time.[2] If we apply the condition $\xi = 1$, we obtain the rule $q_i^e(t+1) = q_i(t)$, which suggests *naive* behavior from both players. Fully *rational* expectations occur when $\xi \to \infty$, so that any mistake made at any period is instantaneously corrected, and we get the rule $q_i^e(t+1) = q_i(t+1)$. We may still have *bounded rational* expectations (see [1]) where firms use information based on the local estimates of the marginal profit function, such that at each time period t, each firm increases (decreases) its production q_i at the period $(t+1)$ if the marginal profit is positive (negative).

In this paper, and for the sake of brevity, we will only illustrate the case when both players choose the strategy defined by naive expectations. The model becomes

$$\begin{cases} q_1(t+1) = f(q_2(t)) \\ q_2(t+1) = g(q_1(t)), \end{cases}$$

which is known as a anti-triangular map, whose dynamics can be studied by using analytical and numerical tools.

§3. The complete model

Let us assume that the market inverse demand function is linear and decreasing:

$$P = p(Q) = a - b(q_1 + q_2), \tag{2}$$

where $Q = q_1 + q_2$ is the industry total output and $a, b > 0$. Following Kopel [5], we consider the case in which production satisfies two conditions. Firstly, there are positive production externalities, and secondly,

[2]The fundamental point in this rule is that palyers update their expectations based on the mistakes they made in the past. Therefore, if no mistakes were done at time t, $q_i(t) = q_i^e(t)$, and so $q_i^e(t+1) = q_i^e(t)$.

for a given level of the rival's production, each firm's individual production shows constant returns to scale. The cost functions of both firms in the market may be represented by

$$C_1(q_1,q_2) = c_1(q_2)\,q_1 \text{ and } C_2(q_1,q_2) = c_2(q_1)\,q_2, \tag{3}$$

with $c_1'(q_2) < 0, c_2'(q_1) < 0$. Notice that each firm has a marginal cost of production $(c_i(\cdot))$ that is constant with respect to its own output but varies negatively with respect to the rival's output. We adopt here the specific asymmetric form for the functions c_i, $i = 1, 2$ as proposed by [9]:

$$\begin{cases} c_1(q_2) = a - bq_2 - 2b(\alpha q_2 - \alpha + 1)^2 \\ c_2(q_1) = a - bq_1 - 2b(\beta q_1 - 1)^2 . \end{cases}$$

Therefore, this particular structure for the marginal cost functions implies that both firms have heterogeneous costs of production and production externalities are nonlinear for both firms.

Combining (1), (2) and (3), we have the following profit functions for both firms:

$$\begin{cases} \Pi_1 = pq_1 - c_1(q_2)\,q_1 \\ \Pi_2 = pq_2 - c_2(q_1)\,q_2 \end{cases}$$

and the reaction functions are given by $f(q_2) = (\alpha q_2 - \alpha + 1)^2$ and $g(q_1) = (\beta q_1 - 1)^2$. Considering the case of naive expectations for both players the model takes the form

$$AT \begin{cases} q_1(t+1) = f(q_2(t)) = (\alpha q_2(t) - \alpha + 1)^2 \\ q_2(t+1) = g(q_1(t)) = (\beta q_1(t) - 1)^2 \end{cases},$$

and was firstly studied by Nonaka [9]. This is an anti-triangular or Cournot map and its dynamics can be rigorously studied by using analytical and numerical tools defined for example in [2], [7], [8], among others. This duopoly model admits multiple Nash equilibria, namely two, three or four, depending on the values of the two parameters α, β, and they can be obtained as the non-negative solutions of the algebraic system $(\alpha q_2 - \alpha + 1)^2 = q_1$, $(\beta q_1 - 1)^2 = q_2$.

The easiest way to study the anti-triangular map is to consider the second order composition of AT, that is,

$$T \begin{cases} (f \circ g)(q_1) = F(q_1) = (\alpha(\beta q_1 - 1)^2 - \alpha + 1)^2 \\ (g \circ f)(q_2) = G(q_2) = (\beta(\alpha q_2 - \alpha + 1)^2 - 1)^2, \end{cases} \tag{4}$$

where variables q_1 and q_2 are acting now independently from each other. This is a product map and we denote it by $T(q_1, q_2) = (F(q_1), G(q_2))$.

Now, the dynamic behavior of the Cournot map can be deduced from the one-dimensional maps $F(q_1)$ and $G(q_2)$, where the basis map $F(q_1) = (\alpha(\beta q_1 - 1)^2 - \alpha + 1)^2$ and the fiber map $G(q_2) = (\beta(\alpha q_2 - \alpha + 1)^2 - 1)^2$ are trimodal with the following critical points

$$q_1^{c_1} = \frac{1}{\beta},\ q_1^{c_2} = \frac{\alpha + \sqrt{\alpha^2 - \alpha}}{\alpha\beta},\ q_1^{c_3} = \frac{\alpha - \sqrt{\alpha^2 - \alpha}}{\alpha\beta}$$

$$q_2^{c_1} = \frac{\alpha - 1}{\alpha},\ q_2^{c_2} = \frac{\beta\alpha - \beta + \sqrt{\beta}}{\beta\alpha},\ q_2^{c_3} = \frac{\beta\alpha - \beta - \sqrt{\beta}}{\beta\alpha}.$$

Since we consider that $1 < \alpha < 2,\ 0 < \beta < 2$, the critical points are always well defined and the dynamics of the map T is fully characterized by the trajectories of the critical points of $F(q_1)$ and $G(q_2)$.

For example, if q_1^* is a fixed point of F, then $(q_1^*, g(q_1^*))$ is a fixed point of the Cournot map with eigenvalues $\lambda_1 = F'(q_1^*)$ and $\lambda_2 = -F'(q_1^*)$ and if the product map T has a period four orbit, then the anti-triangular map has a period eight orbit where the periodic points of AT are obtained by pairs of combinations of the periodic points of T. Moreover by considering tensor products between the one dimensional invariants associated with the critical orbits of F and G, we can compute the topological entropy of the Cournot map (see [7], [6]).

In what follows we will define the tensor product between two matrices and present the main theorem which allows the computation of the topological entropy of triangular and anti-triangular maps.

Definition 1. *Let A, B be two matrices of type $(m \times n)$ and $(p \times q)$. The tensor product of A and B is a matrix C of type $(mp \times nq)$, represented by $C = A \otimes B$ and defined by*

$$C = A \otimes B = \begin{bmatrix} a_{11}B & a_{12}B & \cdots & a_{1n}B \\ \vdots & \vdots & \ddots & \vdots \\ a_{m1}B & a_{m2}B & \cdots & a_{mn}B \end{bmatrix}.$$

It is well known that the class of product maps studied in this paper admits a Markov partition (by rectangles) which is determined by the itineraries of the critical points associated with the basis map and the fiber map. Given a Markov partition $\mathcal{R} = \{R_j\}_{j=1}^m$, the transition matrix $A = (a_{ij})$ of type $(m \times m)$ is defined by $a_{ij} = 1$ if $int(f(R_i)) \cap int(R_j) \neq \emptyset$ and $a_{ij} = 0$ in the other case. The subshift space for A is defined as $\Sigma_A = \{s : \mathbb{N} \to \{1, 2, \ldots, m\} : a_{s_i s_{i+1}} = 1\}$. Letting σ be the shift map on the full $m-$shift, $\Sigma_m = \{1, 2, \ldots, m\}^{\mathbb{N}}$, define $\sigma_A = \sigma_{|\Sigma_A} : \Sigma_A \to \Sigma_A$, that is the subshift of finite type that characterizes the dynamics of the system.

Let us denote by $\mathcal{R}_{q1}, \mathcal{R}_{q2}, A_{q1}, A_{q2}$ the Markov partitions and the transition matrices associated with the critical orbits of the one-dimensional basis map F and the fiber map G. Then the following theorem holds [7], [6], [2]

Theorem 2. *Let I be a compact interval of the real line and let $T(q_1, q_2) = (F(q_1), G(q_2))$ be a continuous triangular map. Suppose that the basis map F admits n critical orbits of finite period $p_1, \ldots, p_n$ and the fiber map G admits m critical orbits of finite periods $q_1, \ldots, q_m$. Then the Markov partition of the map T is given by the Cartesian product $\mathcal{R}_{q_1} \times \mathcal{R}_{q_2}$ and the transition matrix A of T is given by the following tensor product: $A = A_{q_2} \otimes A_{q_1}$.*

Corollary 3. *The topological entropy of a continuous triangular map $T(q_1, q_2) = (F(q_1), G(q_2))$ is given by the sum of the topological entropies of F and G, that is $h(T) = h(F) + h(G)$.*

Now, the topological entropy of a generic Cournot map is formally characterized by:

Corollary 4. *Let I be a compact interval of the real line and let $T : I^2 \to I^2$, $AT(q_1, q_2) = (f(q_2), g(q_1))$ be a continuous Cournot map. We denote by*

$$T(q_1, q_2) = ((f \circ g)(q_1), (g \circ f)(q_2)) = (F(q_1), G(q_2))$$

the second iterate (compose) of the map AT which is by construction a triangular (product) map. Then

$$h(AT) = \frac{h(T)}{2} = \frac{h(F) + h(G)}{2}.$$

The Cournot map shows a very rich dynamic behavior which can be easily observed in Figure 1, where several bifurcation diagrams of the variable q_1 are presented when β is varied between 1 and 2, and α takes the values $\alpha = 1, 1.2, 1.4, 1.8$. We can observe period-doubling routes to chaos with several stability windows and chaotic regions, and some inverse bifurcations when $\alpha = 1.2$. In all of these cases the dynamics of variable q_2 is very similar. We found stable Nash equilibria for values of $\beta < 1$.

Let us consider the map T defined in (4) and fix $\alpha = 1.8$. In order to compute the topological entropy of the Cournot map, we have to find some periodic critical orbits for the basis map F and for the fiber map G. By observing the bifurcation diagram (Figure 1) we decided to consider the period four stability windows located immediately after $\beta = 1$, since $1 < \beta < 2$ presents more complex dynamics. This is a periodic orbit with

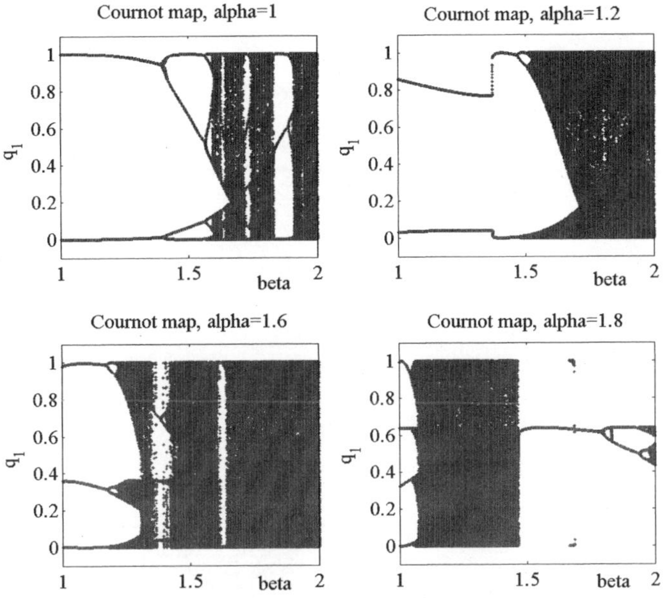

Fig. 1. Several bifurcation diagram for the q_1 variable when the parameter β is varied between 1 and 2 and the parameter α assumes the following values: $1, 1.2, 1.4, 1.8$.

chaos and by solving $(f \circ g)^4 (q_1, \beta) = q_1$ for $q_1^{c_3}$ and $\alpha = 1.8$ we find the exact value of β where the bifurcation takes place. We obtain then, that for $\beta = 1.00991$ the critical orbit of the trimodal map $(f \circ g)$ initiated in $q_1^{c_3}$ is of period four. The critical point $q_1^{c_3}$ is the only significative one, since the diagonal $q_1 = (f \circ g)(q_1)$ intersects only once the trimodal map $(f \circ g)(q_1)$ in the vicinity of this critical point. By the same reason we obtain that $G(q_2) = (g \circ f)(q_2)$ has also a period four critical orbit, which is given by the trajectory of the significant critical point, $q_2^{c_2}$. We may completely ignore the other orbits (critical or non-critical) since the kneading theory assures that the critical orbit is predominant and fully characterizes the dynamics, giving the maximal entropy of the system.

For this parameters setting ($\alpha = 1.8$ and $\beta = 1.00991$) the Cournot map admits 2 unstable nonnegative Nash equilibrium points given by $(0.1753, 0.6771)$ and $(2.0967, 1.2489)$ where $q_1^1 = 0.1753$, $q_1^2 = 2.0967$ and $q_2^1 = 0.6771$, $q_2^2 = 1.2489$ are also the fixed points of the one-dimensional

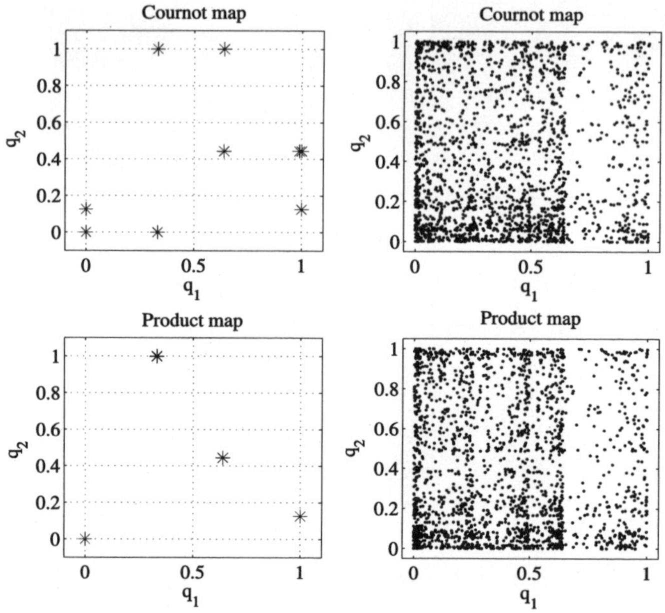

Fig. 2. Period 8 orbit and strange attractor for the antriangular map and period 4 orbit and strange attractor for the product map when $\alpha = 1.8$ and $\beta = 1.00991$, $\beta = 1.2$

trimodal maps $F(q_1)$ and $G(q_2)$. The study of the local stability of these fixed points is based on the localization, on the complex circle, of the eigenvalues of the Jacobian matrix of the two-dimensional map, which is given by

$$J(q_1, q_2) = \begin{bmatrix} 0 & 2\alpha(\alpha q_2 - \alpha + 1) \\ 2\beta(\beta q_1 - 1) & 0 \end{bmatrix}.$$

For the first fixed point we have complex conjugate eigenvalues given by $\lambda = \pm 1.5830i$, with modulus greater than one, while for the second fixed point we have $\lambda = \pm 3.4303$, both displaying real eigenvalues with modulus greater than one, which shows the instability of the Nash equilibria.

The main dynamics of the Cournot map is fully characterized by the dynamics of the critical orbits. We have that the kneading sequences of the two period four critical orbits associated with the maps $(f \circ g)$ and $(g \circ f)$ are given by $(CRLL, crll)$, where $\mathcal{A}_{q_1} = \{L, C, R\}$ and $\mathcal{A}_{q_2} = \{l, c, r\}$ are the alphabets associated with the generated partitions of the phase space of these maps. The transition matrices of the period four critical orbits are given by

$$A_{q_1} = A_{q_2} = \begin{bmatrix} 1 & 1 & 1 \\ 0 & 0 & 1 \\ 0 & 1 & 0 \end{bmatrix},$$

and the transition matrix A of the map T is then

$$A = A_{q_1} \otimes A_{q_2} = \begin{bmatrix} 1 & 1 & 1 & 1 & 1 & 1 & 1 & 1 & 1 \\ 0 & 0 & 1 & 0 & 0 & 1 & 0 & 0 & 1 \\ 0 & 1 & 0 & 0 & 1 & 0 & 0 & 1 & 0 \\ 0 & 0 & 0 & 0 & 0 & 0 & 1 & 1 & 1 \\ 0 & 0 & 0 & 0 & 0 & 0 & 0 & 0 & 1 \\ 0 & 0 & 0 & 0 & 0 & 0 & 0 & 1 & 0 \\ 0 & 0 & 0 & 1 & 1 & 1 & 0 & 0 & 0 \\ 0 & 0 & 0 & 0 & 0 & 1 & 0 & 0 & 0 \\ 0 & 0 & 0 & 0 & 1 & 0 & 0 & 0 & 0 \end{bmatrix}.$$

Since the logarithm of the spectral radius of a Markov transition matrix gives the topological entropy of a multimodal map, we have that $h\,(fog) = h\,(gof) = \log(\lambda_{\max}) = 0.6094$ which means that the topological entropy of the triangular map and of the Cournot map is $h\,(T) = 1.2188$ and $h\,(AT) = 0.6094$.

Figure 2 represents the period 8 orbit of the Cournot map and the period 4 orbit of the product map, for the considered parameters setting. It shows also, as a mere example, some strange attractors simulated for $\beta = 1.2$ and $\alpha = 1.8$.

Acknowledgement. Financial support from the "Fundação Ciência e Tecnologia", Lisbon, is grateful acknowledged, under the contract POCTI/ECO /48628/2002, partially funded by the European Regional Development Fund (ERDF).

References

[1] H. N. Agiza and A. A. Elsadany, Chaotic dynamics in nonlinear duopoly game with heterogeneous players, Appl. Math. Comput., **149** (2004), 843–860.

[2] F. Balibrea, J. S. Cánovas and A. Linero, New results on topological dynamics of antitriangular maps, Appl. Gen. Topol., **2** (2001), 51–61.
[3] G. I. Bischi, C. Chiarella and M. Kopel, The long run outcomes and global dynamics of a duopoly game with misspecified demand functions, Int. Game Theory Rev., **6** (2004), 343–380.
[4] G. I. Bischi and M. Kopel, Equilibrium selection in a nonlinear duopoly game with adaptive expectations, J. Econom. Behavior Organization, **46** (2001), 73–100.
[5] M. Kopel, Simple and complex adjustment dynamics in Cournot duopoly models, Chaos Solitons Fractals, **12** (1996), 2031–2048.
[6] D. A. Mendes and J. S. Ramos, Kneading theory for triangular maps, Int. J. Pure Appl. Math., **10** (2004), 421–450.
[7] D. A. Mendes and J. S. Ramos, Computing the topological entropy of triangular maps of the plane, Grazer Math. Ber., **346** (2004), 283–297.
[8] D. A. Mendes, V. Mendes and O. Gomes, Asymmetric Cournot duopoly games with adaptive and boundedly rational expectations, Working Paper, IBS-ISCTE Business School, Lisbon, Portugal.
[9] Y. Nonaka, Complexity leads to benefits: Pareto-improving chaos in a heterogeneous duopoly market, Nonlinear Dynamics and Heterogenous Interacting Agents, (ed. T. Lux), Lecture Notes in Econom. and Math. Systems, **550**, Springer-Verlag, 2005, pp. 103–122.
[10] T. Puu, Chaos in duopoly pricing, Chaos Solitons Fractals, **1** (1991), 573–581.
[11] X. Vives, Oligopoly Pricing, MIT Press, Cambridge, Mass, 2001.
[12] X. Vives, Information and Learning in Markets, forthcoming, 2007.

D. A. Mendes
Department of Quantitative Methods
IBS–ISCTE Business School and UNIDE

V. M. Mendes
Department of Economics
ISCTE and UNIDE, Lisbon

J. S. Ramos
Department of Mathematics
Instituto Superior Técnico, UTL, Lisbon

O. Gomes
Escola Superior Comunicação Social
IP Lisbon, and UNIDE - ISCTE

E-mail address: `diana.mendes@iscte.pt`
`vivaldo.mendes@iscte.pt`
`sramos@math.ist.utl.pt`
`ogomes@escs.ipl.pt`

Advanced Studies in Pure Mathematics 53, 2009
Advances in Discrete Dynamical Systems
pp. 169–177

Equilibrium dynamics in an overlapping generations economy with endogenous labor supply

Atsue Mizushima [1]

Abstract.

This paper develops a two–period overlapping generation model to examine the behavior of an economy that incorporates endogenous labor–care choice. Assuming a log–linear utility function and a Cobb–Douglas production function, we show that there exists multiple equilibria, comprising a unique trajectory satisfying saddle–path stability and other equilibria, which have an infinite number of converging transition paths.

§1. Introduction

It is wildly recognized that multiple equilibria are a common feature among several dynamic models. Typically, self–fulfilling expectations may yield many perfect foresight paths or sunspot equilibria. In an overlapping generations model, dynamic indeterminacy hinges on complicated preferences or specific restrictions on production technologies (see, for example [5] and [3]). This paper tries to complement the existing literature by studying the labor–care choice by using a two period overlapping generations model.

In the model presented later, we assume that labor supply decisions are endogenous as young agents can choose to work at household on a household–produced health or on the market place to produce market goods. When agents derive utility from the status of their aged parents'

Received December 5, 2006.
2000 *Mathematics Subject Classification.* J22, O40.
Key words and phrases. multiple equilibria, overlapping generations, household production.
[1] I would like to thank Koichi Futagami, Kazuo Mino, and Mitsu-yoshi Yanagihara for helpful comments and suggestions. All remaining errors are naturally my own.

health, working at household are modeled by allowing young agents to participate in the production of household health status of old agents.[1]

Using this model, we show that the equilibrium dynamics can feature multiple equilibria. More specifically, there exist multiple equilibria, comprising a unique trajectory satisfying saddle–path stability and other equilibria, which have an infinite number of converging transition paths in a simple utility function and Cobb–Douglas production technology.

The remainder of this paper is organized as follows. Section 2 sets up the basic model. Section 3 derives the equilibrium dynamics. Section 4 concludes this paper.

§2. The model

Consider an infinite–horizon economy composed of agents and perfectly competitive firms. A new generation, referred to as generation t, is born in each period $t = 1, 2, 3, \cdots$. Generation t is composed of a continuum of $N_t > 0$ units of agents who live for a maximum of two period; young and old age. At each date, new generations, each consisting of a continuum of agents with a unit measure, are born.

Agents have altruism and derive utility from the status of their aged parents' health. The probability that an agent dies at the beginning of old age after he or she has child is $1 - p$, and the probability that he or she lives throughout old period is $p \in (0, 1]$. If an agent alive in his or her old age, he or she also has a probability of being in poor health. The probability that an agent being in good health throughout the old age is ψ, and that has poor health is $1 - \psi$. Therefore, there are three different states in the two periods of life: good health, poor health, and death. A fraction $p\psi$ of young agents are of type g whose parents have good health, a fraction $p(1 - \psi)$ of young agents are of type b whose parents have poor health. Type d agents, who constitute a fraction $1-p$ of young agents, whose parents die. We express the death–illness status of each young agents as index i.

Each young agent of generation t allocates their unit of time between to work at household on a household–produced health $q_{i,t}^{t}$,[2] or on the market place to produce market goods $l_{i,t}^{t}$. He or she earns wage income $w_t l_{i,t}^{t}$, and saves all wage income, where w_t is the real wage rate. Old

[1]Intergenerational transfers of time using household production have been studied by [4].

[2]In what follows we referred to intergenerational transfers of time synonymously with providing care to aged parents and synonymously with care provision.

agents of generation t consume the proceeds of their savings, which we denote by $c_{i,t+1}^{t}$. We assume the existence of actuarially fair insurance in this paper (see [6] and [2]). Thus the rate of return on the annuities is R_{t+1}/p if they are alive and 0 if they die at the end of period t, where R_{t+1} is the real rental rate. Let us assume that the level of health of old agents whose health status is good (g) or poor (b) at time t is produced at household using the following household health production function:

$$h_{g,t} = dq_{g,t}^{t}, \tag{1}$$
$$h_{b,t} = q_{b,t}^{t}, \tag{2}$$

where $d > 1$ is a productivity parameter. Thus, the marginal productivity of care provision of type g young agents is higher than that of type b young agents. Each young agent of generation t solves the following optimization problem for a given level of R_{t+1} and w_t:

$$U_i = \beta \ln h_{i,t} + pc_{i,t+1}^{t} \quad i = g, b, d$$

s.t.

$$\begin{aligned} c_{i,t+1}^{t} &= R_{t+1} w_t l_{i,t}^{t}, \\ q_{i,t}^{t} + l_{i,t}^{t} &= 1, \\ 0 \le q_{i,t}^{t} \le 1, \ 0 &\le l_{i,t}^{t} \le 1, \\ &(1), (2), \end{aligned} \tag{3}$$

where $\beta \in (0,1)$ measures the degree of altruism towards parents.

The optimal care provision is derived as follows:

$$q_{g,t}^{t} = q_{b,t}^{t} = \begin{cases} \dfrac{\beta}{R_{t+1} w_t} & \text{if } \beta \le R_{t+1} w_t, \\ 1 & \text{if } R_{t+1} w_t \le \beta. \end{cases} \tag{4}$$

Due to the quasi-linear utility function, when $R_{t+1} w_t$ is sufficiently large, the opportunity cost of care provision on a household–produced health status is high, then each agent decreases his or her care provision. If parents die, young agents do not derive any utility from the level of parents' health status; thus we have:

$$q_{d,t}^{t} = 0. \tag{5}$$

Using (3), (4), and (5), we have the aggregate labor supply as follows:[3]

$$(6)\qquad L_t \equiv l_t N_t = \begin{cases} \left(1 - \dfrac{p\beta}{R_{t+1}w_t}\right) N_t & \text{if } \beta \leq R_{t+1}w_t, \\ (1-p)N_t & \text{if } R_{t+1}w_t \leq \beta. \end{cases}$$

Firms are perfectly competitive profit maximizers that produce output using a production function of the Cobb–Douglas form $Y_t = AK_t^{\alpha}L_t^{1-\alpha}$, where Y_t is aggregate output, $A > 0$ is a productivity parameter and K_t is the aggregate capital stock. We assume that capital depreciates fully in the process of production. Thus, profit maximization in the competitive market equates the marginal products of private labor and capital to the real wage and the real rental rate, respectively:

$$(7)\qquad w_t = (1-\alpha)Ak_t^{\alpha},\ R_t = \alpha A k_t^{\alpha-1},$$

where $k_t \equiv K_t/L_t$.

§3. Equilibrium dynamics and indeterminacy

The equilibrium condition for the capital market is given by $K_{t+1} = s_t N_t = w_t l_t N_t$, which implies that the savings of young agents in generation t forms the aggregate capital stock in period $t+1$. Dividing both sides by N_t and substituting in (7) yields the following:

$$(8)\qquad k_{t+1} = \frac{(1-\alpha)Ak_t^{\alpha}l_t}{l_{t+1}}.$$

Using (6) through (8), we obtain the following complete dynamic systems:

<Regime I: $\beta \leq R_{t+1}w_t$ >

$$(9)\qquad l_{t+1}^{1-\alpha} = \frac{p\beta l_t^{1-\alpha}}{\alpha A(1-l_t)((1-\alpha)Ak_t^{\alpha})^{\alpha}},$$

$$(10)\qquad k_{t+1}^{1-\alpha} = \frac{(1-\alpha)Ak_t^{\alpha}\alpha A(1-l_t)}{p\beta}.$$

<Regime II: $R_{t+1}w_t \leq \beta$ >

$$(11)\qquad l_{t+1} = 1-p,$$

$$(12)\qquad k_{t+1} = (1-\alpha)Ak_t^{\alpha}.$$

[3] Aggregate labor supply is derived as $L_t \equiv l_t N_t = p\psi(1-q_{g,t}^{\,t}) + p(1-\psi)(1-q_{b,t}^{\,t}) + (1-p)(1-q_{d,t}^{\,t})$.

Before stating the equilibrium, we consider the borderline between the regimes. Substituting (7), (8), and (9) into the borderline, $\beta = R_{t+1}w_t$, we can express the borderline as follows:

$$l_t = 1 - p. \tag{13}$$

Thus, Regime I and Regime II is respectively feasible on the area $1-p \leq l_t \leq 1$ and $l_t \leq 1-p$.

The equations (9) through (12) characterize the economic equilibria that are represented the sequences of $\{k_t, l_t\}_{t=1}^{\infty}$ with an initial condition $(k_1, l_1) \geq 0$. Now let us draw the phase diagram on the (k_t, l_t) plane. We refer to the loci representing $k_{t+1} = k_t$ as KK and that representing $l_{t+1} = l_t$ as LL. We have KK and LL loci from (9) through (12):

<Regime I: $1 - p \leq l_t$ >

$$LL_1 \ : \ l_t = 1 - \frac{p\beta}{\alpha A((1-\alpha)Ak_t^{\alpha})^{\alpha}}, \tag{14}$$

$$KK_1 \ : \ k_t = \Big(\frac{(1-\alpha)A\alpha A(1-l_t)}{p\beta}\Big)^{\frac{1}{1-2\alpha}} \tag{15}$$

LL_1 and KK_1 loci respectively intersect the borderline $l_t = 1-p$ at the points A^{ll} and A^{kk}, where $A^{ll} \equiv (k_t, l_t) = (((\frac{\beta}{\alpha A})^{\frac{1}{\alpha}} \frac{1}{(1-\alpha)A})^{\frac{1}{\alpha}},\ 1-p)$ and $A^{kk} \equiv (k_t, l_t) = ((\frac{(1-\alpha)A\alpha A}{\beta})^{\frac{1}{1-2\alpha}},\ 1-p)$.

<Regime II : $l_t \leq 1 - p$ >

$$LL_2 \ : \ l_t = 1 - p, \tag{16}$$

$$KK_2 \ : \ k_t = ((1-\alpha)A)^{\frac{1}{1-\alpha}}. \tag{17}$$

For simplicity of analysis, we assume the following:

Assumption 1. *(i)* $\alpha < \frac{1}{2}$, *and (ii)* $\Big(\frac{\beta}{\alpha}\Big)^{1-\alpha}\Big(\frac{1}{1-\alpha}\Big)^{\alpha} \leq A.$

Assumption 1-(i) implies that KK locus is decreasing in l_t. Assumption 1-(ii) implies that KK and LL loci intersect in Regime I.[4]

The initial point at which the economy starts can be derived from:

$$l_1 = \frac{K_1}{k_1 N_1}. \tag{18}$$

---

[4]Assumption 1(i): $\frac{\partial k}{\partial l} = -\frac{1}{1-2\alpha}(\frac{(1-\alpha)A\alpha A(1-l_t)}{p\beta})^{\frac{2\alpha}{1-2\alpha}} < 0$, if $\alpha < \frac{1}{2}$.
Assumption 1(ii): We find the value of k becomes $A^{ll} < ((1-\alpha)A)^{\frac{1}{1-\alpha}} < A^{kk}$, if $(\frac{\beta}{\alpha})^{1-\alpha}(\frac{1}{1-\alpha})^{\alpha} < A$.

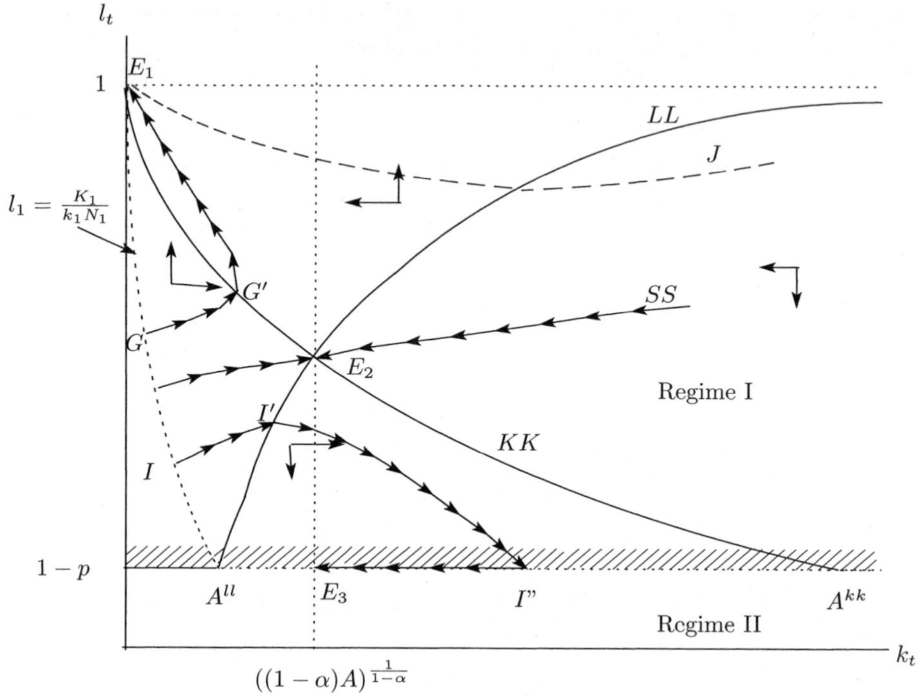

Fig. 1. Phase diagram analysis

The surface representing (18) can be drawn in (k_t, l_t) space as the initial per capita capital stock K_1/N_1 is given exogenously. Therefore, the economy must be initially on the line (18). As can be verified immediately from (18), the contour of (18) when drawn on the (k_t, l_t) plane is downward sloping. The phase diagram of this economy is depicted in Figure 1.

For later reference, we first consider the case in which the initial level of capital is sufficiently large. In this case, the initial contour is drawn in the upper-right corner in Figure 1, and the trajectory is drawn like J. Since any trajectory that is above J does not satisfy the time constraint (see equation (3)), we can exclude these trajectories from the equilibrium. Thus, contour J shows the boundary trajectory in this economy. Next, let us consider the case for example $l_1 = K_1/k_1N_1$ in Figure 1. If the economy initially happens to be on the line SS, it converges to

E_2.[5] If the economy initially happens to be above (below) the line SS, it converges to E_1 (E_3). In these cases, there exists an infinite number of converging transition paths towards E_1 (E_3). Therefore, for a given level of the initial capital stock, the economy converges to one of the equilibria E_1, E_2, or E_3 in the long run.

In order to obtain intuitive implications for these results, let us firstly consider the trajectory which converges to E_1 in Figure 1. On the path towards E_1, any trajectory that starts above the line "SS" for example at point G in Figure 1, initial level of capital is sufficiently high and, thus, so is the opportunity cost of care provision. In the economy both labor supply and the capital–labor ratio initially increase. However, when the KK locus is crossed at point G' in Figure 1, the increased labor supply leads to a decrease in the capital–labor ratio k_t. Because the boundary trajectory of this economy is J, it follows that the trajectory for this regime converges to E_1.

Next, let us consider the equilibrium E_3 in Figure 1. On the equilibrium path towards E_3, any trajectory that starts below the line "SS" for example at point I in Figure 1, the initial level of capital is sufficiently low and, thus, so is the opportunity cost of care provision. In the economy, both labor supply and the capital–labor ratio initially increase. However, when the LL locus is crossed at point I' in Figure 1, the increase in the capital–labor ratio leads to a decrease in labor supply l_t. When the trajectory reaches point I'' in Figure 1, where labor supply is as its lowest, the capital–labor ratio falls and approaches E_3.

§4. Conclusion

We have investigated the steady-state equilibrium dynamics in a model that incorporates endogenous labor–care choice. Care choice is modeled by allowing young agents to participate in the household–produced health. Using a simple utility function and Cobb–Douglas production function, we have shown that the equilibrium dynamics obtained from the model can feature multiple equilibria. More specifically, there exists a unique trajectory satisfying saddle–path stability and other equilibria, which have an infinite number of converging transition paths.

[5]See Appendix A for the conditions for the stability of the steady state.

§ Appendix

Appendix A

We examine the stability of the equilibrium E_2 in this Appendix. To examine the local dynamics of equilibrium E_2, we take a first-order Taylor expansion of the system around the steady state (k^*, l^*). With $\check{l}_t \equiv l_t - l^*$ and $\check{k}_t \equiv k_t - k^*$, the linearization is expressed as:

$$\begin{pmatrix} \check{k}_{t+1} \\ \check{l}_{t+1} \end{pmatrix} = \begin{pmatrix} \dfrac{\alpha}{(1-\alpha)} & -\dfrac{\alpha A}{p\beta} \cdot \dfrac{((1-\alpha)A)^{\frac{1+\alpha}{1-\alpha}}}{1-\alpha} \\ -\dfrac{l^*\alpha^2}{(1-\alpha)((1-\alpha)A)^{\frac{1}{1-\alpha}}} & 1 + \dfrac{((1-\alpha)A)^{\frac{\alpha}{1-\alpha}} l^* \alpha A}{(1-\alpha)p\beta} \end{pmatrix} \begin{pmatrix} \check{k}_t \\ \check{l}_t \end{pmatrix},$$

where $l^* \equiv 1 - \frac{p\beta}{\alpha A((1-\alpha)A)^{\frac{\alpha}{1-\alpha}}}$. The characteristic polynomial becomes:

$$P(\kappa) = \kappa^2 - T\kappa + D,$$

$$\begin{aligned} T &= \frac{\alpha}{1-\alpha} + 1 + \frac{((1-\alpha)A)^{\frac{\alpha}{1-\alpha}} l^* \alpha A}{(1-\alpha)p\beta}, \\ D &= \frac{\alpha}{1-\alpha} + \frac{\alpha}{(1-\alpha)} \cdot \frac{((1-\alpha)A)^{\frac{\alpha}{1-\alpha}} l^* \alpha A}{p\beta}. \end{aligned}$$

[1] checks that the steady state is a saddle point, when $1 - T + D < 0$ holds. It is clear that:

$$1 - T + D = -\frac{((1-\alpha)A)^{\frac{\alpha}{1-\alpha}} l^* \alpha A}{p\beta} < 0.$$

Therefore, E_2 is a saddle point.

References

[1] C. Azariadis, Intertemporal Macroeconomics, Basil Blackwell, Oxford, 1993.

[2] O. J. Blanchard, Debt, deficits, and finite horizons, J. Polit. Econ., **93** (1985), 223–247.

[3] M. Boldrin, Dynamic externalities, multiple equilibria, and growth, J. Econom. Theory, **58** (1992), 198–218.

[4] E. Cardia and P. Michel, Altruism, intergenerational transfers of time and bequests, J. Econom. Dynam. Control, **28** (2004), 1681–1701.

[5] P. Reichlin, Equilibrium cycles in an overlapping generations model with production, J. Econom. Theory, **40** (1986), 89–102.

[6] M. E. Yaari, Uncertain lifetime, life insurance, and the theory of the consumer, Rev. Econom. Stud., **32** (1965), 137–150.

Institute of Economic Research
Kyoto University
Yoshida-honmachi, Sakyo-ku
Kyoto, 606-8501
Japan

E-mail address: mizushima@kier.kyoto-u.ac.jp

Advanced Studies in Pure Mathematics 53, 2009
Advances in Discrete Dynamical Systems
pp. 179–187

Stabilities with respect to a weight function in Volterra difference equations

Satoru Murakami

Abstract.

We establish some results on ρ-stabilities of the zero solution of a system of linear Volterra difference equations. Furthermore, for the equations with nonnegative coefficients we give an easy criterion for the ρ-UAS of the zero solution.

§1. Introduction

In this paper we are concerned with a system of linear Volterra difference equations
(1)

$$x(n+1) = \sum_{k=0}^{n} Q(n-k)x(k) \ \ (= (Q*x)(n)), \qquad n \in \mathbf{Z}^+ := \{0, 1, 2, \cdots\},$$

where $x = \{x(n)\}$ is a sequence of r column vectors, and $Q := \{Q(n)\}$ is a sequence of $r \times r$ matrices which is ρ-summable, that is, $\sum_{n=0}^{\infty} \rho(n)$ $\|Q(n)\| < \infty$, for a weight function $\rho : \mathbf{Z}^+ \to (0, \infty)$. Here and hereafter, we call ρ a weight function if ρ satisfies

$$\rho(0) = 1, \qquad \rho(n+m) \leq \rho(n)\rho(m) \quad \text{for } \forall n, \ m \in \mathbf{Z}^+.$$

The functions $\rho(n) \equiv L^n$, $\rho(n) \equiv L^n(1+n)^p$, $\rho(n) = L^n(1+\log(1+n))^p$ with positive costants L and p are typical ones of weight functions. In this paper, we will introduce the concept of stabilities with respect to ρ (ρ-stabilities, in short), which relates to the decay rate of solutions of Eq. (1) and is exactly same as the usual concept of stabilities in case of $\rho(n) \equiv 1$. As one of our main results in this paper, we will give a result which is a characterization of the ρ-uniform asymptotic stability property of the zero solution of Eq. (1) in connection with the ρ-summability

Received October 12, 2007.
2000 *Mathematics Subject Classification.* 39A11.

of the resolvent matrix of Eq. (1), as well as the invertibility of the characteristic matrix $zE - \sum_{n=0}^{\infty} Q(n)z^{-n}$ for z belonging to the domain $|z| \geq 1/l$, where E denotes the $r \times r$ unit matrix, and l is a number defined by $l := \lim_{n\to\infty} \sqrt[n]{\rho(n)}$ $(= \inf_n \sqrt[n]{\rho(n)})$ and it is assumed to be positive. In fact, in case of $\rho(n) \equiv 1$, the result is identical with the one obtained by Elaydi and Murakami [2]. Furthermore, under the restriction that the matrices $Q(n)$ are all nonegative, we will give an easy criterion for the ρ-UAS of the zero solution of Eq. (1).

The following result is a typical one which is obtainable as an immediate consequence of our main theorems in Sections 2 and 3:

Theorem 1. *Assume that $Q(n) \geq 0, \forall n \geq 0$ and $\sum_{n=0}^{\infty} \|Q(n)\| < \infty$. Then the zero solution of Eq. (1) is UAS, if and only if*

$$\det\left(zE - \sum_{j=0}^{\infty} Q(j)\right) \neq 0 \quad (\forall |z| \geq 1).$$

On the other hand, the $(1+n)^{p+2}$-UAS can **never** *be realized for the zero solution of Eq. (1), when $\sum_{n=0}^{\infty}(n^p\|Q(n)\|) = \infty$ for some positive integer p.*

Indeed, the former part of the theorem is a direct consequence of Theorem 4 (in Section 3) with $\rho(n) \equiv 1$. Also, the latter part of the theorem follows from Theorem 3 (in Section 2), because the $(1+n)^{p+2}$-UAS of the zero solution of Eq. (1) implies that the resolvent $R = \{R(n)\}_{n\in\mathbf{Z}^+}$ of Eq. (1) satisfies $\sup_{n\geq 0}\{(1+n)^{p+2}\|R(n)\|\} < \infty$, and hence $\sum_{n=0}^{\infty}(n^p\|R(n)\|) < \infty$.

§2. Some results on ρ-stabilities

Given $\forall \tau \in \mathbf{Z}^+$ and $\phi : [0, \tau] := \{0, 1, \ldots, \tau\} \to \mathbf{C}^r$. A function $x : \mathbf{Z}^+ \to \mathbf{C}^r$ is called *a solution* of Eq. (1) through (τ, ϕ) if

$$x(n+1) = (Q * x)(n) \quad (\forall n \geq \tau), \quad x(s) = \phi(s) \ (\forall s = 0, 1, \ldots, \tau),$$

and is denoted by the symbol $x(\cdot; \tau, \phi)$.

The zero solution of Eq. (1) is said to be;

(1) *uniformly stable with respect to ρ* (ρ-US) if $\forall\epsilon > 0, \exists\delta > 0$:

$$\|\phi\|_l := \sup_{-\tau\leq\theta\leq 0}(|\phi(\tau+\theta)|l^{\theta}) < \delta \Longrightarrow \rho(n)|x(n+\tau; \tau, \phi)| < \epsilon \ (\forall n \geq 0),$$

(2) *uniformly asymptotically stable with respect to* ρ (ρ-UAS) if the zero solution is ρ-US and if $\exists\delta_0 > 0, \forall\epsilon > 0, \exists N(\epsilon) > 0:$

$$\|\phi\|_l < \delta_0 \Longrightarrow \rho(n)|x(n+\tau;\tau,\phi)| < \epsilon \;\; (\forall n \geq N(\epsilon)).$$

In the above, l is the (nonnegative) constant given by

$$l := \lim_{n\to\infty} \sqrt[n]{\rho(n)} \;\; (= \inf_n \sqrt[n]{\rho(n)}).$$

We remark that if $\rho(n) \equiv 1$, then $l = 1$, and hence the concept of the above ρ-stabilities is exactly same as the concept of the usual stabilities.

In the remainder, we always assume $l > 0$ and the following condition on the weight function ρ:

$$(2) \quad \exists(\text{constant})\ C > 0: \quad \rho(n)l^s \leq C\rho(n+s-1) \quad (\forall n \geq 0,\ s \geq 1).$$

We note that Condition (2) is satisfied (with $C = l$) for the concrete functions $\rho(n) \equiv l^n$, $\rho(n) \equiv l^n(1+n)^p$, and so on.

Notice that the *Z-transform* $\tilde{Q}(z) := \sum_{n=0}^{\infty} Q(n)z^{-n}$ is convergent for any $|z| \geq 1/l$, because $\rho(n) \geq l^n$ and $\sum_{n=0}^{\infty} \|Q(n)\|\rho(n) < \infty$.

A family of $r \times r$ matrices $R = \{R(n)\}_{n\in\mathbf{Z}^+}$ is called the *resolvent* of Eq. (1), if $R(0) = E$ and $R(n+1) = (Q * R)(n)$ for any $n \geq 0$. It is easy to see that the Z-transform $\tilde{R}(z)$ of the resolvent R exists for z with sufficiently large $|z|$. Considering the Z-transform of the both sides of $R(n+1) \equiv (Q * R)(n)$, one can get the relation $z\tilde{R}(z) - zR(0) = \tilde{Q}(z)\tilde{R}(z)$, or $(zE - \tilde{Q}(z))\tilde{R}(z) = zE$.

One of the crucial results in this section is the following:

Theorem 2. *Assume that* $Q = \{Q(n)\}$ *is* ρ*-summable for a weight function* ρ *satisfying Condition (2). Then the following statements are equivalent:*

(1) $\det(zE - \tilde{Q}(z)) \neq 0 \quad (\forall |z| \geq 1/l)$.

(2) R *is* ρ*-summable; that is,* $\displaystyle\sum_{n=0}^{\infty}(\|R(n)\|\rho(n)) < \infty$.

(3) *The zero solution of Eq. (1) is* ρ*-UAS.*

The equivalence (i) $\Longleftrightarrow$ (ii) can be established by applying the theory of Banach algebras [6, Chapter 11] to the space

$$L_\rho(\mathbf{Z}^+;\mathbf{C}) := \{a = \{a(n)\}_{n\in\mathbf{Z}^+} \subset \mathbf{C} \mid \|a\| := \sum_{n=0}^{\infty} \rho(n)|a(n)| < \infty\}.$$

Indeed, the space $L_\rho(\mathbf{Z}^+;\mathbf{C})$ with the convolution product is a commutative Banach algebra, and any nontrivial character χ of $L_\rho(\mathbf{Z}^+;\mathbf{C})$ is given by $\chi = \phi_w$, that is, $\chi(a) \equiv \sum_{n=0}^{\infty} a(n)w^n$, for some $w \in \mathbf{C}$ with $|w| \leq l$. By this fact and the relation $(zE - \tilde{Q}(z))\tilde{R}(z) = zE$, one can verify the equivalence; cf. [3, Proof of Corollary 1 and pp.1213–1214]. Also, the implications (ii)$\Longrightarrow$(iii) and (iii)$\Longrightarrow$(i) can be verified by applying the representation formula of solutions of Eq. (1)

$$x(n+\tau;\tau,\phi) = R(n)\phi(\tau) + \sum_{j=0}^{n-1} R(n-j-1)\sum_{s=1}^{\tau} Q(j+s)\phi(\tau-s),$$

using Condition (2); cf. [2, Proof of Theorem 2].

Applying Theorem 2 with $l = 1$ and $\rho(n) \equiv (1+n)^p$ (p is a nonnegative integer), we get the following corollary:

Corollary 1. Assume that $\sum_{n=0}^{\infty}(\|Q(n)\|n^p) < \infty$. Then the following statements are equivalent:

(1) $\det(zE - \tilde{Q}(z)) \neq 0 \quad (\forall |z| \geq 1)$.

(2) $\sum_{n=0}^{\infty}(\|R(n)\|n^p) < \infty$.

(3) The zero solution of Eq. (1) is $(1+n)^p$-UAS.

In case of $p = 0$, Corollary1 is exactly tha same as [2, Theorem 2]. As shown in the following result, we cannot drop the n^p-summability condition on $Q(n)$ in Corollary1 under *the sign condition* for $Q(n)$.

Theorem 3. *Assume that* $\sum_{n=0}^{\infty}\|Q(n)\| < \infty$, *and that for sufficiently large n, each component $q_{ij}(n)$ of $Q(n)$ has the same sign; that is,*

$$\exists n_0 \geq 0 : \ q_{ij}(n) \geq 0 \ (\forall n \geq n_0, \ i,j = 1,\ldots,r)$$

or

$$q_{ij}(n) \leq 0 \ (\forall n \geq n_0, \ i,j = 1,\ldots,r).$$

If $\sum_{n=0}^{\infty}(\|R(n)\|n^p) < \infty$ *for some positive integer p, then* $\sum_{n=0}^{\infty}(\|Q(n)\|n^p) < \infty$.

Proof. The theorem can be proved by a method similar to the one in [5]; so, we will give only a sketch of the proof. In the following, we may consider the case $p \geq 1$ and assume that the norm of $r \times r$ matrix $A = (a_{ij})$ is given by the l^1 norm, that is, $\|A\| = \sum_{i,j=1}^{r}|a_{ij}|$. Observe that

$\tilde{R}(z) = \sum_{n=0}^{\infty} R(n) z^{-n}$ is convergent on $|z| \geq 1$, and $(zE - \tilde{Q}(z))\tilde{R}(z) = zE$; hence $\tilde{R}(z)$ is invertible and

$$\tilde{Q}(x) \equiv xE - x(\tilde{R}(x))^{-1} \ (=: F(x)) \quad (\forall x \geq 1).$$

It follows that $F \in C^p([1,\infty))$, moreover if $M > n_0$ and $h > 0$, then

$$\begin{aligned} \left\|\frac{F(1+h)-F(1)}{h}\right\| &= \left\|\sum_{n=0}^{\infty} Q(n)\frac{(1+h)^{-n}-1}{h}\right\| \\ &\geq \left(\sum_{n=n_0}^{M} - \sum_{n=0}^{n_0-1}\right)\|Q(n)\|\frac{1-(1+h)^{-n}}{h} \end{aligned}$$

by the sign condition on Q. Noting that

$$\sup_{0\leq n\leq M}\left|\frac{1-(1+h)^{-n}}{h} - n\right| \leq 1$$

for small $h > 0$, we can establish that

$$\sum_{n=0}^{\infty} n\|Q(n)\| < \infty, \quad F'(x) = \frac{d}{dx}\tilde{Q}(x) = \sum_{n=0}^{\infty}(-n)Q(n)x^{-n-1} \quad (\forall x \geq 1)$$

by a contradiction; cf. [5, Proof of Theorem 2]. Repeating the above argument, we can derive that

$$\sum_{n=0}^{\infty} n^k\|Q(n)\| < \infty,$$

$$F^{(k)}(x) = \sum_{n=0}^{\infty}(-1)^k n(n+1)\cdots(n+k-1)Q(n)x^{-n-k} \quad (\forall x \geq 1)$$

for $k = 1, 2, \ldots, p$. Letting $k = p$, we get the required one. Q.E.D.

§3. Stabilities in equations with nonnegative matrices

For any two $l \times q$ matrices $A = (a_{ij})$ and $B = (b_{ij})$, the inequality $A \geq B$ means $a_{ij} \geq b_{ij}$ for all $i = 1, \cdots, l$, $j = 1, \cdots, q$. If $A \geq 0$, then A is called a nonnegative matrix. Also, a (column) vector b is said to be strictly positive, if all components of b are positive. We use the symbol $b \gg 0$ to denote that b is strictly positive.

In this section, we treat Eq. (1) with nonnegative matrices $Q(n)$, $\forall n \in \mathbf{Z}^+$, and establish the following result on the ρ-stabilities.

Theorem 4. *Assume that $Q(n) \geq 0, \forall n \geq 0$ and that $Q = \{Q(n)\}$ is ρ-summable for a weight function ρ satisfying Condition (2). Then the following statements are equivalent:*

(1) $\det(zE - \sum_{n=0}^{\infty} Q(n)l^n) \neq 0 \quad (\forall |z| \geq 1/l)$.
(2) *The zero solution of Eq. (1) is ρ-UAS.*
(3) *For any $b \gg 0$ and any solution y of $y(n+1) = (Q*y)(n)+b/l^n$, $l^n y(n)$ tends to a strictly positive vector as $n \to \infty$.*
(4) *For some $b \gg 0$ and some solution y of $y(n+1) = (Q*y)(n)+b/l^n$, $l^n y(n)$ tends to a strictly positive vector as $n \to \infty$.*

Applying Theorem 4 with $l = 1$ and $\rho(n) \equiv (1+n)^p$ (p is a nonnegative integer), we get the following corollary:

Corollary 2. Assume that $Q(n) \geq 0, \forall n \geq 0$ and that $\sum_{n=0}^{\infty}(\|Q(n)\| n^p) < \infty$. Then the following statements are equivalent:

(1) $\det(zE - \sum_{n=0}^{\infty} Q(n)) \neq 0 \quad (\forall |z| \geq 1)$.
(2) The zero solution of Eq. (1) is $(1+n)^p$-UAS.
(3) For any $b \gg 0$, any solution of $y(n+1) = (Q*y)(n) + b$ tends to a strictly positive vector as $n \to \infty$.
(4) For some $b \gg 0$, some solution of $y(n+1) = (Q*y)(n) + b$ tends to a strictly positive vector as $n \to \infty$.

In Corollary2, (or Theorem 4), we cannot always drop the nonnegativity condition on $Q(n)$ as the following example shows.

Example 1. Let us consider the scalar equation (1) with

$$Q(0) = 0, \qquad Q(n) = (-1)^{n-1}\frac{\alpha}{n^2}(n = 1,\ 2, \ldots),$$

where $\alpha = 6/\pi^2$. Then $\sum_{n=0}^{\infty} Q(n) = \alpha \times (\pi^2/12) = 1/2$; thus, Condition (i) in Corollary2 is satisfied. But the zero solution of Eq. (1) is not UAS, because the equation $\det(zE - \tilde{Q}(z)) = 0$ has a root $z = -1$.

Without assuming the nonnegative condition on $Q(n)$, one can obtain the following result (like a comparison theorem in ODE) on stabilities in Eq. (1) by checking Condition (i) in Theorem 2. Here and hereafter, for any matrix $A = (a_{ij})$ we employ the notation $|A| = (|a_{ij}|)$.

Theorem 5. *Assume that $Q = \{Q(n)\}$ is ρ-summable for a weight function ρ satisfying Condition (2), and that*

$$\det\left(zE - \sum_{n=0}^{\infty}(|Q(n)|l^n)\right) \neq 0 \quad (\forall |z| \geq 1/l).$$

Then the zero solution of Eq. (1) is ρ-UAS.

Example 2. Let $|\alpha| < 1/2$, $|\beta| < 1$, $|\gamma| < 1$, and consider Eq. (1) with

$$Q(n) \equiv \begin{pmatrix} \alpha^{n+1} & 0 \\ \beta^n & \gamma/(n+1)(n+2) \end{pmatrix}.$$

Then

$$\sum_{n=0}^{\infty} |Q(n)| = \begin{pmatrix} |\alpha|/(1-|\alpha|) & 0 \\ 1/(1-|\beta|) & |\gamma| \end{pmatrix}.$$

Observe that $|\alpha|/(1-|\alpha|)$ and $|\gamma|$ are the eigenvalues for the nonnegative matrix $\sum_{n=0}^{\infty} |Q(n)|$, and that they are smaller than 1 by the requirements on α and γ. Therefore, the zero solution of Eq. (1) is UAS by Theorem 5 with $l = 1$.

Before concluding the paper, we will give a sketch of the proof of Theorem 4. To do this, we first recall the following result which is well known as Perron-Frobenius's Theorem [1, 4]

Lemma 1. Let P be a nonnegative square matrix. Then

(1) The spectral radius $r := r_\sigma(P)$ is an eigenvalue of P and there exists a nonnegative eigenvector x, $x \neq 0$ such that $Px = rx$.
(2) $r_\sigma(P) = \sup\{\nu \in \mathbf{R} : \ Px \geq \nu x \quad \text{for some nonzero } x \geq 0\}$.
(3) If $\lambda > r_\sigma(P)$, then $(\lambda E - P)^{-1} \geq 0$.

The following result is an immediate consequence of Perron-Frobenius's Theorem (ii).

Proposition 1. Let P and Q be $r \times r$ matrices such that $|P| \leq Q$. Then $r_\sigma(P) \leq r_\sigma(Q)$.

In the proof of the theorem, the following proposition plays a crucial role.

Proposition 2. Assume that $Q(n) \geq 0, \forall n \geq 0$ and $\sum_{n=0}^{\infty} (\|Q(n)\| \rho(n)) < \infty$. Then the following statements are equivalent:

(1) $\det(zE - \tilde{Q}(z)) \neq 0 \quad (\forall |z| \geq 1/l)$.
(2) $r_\sigma(\sum_{n=0}^{\infty}(Q(n)l^n)) < 1/l$; in other words, $\det(zE - \sum_{n=0}^{\infty} Q(n)l^n) \neq 0 \quad (\forall |z| \geq 1/l)$.
(3) For some $b \gg 0$, there is a $y^* \gg 0$ such that $\left((1/l)E - \sum_{k=0}^{\infty} Q(k)l^k\right) y^* = b$.

Proof. (ii)$\Longrightarrow$(i). This is a direct consequence of Proposition 1, because of the inequality $|\tilde{Q}(z)| \leq \sum_{n=0}^{\infty} Q(n)l^n$ for $|z| \geq 1/l$.

(i)$\Longrightarrow$(ii). Let us consider the continuous function $f(t) = t - r_\sigma(\tilde{Q}(t))$ defined on $[1/l, \infty)$. Assume $f(1/l) \le 0$. Since $f(\infty) = \infty$, there is some $\lambda_1 \ge 1/l$ such that $f(\lambda_1) = 0$; in other words, $\lambda_1 = r_\sigma(\sum_{n=0}^\infty Q(n)\lambda_1^{-n})$. Then λ_1 is an eigenvalue of the matrix $\sum_{n=0}^\infty Q(n)\lambda_1^{-n}$ by Perron-Frobenius's Theorem (i), which contradicts (i). Thus we must get $f(1/l) > 0$, which is equivalent to (ii).

(ii)$\Longrightarrow$(iii). This is a direct consequence of Perron-Frobenius's Theorem (iii).

(iii)$\Longrightarrow$(ii). Since $r := r_\sigma(\sum_{n=0}^\infty Q(n)l^n)$ equals the spectral radius of the transposed matrix of $\sum_{n=0}^\infty Q(n)l^n$, Perron-Frobenius's Theorem (i) implies that there is a nonzero row vector $c \ge 0$ such that $c(\sum_{n=0}^\infty Q(n)l^n) = rc$. Then $rcy^* = c(\sum_{n=0}^\infty Q(n)l^n)y^* = c(y^*/l - b)$, or $cb = (1/l - r)cy^*$. Since $cy^* > 0$ and $cb > 0$, we must get $r < 1/l$, which shows (ii). Q.E.D.

Sketch of Proof of Theorem 4. The equivalence relation (i) $\Longleftrightarrow$ (ii) is an immediate consequence of Theorem 2 and Proposition 2. Also, the implication (iii) $\Longrightarrow$ (iv) is obvious.

(i) $\Longrightarrow$ (iii). Given any $b \gg 0$ and a solution y of $y(n+1) = (Q * y)(n) + b/l^n$ on $[\sigma, \infty)$ such that $y \equiv \phi$ on $[0, \sigma]$. By Proposition 2, there is a $y^* \gg 0$ such that $\left((1/l)E - \sum_{k=0}^\infty Q(k)l^k\right)y^* = b$. We will certify that $\lim_{n\to\infty}\{l^n y(n)\} = y^*$. Observe that the solution y is written by the representation formula

$$y(n) = R(n-\sigma)\phi(\sigma) + \sum_{s=\sigma}^{n-1} R(n-s-1)\left(\sum_{j=-\sigma}^{-1} Q(s-\sigma-j)\phi(\sigma+j) + \frac{b}{l^s}\right).$$

Since $\sum_{n=0}^\infty \|R(n)\|\rho(n) < \infty$ by Theorem 2, one can easily deduce that $l^n\|R(n-\sigma)\phi(\sigma)\| \to 0$ and $l^n\|\sum_{s=\sigma}^{n-1} R(n-s-1)\{\sum_{j=-\sigma}^{-1} Q(s-\sigma-j)\phi(\sigma+j)\}\| \to 0$ as $n \to \infty$. Then

$$\begin{aligned}
\lim_{n\to\infty}\{l^n y(n)\} &= \lim_{n\to\infty}\sum_{s=\sigma}^{n-1}\{R(n-s-1)l^{n-s}b\} \\
&= \lim_{n\to\infty}\sum_{w=0}^{n-\sigma-1}\{R(w)l^{w+1}b\} \\
&= l\tilde{R}(1/l)b \\
&= l\tilde{R}(1/l)\left(\frac{1}{l}E - \sum_{k=0}^\infty Q(k)l^k\right)y^* = y^*,
\end{aligned}$$

because of $z(zE - \tilde{Q}(z))^{-1} = \tilde{R}(z)$ for $|z| \geq 1/l$.

(iv) $\Longrightarrow$ (i). Let y be a solution of $y(n+1) = (Q * y)(n) + b/l^n$ satisfying $\lim_{n\to\infty}(l^n y(n)) = y^* \gg 0$, where $b \gg 0$ is some vector. Since $\sum_{k=0}^{\infty} \|Q(k)\|\rho(k) < \infty$, one can easily check that $\lim_{n\to\infty}(l^n \sum_{s=0}^{\infty} Q(n-s)y(s)) = \{\sum_{k=0}^{\infty} l^k Q(k)\}y^*$. Hence

$$\begin{aligned} y^* = \lim_{n\to\infty}(l^{n+1}y(n+1)) &= \lim_{n\to\infty}\left(l^{n+1}\sum_{s=0}^{n} Q(n-s)y(s) + lb\right) \\ &= l\{\sum_{k=0}^{\infty} l^k Q(k)\}y^* + lb \end{aligned}$$

or

$$\left(\frac{1}{l}E - \sum_{k=0}^{\infty} Q(k)l^k\right)y^* = b;$$

thus Condition (iii) of Proposition 2 is satisfied. Therefore, (i) follows from Proposition 2. Q.E.D.

References

[1] A. Berman and R. J. Plemmons, Nonnegative Matrices in Mathematical Sciences, Academic Press, New York, 1979.

[2] S. Elaydi and S. Murakami, Asymptotic stability versus exponential stability in linear Volterra difference equations of convolution type, J. Differ. Equations Appl., **2** (1996), 401–410.

[3] T. Furumochi, S. Murakami and Y. Nagabuchi, A generalization of Wiener's lemma and its application to Volterra difference equations on a Banach space, J. Differ. Equations Appl., **10** (2004), 1201–1214.

[4] D. G. Luenberger, Introduction to Dynamical Systems, Theory, Models and Applications, J. Wiley, New York, 1979.

[5] S. Murakami, Exponential asymptotic stability for scalar linear Volterra equations, Differential Integral Equations, **4** (1991), 519–525.

[6] W. Rudin, Functional Analysis, McGraw-Hill, New Delhi, 1988.

Department of Applied Mathematics
Okayama University of Science
1-1 Ridaicho, Okayama 700-0005
Japan

E-mail address: `murakami@@youhei.xmath.ous.ac.jp`

Advanced Studies in Pure Mathematics 53, 2009
Advances in Discrete Dynamical Systems
pp. 189–201

A note on asymptotic stability condition for delay difference equations

Piyapong Niamsup

Abstract.

In this paper, we obtain the necessary and sufficient condition for the asymptotic stability of the linear delay difference equation

$$x_{n+1} - x_{n-1} + p\sum_{j=1}^{N} x_{n-k+(j-1)l} = 0$$

where $n = 0, 1, 2, ...$, p is a real number ,and k, l, and N are positive integers such that $k > (N-1)l$.

§1. Introduction

In [5], the asymptotic stability condition for the linear delay difference equation

$$x_{n+1} - x_n + p\sum_{j=1}^{N} x_{n-k+(j-1)l} = 0 \tag{1}$$

where $n \in \mathbb{N}_0 = \mathbb{N} \cup \{0\}$, p is a real number and k, l, and N are positive integers with $k > (N-1)l$, is given as follows.

Theorem A. Let k, l, and N be positive integers with $k > (N-1)l$. Then the zero solution of (1.1) is asymptotically stable if and only if

$$0 < p < \frac{2\sin\left(\frac{\pi}{2M}\right)\sin\left(\frac{l\pi}{2M}\right)}{\sin\left(\frac{Nl\pi}{2M}\right)} \tag{2}$$

where $M = 2k + 1 - (N-1)l$.

Received August 24, 2007.
Revised October 12, 2007.
2000 *Mathematics Subject Classification.* Primary 39A11.
Key words and phrases. Asymptotically stable, delay difference equations, characteristic equations.

Theorem A. generalizes asymptotic stability conditions given in [1 p.87, 2-3, 5, 6 p.65]. Theorem A. is proved using the fact that the zero solution of a linear difference equation is asymptotically stable if and only if all the roots of its characteristic equation lie inside the unit disk. In [4], we give necessary and sufficient conditions for the asymptotic stability of the following linear difference equation

$$x_{n+1} - a^2 x_{n-1} + b x_{n-k} = 0.$$

Motivated by these results, we are interested in the asymptotic stability of the linear delay difference equation of higher order which is similar to (1.1) as follows:

$$x_{n+1} - x_{n-1} + p \sum_{j=1}^{N} x_{n-k+(j-1)l} = 0 \tag{3}$$

where $n \in \mathbb{N}_0 = \mathbb{N} \cup \{0\}$, p is a real number, and k, l, and N are positive integers with $k > (N-1)l$. These linear difference equations may be used as discrete models of population dynamics of Baleen whales, [2]. Our main theorem is the following.

Theorem 1.1. Let k, l, and N be positive integers with k odd, l even and $k > (N-1)l$. Then the zero solution of (1.3) is asymptotically stable if and only if

$$0 < p < \frac{2\sin\left(\frac{\pi}{M}\right)\sin\left(\frac{l\pi}{2M}\right)}{\sin\left(\frac{Nl\pi}{2M}\right)} \tag{4}$$

where $M = 2k - (N-1)l$.

Remark 1.1. For $p > 0$ and k is even, we have $F(-1) = pN > 0$ and $\lim_{z \to -\infty} F(z) = -\infty$; hence F has a root which lies outside the unit disk and the zero solution of (1.4) is not asymptotically stable.

§2. Proof of Theorem

The characteristic equation of (1.1) is given by

$$F(z) = z^{k+1} - z^{k-1} + p\left(z^{(N-1)l} + \cdots + z^l + 1\right) = 0. \tag{5}$$

For $p = 0$, $F(z)$ has simple roots at 1 and -1 and root at 0 of multiplicity $k-1$. We first consider the location of the roots of (2.1) as p varies. Throughout the paper, we denote the unit circle by C and let $M = 2k - (N-1)l$.

Proposition 2.1. Let z be a root of (2.1) which lies on C. Then the roots z and p are of the form

$$z = e^{w_m i}, \text{ and} \tag{6}$$

$$p = 2\,(-1)^m \frac{\sin w_m \sin \frac{l w_m}{2}}{\sin \frac{N l w_m}{2}} \equiv p_m \tag{7}$$

for some $m = 0, 1, ..., M-1$ where $w_m = \frac{2m+1}{M}\pi$. Conversely, if p is given by (2.3), then $z = e^{w_m i}$ is a root of (2.1).

Proof. We consider roots of (2.1) which lie on C except the roots $z = 1$ and $z = -1$. Suppose that the value z satisfies $z^{Nl} = 1$ and $z^l \neq 1$. Then $z^{(N-1)l} + \cdots + z^l + 1 = 0$ and z is not a root of (2.1) which lies on C and we shall consider only the value z such that $z^{Nl} \neq 1$ or $z^l = 1$. Thus (2.1) can be written as

$$p = -\frac{z^{k-1}(z^2-1)}{z^{(N-1)l} + \cdots + z^l + 1}. \tag{8}$$

Since p is real, we have

$$\begin{aligned} p &= -\frac{\overline{z}^{k-1}(\overline{z}^2-1)}{\overline{z}^{(N-1)l} + \cdots + \overline{z}^l + 1} \\ &= -\frac{(z^2-1)z^{-k-1+(N-1)l}}{z^{(N-1)l} + \cdots + z^l + 1} \end{aligned} \tag{9}$$

where $\overline{z}$ denotes the conjugate of z. It follows from (2.4) and (2.5) that

$$z^{2k-(N-1)l} = -1$$

which implies that (2.2) is valid for $m = 0, 1, ..., M-1$ except for those integers m such that $e^{Nlw_m i} = 1$ and $e^{lw_m i} \neq 1$. We now show that p is of the form stated in (2.3). There are two cases to be considered as follows.

Case 1. z is of the form $e^{w_m i}$ for some $m = 1, 2, ..., M-1$ and $z^{Nl} \neq 1$.

From (2.4) we have

$$
\begin{aligned}
p &= -\frac{z^{k-1}(z^2-1)\left(z^l-1\right)}{z^{Nl}-1} \\
&= -\frac{e^{(k-1)w_m i}\left(e^{2w_m i}-1\right)\left(e^{lw_m i}-1\right)}{e^{Nlw_m i}-1} \\
&= -\frac{e^{(k-(N-1)\frac{l}{2})w_m i}\left(e^{w_m i}-e^{-w_m i}\right)\left(e^{\frac{lw_m i}{2}}-e^{-\frac{lw_m i}{2}}\right)}{e^{\frac{Nlw_m i}{2}}-e^{-\frac{Nlw_m i}{2}}} \\
&= -e^{(k-(N-1)\frac{l}{2})w_m i}\frac{2i\sin(w_m)\sin\left(\frac{lw_m}{2}\right)}{\sin\left(\frac{Nlw_m}{2}\right)} \\
&= -e^{\frac{(2m+1)}{2}\pi i}\frac{2i\sin(w_m)\sin\left(\frac{lw_m}{2}\right)}{\sin\left(\frac{Nlw_m}{2}\right)} \\
&= 2(-1)^m\frac{\sin(w_m)\sin\left(\frac{lw_m}{2}\right)}{\sin\left(\frac{Nlw_m}{2}\right)} \equiv p_m.
\end{aligned}
$$

Case 2. z is of the form $e^{w_m i}$ for some $m = 1, 2, ..., M-1$ and $z^{Nl} = 1$.

In this case, we have $lw_m = 2q\pi$ for some positive integer q. Then taking the limit as $lw_m \to 2q\pi$ we obtain

$$
p = \frac{2(-1)^{m+q(N-1)}\sin(w_m)}{N}. \tag{10}
$$

From these two cases, we conclude that p is of the form in (2.3) for $m = 1, 2, ..., M-1$ except for those m such that $e^{Nlw_m i} = 1$ and $e^{lw_m i} \neq 1$.

Conversely, if p is given by (2.3), then it is obvious that $z = e^{w_m i}$ is a root of (2.1). This completes the proof of the proposition. Q.E.D.

We now consider p as a function of z:

$$
p(z) = -\frac{z^{k-1}(z^2-1)}{z^{(N-1)l}+...+z^l+1}.
$$

Then, we have

$$
\begin{aligned}
\frac{dp(z)}{dz} = &- \frac{z^{k-2}\left(2z^2+(k-1)(z^2-1)\right)}{z^{(N-1)l}+...+z^l+1} \\
&+ \frac{z^{k-2}(z^2-1)\left\{(N-1)lz^{(N-1)l}+...+lz^l\right\}}{\left(z^{(N-1)l}+...+z^l+1\right)^2}.
\end{aligned} \tag{11}
$$

From this we have

Lemma 2.1. $\left.\frac{dp}{dz}\right|_{z=e^{w_m i}} \neq 0$. In particular, the roots of (2.1) which lie on C are simple.

Proof. Suppose on the contrary that $\left.\frac{dp}{dz}\right|_{z=e^{w_m i}} = 0$. We divide (2.7) by $\frac{p(z)}{z}$ to obtain

$$\text{(12)} \qquad \frac{2z^2+(k-1)(z^2-1)}{z^2-1} - \frac{l\{(N-1)z^{(N-1)l}+...+z^l\}}{z^{(N-1)l}+...+z^l+1} = 0.$$

Substituting z by $\frac{1}{z}$ in (2.8) we obtain

(13)

$$\frac{2+(k-1)(1-z^2)}{1-z^2} - \frac{l\{(N-1)+(N-2)z^l+...+z^{(N-2)l}\}}{z^{(N-1)l}+...+z^l+1} = 0.$$

By adding (2.8) and (2.9), we obtain

$$2k-(N-1)l=0$$

which contradicts $k \geq (N-1)l$. This completes the proof. Q.E.D.

From Lemma 2.1, there exists a neighborhood of $z = e^{w_m i}$ such that the mapping $p(z)$ is one-to-one and the inverse of $p(z)$ exists locally. Now, let z be expressed as $z = re^{i\theta}$. Then we have

$$\frac{dz}{dp} = \frac{z}{r}\left\{\frac{dr}{dp} + ir\frac{d\theta}{dp}\right\}$$

which implies that

$$\frac{dr}{dp} = Re\left\{\frac{r}{z}\frac{dz}{dp}\right\}$$

as p varies and remaining real. The following result describes the behavior of the roots of (2.1) as p varies.

Proposition 2.2. The moduli of the roots of (2.1) on C increases as $|p|$ increases.

Proof. Let r be the modulus of z. Let $z = e^{w_m i}$ be a root of C. To prove this proposition, it suffices to show that

$$\text{(14)} \qquad \left.\frac{dr}{dp}\cdot p\right|_{z=e^{w_m i}} > 0.$$

There are two cases to be considered.

Case 1. $z^{Nl} \neq 1$. In this case we have

$$p(z) = -\frac{z^{k-1}(z^2-1)(z^l-1)}{z^{Nl}-1} = -\frac{z^{k-1}f(z)}{z^{Nl}-1}$$

where $f(z) = z\left(z^l - 1\right)$. Then

$$\frac{dp}{dz} = -\frac{z^{k-2}g(z)}{\left(z^{Nl}-1\right)^2}$$

where $g(z) = ((k-1)f(z) + zf'(z))\left(z^{Nl}-1\right) - Nlz^{Nl}f(z)$. Letting $w(z) = -\frac{\left(z^{Nl}-1\right)^2}{z^{k-1}g(z)}$, we obtain

$$\frac{dr}{dp} = Re\left(\frac{r}{z}\frac{dz}{dp}\right) = rRe(w).$$

We now compute $Re(w)$. We note that

$$\begin{aligned} f\left(\overline{z}\right) &= \frac{f(z)}{z^{l+2}} \text{ and} \\ f'\left(\overline{z}\right) &= \frac{h(z)}{z^{l+1}} \end{aligned}$$

where $h(z) = l(1-z^2)+2(1-z^l)$. From the above relation and $z^M = -1$, we have

$$\begin{aligned} \overline{z}^{k-1}g(\overline{z}) &= \frac{1}{z^{k-1}}\left\{\left((k-1)f\left(\overline{z}\right) + \frac{1}{z}f'\left(\overline{z}\right)\right)\left(\frac{1}{z^{Nl}}-1\right) - \frac{Nl}{z^{Nl}}f\left(\overline{z}\right)\right\} \\ &= \frac{((k-1)f(z)+h(z))\left(1-z^{Nl}\right) - Nlf(z)}{z^{Nl+l+1+k}} \\ &= -\frac{((k-1)f(z)+h(z))\left(1-z^{Nl}\right) - Nlf(z)}{z^{2Nl-k+1}}. \end{aligned}$$

It follows that

$$\begin{aligned}
Re(w) &= \frac{w+\overline{w}}{2} \\
&= -\frac{1}{2}\left\{\frac{\left(z^{Nl}-1\right)^2}{z^{k-1}g(z)}+\frac{\left(\overline{z}^{Nl}-1\right)^2}{\overline{z}^{k-1}g(\overline{z})}\right\} \\
&= -\frac{1}{2}\left\{\frac{\overline{z}^{k-1}g(\overline{z})\left(z^{Nl}-1\right)^2+z^{k-1}g(z)\left(\overline{z}^{Nl}-1\right)^2}{|g(z)|^2}\right\} \\
&= -\frac{1}{2\,|g(z)|^2}\left\{\begin{array}{l} \frac{((k-1)f(z)+h(z))\left(z^{Nl}-1\right)+Nlf(z)}{z^{2Nl-k+1}}\cdot\left(z^{Nl-1}\right)^2+ \\ z^{k-1}\left(((k-1)f(z)+zf'(z))\left(z^{Nl}-1\right)\right. \\ \left.-Nlz^{Nl}f(z)\right)\left(\frac{1}{z^{Nl}}-1\right)^2 \end{array}\right\} \\
&= -\frac{\left(z^{Nl}-1\right)^2 z^{k-1}}{2z^{2Nl}\,|g(z)|^2}\left\{\begin{array}{l} ((k-1)f(z)+h(z))\left(z^{Nl}-1\right) \\ +Nlf(z)+(((k-1)f(z)+zf'(z)) \\ \left(z^{Nl}-1\right))-Nlz^{Nl}f(z) \end{array}\right\} \\
&= -\frac{\left(z^{Nl}-1\right)^3 z^{k-1}}{2z^{2Nl}\,|g(z)|^2}\left\{h(z)+zf'(z)+(2(k-1)-Nl)f(z)\right\}.
\end{aligned}$$

Since

$$h(z)+zf'(z)+(2(k-1)-Nl)f(z)=Mf(z)$$

we obtain

$$Re(w)=\frac{\left(z^{Nl}-1\right)^4 M}{2z^{2Nl}\,|g(z)|^2}\cdot\frac{-z^{k-1}f(z)}{z^{Nl}-1}=\frac{\left(z^{Nl}-1\right)^4 Mp}{2z^{2Nl}\,|g(z)|^2}.$$

The value of $Re(w)$ at $z=e^{w_m i}$ is

$$\begin{aligned}
Re(w) &= \frac{\left(z^{Nl}-1\right)^4}{z^{2Nl}}\cdot\frac{Mp}{2\,|g(z)|^2} \\
&= (2\cos Nlw_m-2)^2\cdot\frac{Mp}{2\,|g(z)|^2}.
\end{aligned}$$

Therefore,

$$\frac{dr}{dp}=\frac{2r\left(\cos Nlw_m-1\right)^2 Mp}{|g(z)|^2}>0$$

and it follows that (2.10) holds at $z=e^{w_m i}$.

Case 2. $z^l=1$. With an argument similar to Case 1., we obtain

$$\frac{dr}{dp}=\frac{2rN^2Mp}{|(M+1)\,z-M+1|^2}$$

which implies that (2.10) is valid for $z = e^{w_m i}$.

This completes the proof. □

We now determine the minimum of the absolute values of p_m given by (2.3). We have the following result.

Proposition 2.3. $p_0 = \min\{|p_m| : m = 0, 1, ..., M-1\}$

To prove Proposition 2.3, we need the following lemmas.

Lemma 2.2. [5] Let N be a positive integer, then

$$\left|\frac{\sin Nt}{\sin t}\right| \leq N$$

holds for all $t \in \mathbb{R}$.

Lemma 2.3. [5] Let $0 < \theta < \frac{\pi}{2}$, then the inequality

$$\sin x\theta \, \sin y\theta \leq \sin\theta \, \sin xy\theta$$

holds for all $x, y \in \left(1, \frac{\pi}{2\theta}\right)$.

Proof of Proposition 2.3. By assumption, l is even which implies that M is also even. It is clear that $p_0 > 0$. Since each p_m is corresponded to $e^{w_m i}$ and its conjugate $\overline{e^{w_m i}}$, it is sufficient to consider p_m for $m = 0, 1, ..., \left[\frac{M-1}{2}\right] = \frac{M}{2} - 1$. To this end, we consider the following three cases.

Case I. $N = 1$. In this case, we have

$$p_m = 2(-1)^m \sin\frac{(2m+1)\pi}{2k}.$$

It follows immediately that $p_m \geq p_0$.

Case II. $N = 2$. It suffices to show that $\frac{1}{p_m} \leq \frac{1}{p_0}$ for $m = 1, 2, ..., \frac{M}{2} - 1$. Since $z^l = -z^{2k}$ and $z = e^{w_m i}$, we get

$$\begin{aligned}
p_m &= -\frac{z^{k-1}(z^2-1)\left(-z^{2k}-1\right)}{z^{4k}-1} \\
&= \frac{z^{k-1}\left(z^2-1\right)}{z^{2k}-1} \\
&= \frac{z - z^{-1}}{z^k - z^{-k}} \\
&= \frac{e^{w_m i} - e^{-w_m i}}{e^{k w_m i} - e^{-k w_m i}} \\
&= \frac{\sin w_m}{\sin k w_m}.
\end{aligned}$$

We observe that

$$\begin{aligned} p_{\frac{M}{2}-i} &= \frac{\sin\frac{2(\frac{M}{2}-i)+1}{M}\pi}{\sin\frac{2(\frac{M}{2}-i)+1}{M}k\pi} \\ &= \frac{\sin\frac{M-(2i-1)}{M}\pi}{\sin\frac{M-(2i-1)}{M}k\pi} \\ &= \frac{\sin\left(\pi-\frac{(2i-1)}{M}\pi\right)}{\sin\left(k\pi-\frac{(2i-1)}{M}k\pi\right)} \\ &= \frac{\sin\frac{(2i-1)}{M}\pi}{\sin\frac{(2i-1)}{M}k\pi} \\ &= p_{i-1}. \end{aligned}$$

Therefore, it suffices to show that

$$\frac{1}{p_m} \le \frac{1}{p_0} \tag{15}$$

for $m = 1, 2, ..., \left[\frac{M}{4} - \frac{1}{2}\right]$. Note that when $M = 4j$ then $\left[\frac{M}{4} - \frac{1}{2}\right] = \frac{M}{4}$ and when $M = 2j$ for an odd number j, then $\left[\frac{M}{4} - \frac{1}{2}\right] = \frac{M}{4} - \frac{1}{2}$. Let $\theta = \frac{\pi}{M}$. Then we have

$$\frac{1}{p_0} = \frac{\sin k\theta}{\sin\theta} \text{ and } \frac{1}{p_m} = \frac{\sin k(2m+1)\theta}{\sin(2m+1)\theta}.$$

Note that $0 < \theta < \frac{\pi}{2}$ and

$$1 \le M - k \le \frac{\pi}{2\theta},\ 1 \le 2m+1 \le \frac{\pi}{2\theta},$$

since $k > l$. It follows from Lemma 2.2 that

$$\sin(M-k)\,\theta\,\sin(2m+1)\,\theta \ge \sin\theta\,\sin(M-k)\,(2m+1)\,\theta.$$

Taking into account that $(M-k)\,\theta = \pi - k\theta$, we obtain (2.8) for $m = 1, 2, ..., \left[\frac{M}{4} - \frac{1}{2}\right]$.

Case III. $N \ge 3$. We will show that

$$|p_m| \ge p_0 \tag{16}$$

for $m = 0, 1, ..., \left[\frac{M-1}{2}\right]$. With the same argument as in Case II, it suffices to show (2.9) for $m = 0, 1, ..., \left[\frac{M}{4} - \frac{1}{2}\right]$. Let $\theta = \frac{\pi}{M}$. Then

$0 < (2m+1)\,\theta \leq \frac{\pi}{2}$ and

$$|p_m| = 2\sin{(2m+1)\,\theta}\left|\frac{\sin\frac{(2m+1)l\theta}{2}}{\sin\frac{(2m+1)Nl\theta}{2}}\right|.$$

By Lemma 2.3 and Jordan's inequality, namely $\frac{2}{\pi} \leq \frac{\sin\theta}{\theta} \leq 1$ for $0 \leq \theta \leq \frac{\pi}{2}$, we obtain

$$|p_m| \geq 2 \cdot \frac{2}{\pi}\,(2m+1)\,\theta \cdot \frac{1}{N} = \frac{4\,(2m+1)\,\theta}{\pi N}. \tag{17}$$

We will show that (2.13) holds in the following three subcases:

Subcase (IIIa): $\frac{Nl\theta}{2} \leq \frac{\pi}{2}$. In this subcase we have

$$p_0 = \frac{2\sin\theta\,\sin\frac{l\theta}{2}}{\sin\frac{Nl\theta}{2}} \leq \frac{2\cdot\theta\cdot\frac{l\theta}{2}}{\frac{2}{\pi}\cdot\frac{Nl\theta}{2}} = \frac{\pi\theta}{N}. \tag{18}$$

Inequalities (2.14) and (2.15) imply that (2.13) holds for $m = 0, 1, ...,$ $\left[\frac{M}{4} - \frac{1}{2}\right]$.

Subcase (IIIb): $\frac{Nl\theta}{2} > \frac{\pi}{2}$. In this subcase we have

$$\frac{Nl\theta}{2} = \frac{Nl\pi}{2M} < \frac{\pi}{2}\cdot\frac{Nl}{(N-1)\,l} = \frac{\pi}{2}\cdot\frac{N}{(N-1)}$$

since $k > (N-1)l$ and $M = 2k - (N-1)\,l > 2(N-1)l - (N-1)l = (N-1)l$. By using $\sin\frac{Nl\theta}{2} = \sin\left(\pi - \frac{Nl\theta}{2}\right)$, we get

$$p_0 = \frac{2\sin\theta\,\sin\frac{l\theta}{2}}{\sin\frac{Nl\theta}{2}} \leq \frac{2\cdot\theta\cdot\frac{l\theta}{2}}{\frac{2}{\pi}\cdot\left(\pi - \frac{Nl\theta}{2}\right)} = \frac{\pi l\theta^2}{2\pi - Nl\theta}.$$

It follows from (2.14), (2.15), and (2.16) that

$$\begin{aligned}
\frac{|p_m|}{p_0} &\geq \frac{4(2m+1)\theta}{\pi N}\cdot\frac{2\pi - Nl\theta}{\pi l\theta^2} \\
&= \frac{4(2m+1)}{\pi^2}\left(\frac{2\pi}{Nl\theta} - 1\right) \\
&> \frac{4(2m+1)}{\pi^2}\left(\frac{2\,(N-1)}{N} - 1\right) \\
&= \frac{4(2m+1)}{\pi^2}\left(1 - \frac{2}{N}\right).
\end{aligned}$$

From the above we have the following:

(i) If $N \geq 12$ and $m \geq 1$, then (2.13) holds.

(ii) If $N \geq 4$ and $m \geq 2$, then (2.13) holds.
(iii) If $N = 3$ and $m \geq 4$, then (2.13) holds.
We now consider the remaining cases.
(iv) $N \geq 4$ and $m = 1$. In this case it follows from (2.15) that $l\theta < \frac{\pi}{3}$ which implies that

$$|p_1| = \left| \frac{2\sin 3\theta \sin \frac{3l\theta}{2}}{\sin \frac{3Nl\theta}{2}} \right| \geq 2 \cdot \frac{2}{\pi} \cdot 3\theta \cdot \frac{2}{\pi} \cdot \frac{3l\theta}{2} = \frac{36l\theta^2}{\pi^2}. \tag{19}$$

It follows from (2.15), (2.16), and (2.17) that

$$\begin{aligned} \frac{|p_1|}{p_0} &\geq \frac{36l\theta^2}{\pi^2} \cdot \frac{2\pi - Nl\theta}{\pi l\theta^2} > \frac{36}{\pi^3}(2\pi - Nl\theta) \\ &> \frac{36}{\pi^3}\left(2\pi - \frac{\pi N}{N-1}\right) = \frac{72}{\pi^3}\left(\pi - \frac{\pi N}{2(N-1)}\right) \\ &\geq \frac{24}{\pi^2} > 1. \end{aligned}$$

(v) $N = 3$ and $1 \leq m \leq 3$. By (2.15) and the assumption of Subcase (IIIb) it follows that $\frac{\pi}{6} < \frac{l\theta}{2} < \frac{\pi}{4}$ and we have

$$\frac{|p_m|}{p_0} = \left| \frac{\sin(2m+1)\theta \sin \frac{3l\theta}{2}}{\sin \frac{3(2m+1)l\theta}{2} \sin \frac{l\theta}{2}} \right| \frac{\sin \frac{(2m+1)l\theta}{2}}{\sin\theta}. \tag{20}$$

By Lemma 2.3, we get

$$\left| \frac{\sin(2m+1)\theta \sin \frac{3l\theta}{2}}{\sin \frac{3(2m+1)l\theta}{2} \sin \frac{l\theta}{2}} \right| \geq \frac{1}{3} \left| \frac{\sin \frac{3l\theta}{2}}{\sin \frac{l\theta}{2}} \right| = \frac{1}{3}\left| 3 - 4\sin^2 \frac{l\theta}{2} \right| > \frac{1}{3}.$$

By Jordan's inequality we have

$$\frac{\sin \frac{(2m+1)l\theta}{2}}{\sin\theta} > \frac{\frac{2}{\pi} \cdot (2m+1)\theta}{\theta} = \frac{2(2m+1)}{\pi}.$$

Therefore,

$$\frac{|p_m|}{p_0} > \frac{2(2m+1)}{3\pi} > 1 \qquad \text{for } m = 2, 3.$$

If $m = 1$ and $p_1 > 0$, using $\frac{\pi}{6} < \frac{l\theta}{2} < \frac{\pi}{4}$, we obtain

$$\frac{p_1}{p_0} = \frac{3 - 4\sin^2 \frac{l\theta}{2}}{4\sin^3 \frac{3l\theta}{2} - 3} \frac{\sin 3\theta}{\sin\theta} > 1 \cdot \frac{1}{\theta} \cdot \frac{2}{\pi} \cdot 3\theta = \frac{6}{\pi} > 1.$$

This completes the proof of Proposition 2.3. Q.E.D.

We are now ready to prove Theorem 1.1.

Proof of Theorem 1.1. Note that when $p < 0$ we have $F(1) = -pN < 0$ and $\lim_{z\to+\infty} F(z) = +\infty$. Thus F has a root which lies outside the unit disk. For $p = 0$, $F(z)$ has simple roots at 1 and -1 and root at 0 of multiplicity $k-1$. Let $z_1(p)$ be the branch of the root of (2.1) with $z_1(0) = 1$. Then it follows from (2.7) that

$$\left.\frac{dz_1}{dp}\right|_{p=0} = -\frac{N}{2} < 0.$$

By the continuity of the roots with respect to p, this implies that if $p > 0$ is sufficiently small then all the roots of (2.1) lie inside the unit disk. Next, Proposition 2.3 shows that p_0 is a positive minimum value of p such that a root of (2.1) intersects C as p increases from 0. Then by Proposition 2.2, if $p \geq p_0$, then there exists a root of (2.1) which lies outside the unit disk. From these arguments, we conclude that all the roots of (2.1) lie inside the unit disk if and only if $0 < p < p_0$. Therefore, the zero zolution of (1.3) is asymptotically stable if and only the condition (1.4) holds. Q.E.D.

Remark 2.1. For the case k and l are odd positive integers, N must also be odd (otherwise, $F(z)$ will have a root at -1 so that the zero solution of (1.3) is not asymptotically stable). Note that M is still an even integer. When $N = 1$ the same argument as in *Case I* of the proof of Proposition 2.3 shows that p_0 is the positive minimum of $|p_m|$ for $m = 0, 1, ..., \frac{M}{2} - 1$. When $N = 3$, the same argument as in *Case III* of the proof of Proposition 2.3 shows that p_0 is the positive minimum of $|p_m|$ for $m = 0, 1, ..., \left[\frac{M}{4} - \frac{1}{2}\right]$. However, we can not conclude from the proof in *Case III* of Proposition 2.3 that p_0 is the positive minimum of $|p_m|$ for $m = 0, 1, ..., \frac{M}{2} - 1$. We then have the following conclusion:

Theorem 2.4. *Let k, l, and N be positive integers with k and l odd and $k > (N-1)l$. Then the zero solution of (1.3) is asymptotically stable if and only if*

$$0 < p < p_0^*$$

where $M = 2k - (N-1)l$, $p_0^* = \min\{p_0, p^*\}$, $p_0 = \frac{2\sin\left(\frac{\pi}{M}\right)\sin\left(\frac{l\pi}{2M}\right)}{\sin\left(\frac{Nl\pi}{2M}\right)}$ and

$$p^* = \min\left\{p_m : m = \left[\frac{M}{4} - \frac{1}{2}\right] + 1, \left[\frac{M}{4} - \frac{1}{2}\right] + 2, ..., \frac{M}{2} - 1\right\}.$$

Acknowledgment. This research is supported by the Thailand Research Fund.

References

[1] Y. Kuang, Delay Differential Equations with Applications in Population Dynamics, Academic Press, San Diego, 1993.

[2] S. A. Kuruklis, The asymptotic stability of $x_{n+1} - ax_n + bx_{n-k} = 0$, J. Math. Anal. Appl., **188** (1994), 719–731.

[3] S. A. Levin and R. M. May, A note on difference-delay equations, Theoret. Population Biology, **9** (1976), 178–187.

[4] P. Niamsup and Y. Lenbury, The asymptotic stability of $x_{n+1} - a^2 x_{n-1} + bx_{n-k} = 0$, Kyungpook Math. J., **48** (2008), 173–181.

[5] R. Ogita, H. Matsunaga and T. Hara, Asymptotic stability for a class of linear delay difference equations of higher order, J. Math. Anal. Appl., **248** (2000), 83–96.

[6] V. G. Papanicolaou, On the asymptotic stability of class of linear difference equations, Math. Magazine, **69** (1996), 34–43.

[7] G. Stépán, Retarded Systems: Stability and Characteristic Functions, Longman, Harlow, UK, 1989.

Department of Mathematics
Faculty of Science
Chiangmai University
Chiangmai, 50200
Thailand

E-mail address: `scipnmsp@chiangmai.ac.th`

Advanced Studies in Pure Mathematics 53, 2009
Advances in Discrete Dynamical Systems
pp. 203–214

Global externalities, endogenous growth and sunspot fluctuations

Kazuo Nishimura, Harutaka Takahashi and Alain Venditti

Abstract.

We consider a two-sector economy with Cobb-Douglas technologies, labor-augmenting global external effects and increasing social returns. We prove the existence of a normalized balanced growth path and we give conditions for the occurrence of sunspot fluctuations that are compatible with both types of capital intensity configuration at the private level provided the elasticity of intertemporal substitution in consumption admits intermediary values. We finally show that the existence of period-two cycles requires the consumption good to be physical capital intensive at the private level.

§1. Introduction

Over the last decade, a large literature has focussed on the existence of local indeterminacy, i.e. a continuum of equilibrium paths converging toward a steady state, and sunspot fluctuations in endogenous growth models. These studies provide a possible explanation of the facts that economic growth rates are volatile over time and dispersed across countries. Within infinite-horizon two-sector models, the occurrence of multiple equilibrium paths is related to the existence of market imperfections such that productive externalities. Two different strategies have been followed to modelize these learning-by-doing effects. A first set of contributions builds upon Benhabib and Farmer [1] and consider sector-specific externalities with constant returns at the social level. Most of the papers are based on continuous-time models and local indeterminacy is shown to arise if the final good sector is human capital (labor) intensive at the private level but physical capital intensive at the social

Received November 14, 2006.
2000 *Mathematics Subject Classification.* C62, E32, O41.
Key words and phrases. Global externalities, increasing returns, endogenous growth, intertemporal substitution in consumption, indeterminacy, sunspot fluctuations, period-two cycles.

level.[1] It is worth noticing however that Mino *et al.* [14] have recently considered a discrete-time two-sector analogue model and show that the existence of local indeterminacy follows from a more complex set of conditions based on capital intensities differences at the private and social level, the rate of depreciation of capital and the discount factor.

A second set of contributions build upon the initial endogenous growth framework developed by Lucas [12] and Romer [17]. Local indeterminacy is derived from the consideration of global externalities and increasing returns at the social level. Most of the papers are however based on specific functional forms with a Leontief technology in the investment good sector, and/or assume a degenerate allocation of human capital (labor) across sectors.[2] In such a framework, there is no clear condition in terms of capital intensities differences to ensure the occurrence of multiple equilibria. Considering a non-trivial allocation mechanism of labor between the two sectors, Drugeon *et al.* [10] have shown within a general continuous-time model with labor-augmenting global external effects borrowed from Boldrin and Rustichini [6] that the occurrence of local indeterminacy necessarily requires a physical capital intensive investment good sector at the private level. This condition appears to be the complete opposite to the one derived with sector-specific externalities. The aim of this paper is then to explore the robustness of this result by considering a discrete-time formulation.

Considering a discrete-time analogue of the Drugeon *et al.* [10] model, Goenka and Poulsen [11] suggest that local indeterminacy may arise under both configurations for the capital intensity difference at the private level. However, their conditions are quite complex as they involve the endogenous growth rate, and they do not show that a non-empty set of economies may satisfy these conditions. We then consider a discrete-time model with Cobb-Douglas technologies and labor-augmenting global externalities. We prove the existence of a balanced growth path and we give conditions on the Cobb-Douglas coefficients for the occurrence of sunspot fluctuations that are compatible with both types of capital intensity configuration at the private level provided the elasticity of intertemporal substitution in consumption admits intermediary values. However, the occurrence of period-two cycles requires the consumption good to be physical capital intensive at the private level. All these results are finally illustrated through numerical examples.

---

[1] Benhabib *et al.* [2], Mino [13]. See also Bond *et al.* [7] in which similar results are obtained from the consideration of distortionary factor taxes.

[2] See Benhabib and Perli [4], Boldrin and Rustichini [6], Boldrin *et al.* [5], Xie [19].

The rest of the paper is organized as follows: Section 2 sets up the basic model. In Section 3 we study the existence of a balanced growth path. Section 4 provides the main results on local indeterminacy. Section 5 contains concluding comments. The proofs are gathered in a final Appendix.

§2. The model

We consider a discrete-time two-sector economy having an infinitely-lived representative agent with a single period utility function given by

$$u(c) = \tfrac{c^{1-\sigma}}{1-\sigma}$$

where c is consumption and $\sigma > 0$ is the inverse of the elasticity of intertemporal substitution in consumption. The labor supply is inelastic. There are two goods: the consumption good, c, and the capital good, k. Each good is produced with a Cobb-Douglas technology. We assume that the production functions contain positive global externalities given by the average capital stock in the economy and which can be interpreted as a labor augmenting technical progress. We denote by y and c the outputs of sectors k and c:

$$c = K_c^{\alpha_1}(eL_c)^{\alpha_2}, \qquad y = AK_y^{\beta_1}(eL_y)^{\beta_2}$$

with $e = \bar{k}$ and $A > 0$ a normalization constant. Labor is normalized to one, $L_c + L_y = 1$ and the total stock of capital is $K_c + K_y = k$. We assume that the economy-wide average $\bar{k}$ is taken as given by individual firms. At the equilibrium, all firms of sector $i = c, y$ being identical, we have $\bar{k} = k$.

We assume constant returns to scale at the private level, i.e. $\alpha_1 + \alpha_2 = \beta_1 + \beta_2 = 1$. At the social level the returns to scale are therefore increasing. It can be easily shown that if $\beta_1/\beta_2 > (<)\alpha_1/\alpha_2$ the investment (consumption) good sector is capital intensive from the private perspective.

We assume complete depreciation of capital in one period so that the capital accumulation equation is $y_t = k_{t+1}$. Optimal factor allocations across sectors are obtained by solving the following program:

$$(1) \qquad \begin{array}{cl} \max\limits_{K_{ct},L_{ct},K_{yt},L_{yt}} & K_{ct}^{\alpha_1}(e_tL_{ct})^{\alpha_2} \\ s.t. & k_{t+1} = AK_{yt}^{\beta_1}(e_tL_{yt})^{\beta_2} \\ & 1 = L_{ct} + L_{yt} \\ & k_t = K_{ct} + K_{yt} \\ & e_t \ \ given \end{array}$$

Denote by p_t, w_t and r_t respectively the price of the capital good, the wage rate of labor and the rental rate of capital, all in terms of the price of the consumption good. For any (k_t, k_{t+1}, e_t), solving the associated first order conditions gives inputs as $\tilde{K}_c(k_t, k_{t+1}, e_t)$, $\tilde{L}_c(k_t, k_{t+1}, e_t)$, $\tilde{K}_y(k_t, k_{t+1}, e_t)$ and $\tilde{L}_y(k_t, k_{t+1}, e_t)$. We then define the efficient production frontier as

$$T(k_t, k_{t+1}, e_t) = \tilde{K}_c(k_t, k_{t+1}, e_t)^{\alpha_1}[e_t\tilde{L}_c(k_t, k_{t+1}, e_t)]^{\alpha_2}$$

which describes the standard trade-off between consumption and investment: for a given e_t, $T(k_t, k_{t+1}, e_t)$ is increasing with respect to k_t and decreasing with respect to k_{t+1}. Using the envelope theorem we then derive the equilibrium prices

$$(2) \qquad p_t = -T_2\big(k_t, k_{t+1}, e_t\big), \quad r_t = T_1\big(k_t, k_{t+1}, e_t\big)$$

where $T_1 = \frac{\partial T}{\partial k}$ and $T_2 = \frac{\partial T}{\partial y}$. The representative consumer's optimization program is finally given by

$$(3) \qquad \begin{array}{cc} \max\limits_{\{k_t\}_{t=0}^{+\infty}} & \sum\limits_{t=0}^{\infty} \delta^t \dfrac{[T(k_t, k_{t+1}, e_t)]^{1-\sigma}}{1-\sigma} \\ s.t. & k_0 = \hat{k}_0, \{e_t\}_{t=0}^{+\infty} \;\; given \end{array}$$

with $\delta \in (0,1)$ the discount factor. The corresponding Euler equation is

$$(4) \qquad -c_t^{-\sigma}p_t + \delta c_{t+1}^{-\sigma}r_{t+1} = 0$$

Let $\{k_t\}_{t=0}^{+\infty}$ denote a solution of (4) which obviously depends on $\{e_t\}_{t=0}^{\infty}$, i.e. $k_t = k(t, \{e_t\}_{t=0}^{\infty})$ for all $t \geq 0$. As we have assumed that $e_t = \bar{k}_t$ with $\bar{k}_t$ the economy-wide average capital stock, expectations are realized if there exists a solution of an infinite-dimensional fixed-point problem such that $e_t = k(t, \{e_t\}_{t=0}^{\infty})$ for any $t = 0, 1, 2, ...$ Assuming that such a solution exists,[3] prices may now be written as

$$(5) \qquad r(k_t, k_{t+1}) = T_1(k_t, k_{t+1}, k_t), \quad p(k_t, k_{t+1}) = -T_2(k_t, k_{t+1}, k_t)$$

and consumption at time t is given by a linear homogeneous function.[4]

$$(6) \qquad c(k_t, k_{t+1}) = T(k_t, k_{t+1}, k_t)$$

Equation (4) finally becomes:

$$(7) \qquad p(k_t, k_{t+1},)c(k_t, k_{t+1})^{-\sigma} = \delta r(k_{t+1}, k_{t+2})c(k_{t+1}, k_{t+2})^{-\sigma}$$

---

[3]A detailed treatment of the existence of such a solution within a discrete-time version of the Lucas [12] model is provided in Mitra [15].

[4]See Proposition 1 in Drugeon and Venditti [9] or Lemma 1 in Drugeon *et al.* [10].

Any solution $\{k_t\}_{t=0}^{+\infty}$ of (7) which also satisfies the transversality condition

(8) $$\lim_{t\to+\infty} \delta^t c(k_t, k_{t+1})^{-\sigma} p(k_t, k_{t+1}) k_{t+1} = 0$$

and the summability condition

(9) $$\sum_{t=0}^{\infty} \delta^t \frac{[c(k_t,k_{t+1})]^{1-\sigma}}{1-\sigma} < +\infty$$

is called an equilibrium path.[5]

§3. Balanced growth path

We call a path $\{k_t\}_{t=0}^{+\infty}$ satisfying $k_0 = \hat{k}_0$ and $k_t \le k_{t+1} \le Ak_t$, a feasible path (from $k_0 = \hat{k}_0$). We call $\{k_t\}_{t=0}^{+\infty}$ a balanced growth path if it is in equilibrium and $k_{t+1}/k_t = \theta$ for $t = 0, 1, \ldots$.

Lemma 1. *Along an equilibrium path, prices satisfy*

(10) $$r(k_t, k_{t+1}) = A\alpha_1 \left[\frac{\alpha_2\beta_1}{\alpha_1\beta_2}\left(\frac{k_{t+1}}{A\bar{g}(k_t,k_{t+1})}\right)^{1/\beta_2}\right]^{\alpha_2},$$
$$p(k_t, k_{t+1}) = \frac{r_t \bar{g}(k_t,k_{t+1})}{\beta_1 k_{t+1}}$$

with

$$\bar{g}(k_t, k_{t+1}) = \left\{K_y \in (0, Ak)/\frac{\alpha_1\beta_2}{\alpha_2\beta_1} = \frac{[k_t - K_y][(k_{t+1}/A)^{1/\beta_2} k_t^{-1} K_y^{-\beta_1/\beta_2}]}{[1-(k_{t+1}/A)^{1/\beta_2} k_t^{-1} K_y^{-\beta_1/\beta_2}]K_y}\right\}$$

a linear homogeneous function.

Define $\theta_t = k_{t+1}/k_t (= c_{t+1}/c_t)$ the growth factor of capital (and thus consumption) at time t. Notice that $ln\theta_t$ is the growth rate of capital. By the feasibility condition $k_t \le k_{t+1} \le Ak_t$, it holds that $\theta_t \in (0, A)$. Denoting $g(\theta_t^{-1}) = \bar{g}(\theta_t^{-1}, 1)$ and using (10), the Euler equation (7) can be transformed into an implicit recursive equation as follows

(11) $$\delta\beta_1 g(\theta_{t+1}^{-1})^{-\alpha_2/\beta_2} = g(\theta_t^{-1})^{1-(\alpha_2/\beta_2)}\theta_t^{\sigma}$$

A balanced growth factor is defined by $\theta_t = \theta_{t+1} = \theta$ and satisfies equation (11) together with conditions (8)-(9). Using the normalization constant A we will show that there exists such a normalized balanced growth factor.

Proposition 1. *Let $\tilde{\sigma} = ln\beta_1/ln(\delta\beta_1) > 0$. There exists a normalized balanced growth factor (NBGF) $\theta^* = (\delta\beta_1)^{-1}$ solution of the Euler equation (11) if $\sigma > \tilde{\sigma}$ and the normalization constant A is set at the following value*

(12) $$A^* = \left[\frac{\alpha_2\beta_1}{\alpha_1\beta_2}\frac{1-(\delta\beta_1)^{\sigma}\left(1-\frac{\alpha_1\beta_2}{\alpha_2\beta_1}\right)}{(\delta\beta_1)^{(1+\sigma)/\beta_2}}\right]^{\beta_2}$$

[5]See Lemma 1 and Corollary 1 in Boldrin *et al.* [5].

Proposition 1 establishes the existence of a normalized balanced growth rate (NBGR) $ln\theta^*$ from which we define a normalized balanced growth path (NBGP), namely $k_t = \hat{k}_0\theta^{*t} = \hat{k}_0(\delta\beta_1)^{-t}$.

§4. Local indeterminacy

The local stability properties of the NBGP are obtained from the linearization of the Euler equation (11) around θ^*.

Assumption 1. $\alpha_1\beta_2 \neq \alpha_2\beta_1$

Assumption 1, which is equivalent to a non-zero capital intensity difference at the private level, implies that the technologies are not identical.

Lemma 2. *Under Assumption 1, if $\sigma > \tilde{\sigma}$ and $A = A^*$ as defined in Proposition 1, the linearization of the Euler equation (11) around θ^* gives*

$$\frac{d\theta_{t+1}}{d\theta_t} = \sigma\beta_2 \frac{1-\frac{\alpha_1\beta_2}{\alpha_2\beta_1}+\frac{1-(\delta\beta_1)^\sigma\left(1-\frac{\alpha_1\beta_2}{\alpha_2\beta_1}\right)}{\beta_2(\delta\beta_1)^\sigma}}{\alpha_2\left(1-\frac{\alpha_1\beta_2}{\alpha_2\beta_1}\right)} - \beta_1\left(1-\frac{\alpha_1\beta_2}{\alpha_2\beta_1}\right) \tag{13}$$

Notice that the Euler equation (11) is forward looking since the initial growth rate $\theta_0 = k_1/\hat{k}_0$ is not pre-determined. As a result, if $d\theta_{t+1}/d\theta_t \in (-1,1)$, then every initial point in the neighborhood of the NBGR has indeterminate equilibrium paths satisfying (8) and (9), and we say that the NBGP is locally indeterminate. If on the contrary $|d\theta_{t+1}/d\theta_t| > 1$, the NBGR is locally unstable and starting from $\hat{k}_0$ one possible equilibrium consists in jumping at $t = 1$ on the NBGP. In such a case the NBGP is locally determinate.

Proposition 2. *Let $\tilde{\sigma} = ln\beta_1/ln(\delta\beta_1)$. Under Assumption 1, if one of the following set of conditions is satisfied:*

i) the investment good is capital intensive at the private level with

$$\frac{\tilde{\sigma}}{\beta_1}\left[1-\beta_1^2\left(1-\frac{\alpha_1\beta_2}{\alpha_2\beta_1}\right)\right] - \alpha_2\left(1-\frac{\alpha_1\beta_2}{\alpha_2\beta_1}\right)\left[1+\beta_1\left(1-\frac{\alpha_1\beta_2}{\alpha_2\beta_1}\right)\right] < 0 \tag{14}$$

ii) the consumption good is capital intensive at the private level with

$$\frac{\tilde{\sigma}}{\beta_1}\left[1-\beta_1^2\left(1-\frac{\alpha_1\beta_2}{\alpha_2\beta_1}\right)\right] + \alpha_2\left(1-\frac{\alpha_1\beta_2}{\alpha_2\beta_1}\right)\left[1-\beta_1\left(1-\frac{\alpha_1\beta_2}{\alpha_2\beta_1}\right)\right] < 0 \tag{15}$$

then there exist $\underline{\sigma} \geq \tilde{\sigma}$ and $\bar{\sigma} > \underline{\sigma}$ such that the NBGP $k_t = \hat{k}_0(\delta\beta_1)^{-t}$ is locally indeterminate when $\sigma \in (\underline{\sigma}, \bar{\sigma})$.

We easily derive from Proposition 2 that there exists some case in which the growth rate of capital exhibits period-two cycles.

Corollary 1. *Let* $\tilde{\sigma} = ln\beta_1/ln(\delta\beta_1)$. *Under Assumption 1, if the consumption good is capital intensive at the private level and condition (15) holds, then there exist* $\underline{\sigma} \geq \tilde{\sigma}$ *and* $\bar{\sigma} > \underline{\sigma}$ *such that the NBGP is locally indeterminate when* $\sigma \in (\underline{\sigma}, \bar{\sigma})$. *Moreover, when* σ *crosses* $\bar{\sigma}$ *from below the NBGP becomes locally determinate and there exist locally indeterminate (determinate) period-two growth cycles in a right (left) neighborhood of* $\bar{\sigma}$.

Corollary 1 shows that the existence of endogenous fluctuations requires a capital intensive consumption good. The intuition for this result, initially provided by Benhabib and Nishimura [3], may be summarized as follows. Consider an instantaneous increase in the capital stock k_t. This results in two opposing forces:

- Since the consumption good is more capital intensive than the investment good, the trade-off in production becomes more favorable to the consumption good. The Rybczinsky theorem thus implies a decrease of the output of the capital good y_t. This tends to lower the investment and the capital stock in the next period k_{t+1}, and thus implies a decrease of the balanced growth factor $\theta_t = k_{t+1}/k_t$.

- In the next period the decrease of k_{t+1} implies again through the Rybczinsky effect an increase of the output of the capital good y_{t+1}. Indeed, the decrease of k_{t+1} improves the trade-off in production in favor of the investment good which is relatively less intensive in capital. Therefore this tends to lower the investment and the capital stock in period $t+2$, k_{t+2}, and thus implies an increase of the balanced growth factor $\theta_{t+1} = k_{t+2}/k_{t+1}$.

Boldrin *et al.* [5] consider a similar two-sector model but assume a Leontief technology in the investment good sector with a degenerate allocation of labor across sectors as the investment good is only produced from physical capital. They prove the existence of global indeterminacy of equilibria and construct robust examples of both topological and ergodic chaos for the dynamics of balanced growth paths. We may expect similar results from Corollary 1. However, showing the existence of chaotic dynamics with a Cobb-Douglas technology in both sector is beyond the goal of the current paper and is left for future research.

Remark. Considering a discrete-time two-sector optimal endogenous growth model with two capital goods, one being consumable while the other is not, Drugeon [8] shows that the occurrence of period-two cycles requires a non-unitary rate of capital depreciation in one sector at least.

In a similar formulation but extended to include sector-specific externalities, Mino *et al.* [14] show that the existence of local indeterminacy is based on the same necessary condition. We prove that the consideration of global externalities allows to get sunspot fluctuations even under full depreciation of capital.

Numerical illustration:
In order to check whether all the conditions of Proposition 2 can be satisfied simultaneously, we perform some numerical simulations.

i) Let $\delta = 0.98$, $\beta_1 = 0.85$ and $\alpha_1 = 0.55$. The investment good is thus capital intensive at the private level and the NBGF is equal to $\theta^* \approx 1.2$. The NBGP is locally indeterminate for any $\sigma \in (\tilde{\sigma}, \bar{\sigma})$ with $\tilde{\sigma} \approx 0.8894345$ and $\bar{\sigma} \approx 1.07066$.

ii) Let $\delta = 0.95$, $\beta_1 = 0.35$ and $\alpha_1 = 0.88$. The consumption good is thus capital intensive at the private level and $\theta^* \approx 3.0075$. The NBGP is locally indeterminate for any $\sigma \in (\tilde{\sigma}, \bar{\sigma})$ with $\tilde{\sigma} \approx 0.953417$ and $\bar{\sigma} \approx 1.07$. Moreover, $\bar{\sigma}$ is a flip bifurcation value so that there exist locally indeterminate (resp. determinate) period-two cycles in a right (resp. left) neighborhood of $\bar{\sigma}$.

§5. Concluding comments

We have considered a discrete-time two-sector endogenous growth model with Cobb-Douglas technologies augmented to include labor-augmenting global externalities. We have proved the existence of a normalized balanced growth path and we have shown that sunspot fluctuations arise under both types of capital intensity configuration at the private level provided the elasticity of intertemporal substitution in consumption admits intermediary values. Moreover, the dynamics of growth rates exhibits period-two cycles if the consumption good is capital intensive at the private level.

§6. Appendix

6.1. Proof of Lemma 1

The Lagrangian associated with program (1) is:

$$\begin{aligned}\mathcal{L} &= K_{ct}^{\alpha_1}(e_tL_{ct})^{\alpha_2} + p_t[AK_{yt}^{\beta_1}(e_tL_{yt})^{\beta_2} - k_{t+1}] + \omega_t[1 - L_{ct} - L_{yt}] \\ &+ r_t[k_t - K_{ct} - K_{yt}]\end{aligned}$$

The first order conditions are:

$$\alpha_1 c_t/K_{cc} = p_t\beta_1 y_t/K_{yt} = r_t \tag{16}$$

$$\alpha_2 c_t/L_{ct} = p_t\beta_2 y_t/L_{yt} = w_t \tag{17}$$

Solving $y_t = AK_{yt}^{\beta_1}(k_t L_{yt})^{\beta_2}$ with respect to L_{yt} gives

(18) $$L_{yt} = (y_t/A)^{1/\beta_2} k_t^{-1} K_{yt}^{-\beta_1/\beta_2}$$

Using $K_{ct} = k_0 - K_{yt}$, $L_{yt} = 1 - L_{ct}$, and merging (16)-(18) we get:

(19) $$\frac{\alpha_1\beta_2}{\alpha_2\beta_1} = \frac{K_{ct}}{L_{ct}}\frac{L_{yt}}{K_{yt}} = \frac{k_t - K_{yt}}{1-(y_t/A)^{1/\beta_2}k_t^{-1}K_{yt}^{-\beta_1/\beta_2}} \frac{(y_t/A)^{1/\beta_2}k_t^{-1}K_{yt}^{-\beta_1/\beta_2}}{K_{yt}}$$

The solution K_{yt}^* is obtained as an implicit linear homogeneous function $\bar{g}(k_t, y_t)$ as shown in Drugeon and Venditti [9].[6] We then derive from (16)

$$r_t = A\alpha_1\left(\frac{k_t L_{ct}}{K_{ct}}\right)^{\alpha_2}, \quad p_t = \frac{r_t K_{yt}}{\beta_1 y_t}$$

Using (18) and (19) with $y_t = k_{t+1}$ gives the final results.

6.2. Proof of Proposition 1

Consider the Euler equation (11):

$$\delta\beta_1 g(\theta_{t+1}^{-1})^{-\alpha_2/\beta_2} = g(\theta_t^{-1})^{1-(\alpha_2/\beta_2)}\theta_t^{\sigma}$$

Along a balanced growth path with $\theta_t = \theta_{t+1} = \theta$, we get $\delta\beta_1 = g(\theta^{-1})\theta^{\sigma}$. Now consider equation (19) with $K_y = \bar{g}(k_t, k_{t+1}) = k_{t+1}\bar{g}(k_t/k_{t+1}, 1) \equiv k_{t+1}g(\theta_t^{-1})$. We derive

(20) $$\frac{\alpha_1\beta_2}{\alpha_2\beta_1} = \frac{\theta_t^{-1} - g(\theta_t^{-1})}{A^{1\beta_2}\theta_t^{-1}g(\theta_t^{-1})^{1/\beta_2} - g(\theta_t^{-1})}$$

If $\theta_t = \theta$ and thus $g(\theta^{-1}) = \delta\beta_1\theta^{-\sigma}$ we get after simplifications

$$\frac{\alpha_1\beta_2}{\alpha_2\beta_1} = \frac{1-\delta\beta_1\theta^{1-\sigma}}{A^{1\beta_2}(\delta\beta_1\theta_t^{-\sigma})^{1/\beta_2} - \delta\beta_1\theta^{1-\sigma}}$$

It follows that $\theta = (\delta\beta_1)^{-1} \equiv \theta^*$ is a solution of this equation if and only if the normalization constant A satisfies $A = A^*$ as defined by (12). Along the stationary balanced growth path $k_t = (\delta\beta_1)^{-t}\hat{k}_0$, using the fact that $c(k, y)$ is homogeneous of degree one and $p(k, y)$ is homogeneous of degree zero, the transversality condition (8) becomes

$$\delta^{-1}\hat{k}_0^{1-\sigma}c(\delta\beta_1, 1)^{-\sigma}p(\delta\beta_1, 1)\lim_{t\to+\infty}(\delta^{\sigma}\beta_1^{\sigma-1})^{t+1} = 0$$

It will be satisfied if $\delta^{\sigma} < \beta_1^{1-\sigma}$ or equivalently $\sigma > \tilde{\sigma} = ln\beta_1/ln(\delta\beta_1) > 0$. Similarly the summability condition (9) becomes

$$\delta^{-1}\hat{k}_0^{1-\sigma}\frac{[c(\delta\beta_1, 1)]^{1-\sigma}}{1-\sigma}\sum_{t=0}^{\infty}(\delta^{\sigma}\beta_1^{\sigma-1})^{t+1} < +\infty$$

and is satisfied if $\delta^{\sigma} < \beta_1^{1-\sigma}$ or equivalently $\sigma > \tilde{\sigma} = ln\beta_1/ln(\delta\beta_1) > 0$. Finally, the feasibility condition requires $(\delta\beta_1)^{-1} \leq A^*$ or equivalently

$$\frac{\alpha_2\beta_1}{\alpha_1\beta_2}\frac{1-(\delta\beta_1)^{\sigma}\left(1-\frac{\alpha_1\beta_2}{\alpha_2\beta_1}\right)}{(\delta\beta_1)^{\sigma/\beta_2}} \geq 1 \quad \Leftrightarrow \quad 1-(\delta\beta_1)^{\sigma} + (\delta\beta_1)^{\sigma}\frac{\alpha_1\beta_2}{\alpha_2\beta_1}\left[1-(\delta\beta_1)^{(\sigma\beta_1)/\beta_2}\right] \geq 0$$

This inequality is always satisfied.

[6] See also Drugeon *et al.* [10].

6.3. Proof of Lemma 2

Notice first that $g(\theta^{*-1}) = g(\delta\beta_1) = (\delta\beta_1)^{1+\sigma}$. Total differentiation of the Euler equation (11) around $\theta = \theta^*$ with $A = A^*$ gives

(21) $$\frac{d\theta_{t+1}}{d\theta_t} = \frac{\sigma\beta_2(\delta\beta_1)^\sigma}{\alpha_2 g'(\theta^{*-1})} + \frac{\beta_2-\alpha_2}{\alpha_2}$$

Consider now equation (20) and let us denote $X = \theta^{-1}$. We get:

$$\frac{\alpha_1\beta_2}{\alpha_2\beta_1} = \frac{X-g(X)}{A^{1\beta_2}Xg(X)^{1/\beta_2}-g(X)}$$

Total differenciation with respect to X gives

$$\frac{dg}{dX} = g'(X) = \frac{(\delta\beta_1)^\sigma\left(1-\frac{\alpha_1\beta_2}{\alpha_2\beta_1}\right)}{1-\frac{\alpha_1\beta_2}{\alpha_2\beta_1}+\frac{1-(\delta\beta_1)^\sigma\left(1-\frac{\alpha_1\beta_2}{\alpha_2\beta_1}\right)}{\beta_2(\delta\beta_1)^\sigma}}$$

Substituting this expression in (21) and using the fact that $\beta_2 - \alpha_2 = -\alpha_2\beta_1(1 - \alpha_1\beta_2/\alpha_2\beta_1)$ give the final result.

6.4. Proof of Proposition 2

Consider the expression (13) which may be expressed as follows

(22) $$\frac{d\theta_{t+1}}{d\theta_t} = \frac{\sigma}{\alpha_2}\left[\frac{\left(1-\frac{\alpha_1\beta_2}{\alpha_2\beta_1}\right)^{-1}}{(\delta\beta_1)^\sigma} - \beta_1\right] - \beta_1\left(1 - \frac{\alpha_1\beta_2}{\alpha_2\beta_1}\right)$$

Local indeterminacy of the NBGP will be obtained if and only if $d\theta_{t+1}/d\theta_t \in (-1, 1)$. We first obtain that

(23) $$\begin{aligned}\lim_{\sigma\to+\infty} \frac{d\theta_{t+1}}{d\theta_t} &= +\infty \Leftrightarrow 1 - \frac{\alpha_1\beta_2}{\alpha_2\beta_1} > 0\\ &= -\infty \Leftrightarrow 1 - \frac{\alpha_1\beta_2}{\alpha_2\beta_1} < 0\end{aligned}$$

Let $\sigma = \tilde{\sigma}$, or equivalently $(\delta\beta_1)^{\tilde{\sigma}} = \beta_1$, so that (13) becomes

$$\left.\frac{d\theta_{t+1}}{d\theta_t}\right|_{\sigma=\tilde{\sigma}} = \frac{1}{\alpha_2\left(1-\frac{\alpha_1\beta_2}{\alpha_2\beta_1}\right)}\left\{\frac{\tilde{\sigma}}{\beta_1}\left[1 - \beta_1^2\left(1 - \frac{\alpha_1\beta_2}{\alpha_2\beta_1}\right)\right] - \alpha_2\beta_1\left(1 - \frac{\alpha_1\beta_2}{\alpha_2\beta_1}\right)^2\right\}$$

i) Consider first the case in which the investment good is capital intensive at the private level, i.e. $1 - \alpha_1\beta_2/\alpha_2\beta_1 > 0$. We get

$$\left.\frac{d\theta_{t+1}}{d\theta_t}\right|_{\sigma=\tilde{\sigma}} > -1$$

$$\Leftrightarrow \frac{\tilde{\sigma}}{\beta_1}\left[1 - \beta_1^2\left(1 - \frac{\alpha_1\beta_2}{\alpha_2\beta_1}\right)\right] + \alpha_2\left(1 - \frac{\alpha_1\beta_2}{\alpha_2\beta_1}\right)\left[1 - \beta_1\left(1 - \frac{\alpha_1\beta_2}{\alpha_2\beta_1}\right)\right] > 0$$

This inequality is always satisfied since $1 - \alpha_1\beta_2/\alpha_2\beta_1 < 1$. We also have

$$\left.\frac{d\theta_{t+1}}{d\theta_t}\right|_{\sigma=\tilde{\sigma}} < 1$$

$$\Leftrightarrow \frac{\tilde{\sigma}}{\beta_1}\left[1 - \beta_1^2\left(1 - \frac{\alpha_1\beta_2}{\alpha_2\beta_1}\right)\right] - \alpha_2\left(1 + \frac{\alpha_1\beta_2}{\alpha_2\beta_1}\right)\left[1 - \beta_1\left(1 - \frac{\alpha_1\beta_2}{\alpha_2\beta_1}\right)\right] < 0$$

This last inequality corresponds to condition (14). If it holds we conclude from (23) that there exists $\bar{\sigma} > \tilde{\sigma}$ such that the NBGP is locally indeterminate when $\sigma \in (\tilde{\sigma}, \bar{\sigma})$.

ii) Consider now the case in which the consumption good is capital intensive at the private level, i.e. $1 - \alpha_1\beta_2/\alpha_2\beta_1 < 0$. We get

$$\left.\frac{d\theta_{t+1}}{d\theta_t}\right|_{\sigma=\tilde{\sigma}} > -1$$

$$\Leftrightarrow \quad \frac{\tilde{\sigma}}{\beta_1}\left[1 - \beta_1^2\left(1 - \frac{\alpha_1\beta_2}{\alpha_2\beta_1}\right)\right] + \alpha_2\left(1 - \frac{\alpha_1\beta_2}{\alpha_2\beta_1}\right)\left[1 - \beta_1\left(1 - \frac{\alpha_1\beta_2}{\alpha_2\beta_1}\right)\right] < 0$$

This last inequality corresponds to condition (15). If it holds we conclude from (23) that there exist $\underline{\sigma} \geq \tilde{\sigma}$ and $\bar{\sigma} > \underline{\sigma}$ such that the NBGP is locally indeterminate when $\sigma \in (\underline{\sigma}, \bar{\sigma})$. We may indeed have $\underline{\sigma} > \tilde{\sigma}$ since it may be the case that $d\theta_{t+1}/d\theta_t|_{\sigma=\tilde{\sigma}} > 1$.

6.5. Proof of Corollary 1

Let the consumption good be capital intensive at the private level and condition (15) holds. Then $d\theta_{t+1}/d\theta_t|_{\sigma=\tilde{\sigma}} > -1$. Considering (23) we derive from Proposition 2 that $d\theta_{t+1}/d\theta_t|_{\sigma=\bar{\sigma}} = -1$ and $d\theta_{t+1}/d\theta_t < -1$ when $\sigma > \bar{\sigma}$. Therefore $\bar{\sigma}$ is a flip bifurcation value (see Ruelle [18]) and the result follows.

References

[1] J. Benhabib and R. Farmer, Indeterminacy and sector-specific externalities, J. Monetary Econom., **37** (1996), 397–419.

[2] J. Benhabib, Q. Meng and K. Nishimura, Indeterminacy under constant returns to scale in multisector economies, Econometrica, **68** (2000), 1541–1548.

[3] J. Benhabib and K. Nishimura, Competitive equilibrium cycles, J. Econom. Theory, **35** (1985), 284–306.

[4] J. Benhabib and R. Perli, Uniqueness and indeterminacy: transitional dynamics in a model of endogenous growth, J. Econom. Theory, **63** (1994), 113–142.

[5] M. Boldrin, K. Nishimura, T. Shigoka and M. Yano, Chaotic equilibrium dynamics in endogenous growth models, J. Econom. Theory, **96** (2001), 97–132.

[6] M. Boldrin and A. Rustichini, Growth and indeterminacy in dynamic models with externalities, Econometrica, **62** (1994), 323–342.

[7] E. Bond, P. Wang and C. Yip, A general two-sector model of endogenous growth with human and physical capital: balanced growth and transitional dynamics, J. Econom. Theory, **68** (1996), 149–173.

[8] J. P. Drugeon, Optimal cycles of unbounded growth, mimeo EUREQua, 2004.
[9] J. P. Drugeon and A. Venditti, Intersectoral external effects, multiplicities and indeterminacies II. The long-run growth case, GREQAM Working Paper 98A20, 1998.
[10] J. P. Drugeon, O. Poulsen and A. Venditti, On intersectoral allocations, factors substitutability and multiple long-run growth paths, Econom. Theory, **21** (2003), 175–183.
[11] A. Goenka and O. Poulsen, Indeterminacy and labor augmenting externalities, J. Econom., **10** (2004), 61–82.
[12] R. Lucas, On the mechanics of economic development, J. Monetary Econom., **22** (1988), 3–42.
[13] K. Mino, Indeterminacy and endogenous growth with social constant returns, J. Econom. Theory, **97** (2001), 203–222.
[14] K. Mino, K. Nishimura, K. Shimomura and P. Wang, Equilibrium dynamics in discrete-time endogenous growth models with social constant returns, Econom. Theory, **34** (2008), 1–23.
[15] T. Mitra, On equilibrium dynamics under externalities in a model of economic development, Jpn Econ. Rev., **49** (1998), 85–107.
[16] K. Nishimura and A. Venditti, Indeterminacy and the role of factor substitutability, Macroecon. Dyn., **8** (2004), 436–465.
[17] P. Romer, Increasing returns and long-run growth, J. Polit. Econom., **94** (1986), 1002–1037.
[18] D. Ruelle, Elements of Differentiable Dynamics and Bifurcation Theory, Academic Press, San Diego, 1989.
[19] D. Xie, Divergence in economic performance: transitional dynamics with multiple equilibria, J. Econom. Theory, **63** (1994), 97–112.

Kazuo Nishimura
Institute of Economic Research
Kyoto University
Japan

Harutaka Takahashi
Department of Economics
Meiji Gakuin University, Tokyo
Japan

Alain Venditti
CNRS - GREQAM, Marseille
France

E-mail address: nishimura@kier.kyoto-u.ac.jp
haru@eco.meijigakuin.ac.jp
alain.venditti@univmed.fr

Advanced Studies in Pure Mathematics 53, 2009
Advances in Discrete Dynamical Systems
pp. 215–223

Some oscillation results for second order linear delay dynamic equations

Lynn Erbe and Allan Peterson

Abstract.

We obtain some oscillation theorems for linear delay dynamic equations on a time scale. We illustrate the results by a number of examples.

§1. Preliminary results

Consider the second order linear delay dynamic equation

$$L[x](t) := (r(t)x^{\Delta}(t))^{\Delta} + \sum_{i=1}^{n} q_i(t)x(\tau_i(t)) = 0. \tag{1}$$

We will be interested in obtaining oscillation theorems for (1) by comparing the solutions to a related equation without delay of the form

$$(r(t)x^{\Delta})^{\Delta} + \sum_{i=1}^{n} Q_i(t)x^{\sigma} = 0, \tag{2}$$

for which many oscillation results are known. We recall that a solution of (1) or (2) is nonoscillatory if it is eventually of one sign. If a solution changes sign infinitely often it is said to be oscillatory.

Let $\mathbb{T}$ be a time scale (nonempty closed subset of the reals $\mathbb{R}$) which is unbounded above. We assume that the coefficient functions $q_i(t) \geq 0$, $i = 1, 2, \cdots, n$, and $r(t) > 0$ are rd-continuous on the time scale interval $[a, \infty)_{\mathbb{T}} := [a, \infty) \cap \mathbb{T}$, (i.e., $r, q_i \in C_{rd}([a, \infty)_{\mathbb{T}})$. Furthermore, we will assume that $\sum_{i=1}^{n} q_i(t) \not\equiv 0$ (for all large t). We will also assume that the delay functions $\tau_i : [a, \infty)_{\mathbb{T}} \to \mathbb{T}$ are rd-continuous, $\tau_i(t) \leq t$, and $\tau_i(t) \to \infty$ as $t \to \infty$, $i = 1, 2, \cdots, n$. For details concerning calculus on

---

Received October 12, 2007.
2000 *Mathematics Subject Classification.* 39A10.
Key words and phrases. comparison theorems, linear oscillation, delay equations, time scale.

time scales and other pertinent definitions, we refer to the books [3], [4], and [11]. Stability and oscillation questions for certain first order delay dynamic equations have been considered in [1] for example.

We start with several auxiliary lemmas which are crucial in the proof of the main results. The first lemma is usually referred to as the Riccati technique. Denote

$$S[z] = \frac{z^2}{r(t) + \mu(t)z}.$$

Lemma 1 ([12], [6]). *The equation*

$$L_{rq}[x] := (r(t)x^\Delta)^\Delta + q(t)x^\sigma = 0 \tag{3}$$

is nonoscillatory if and only if there is a function z satisfying the Riccati dynamic inequality

$$z^\Delta(t) + q(t) + S[z](t) \leq 0 \tag{4}$$

with $r(t) + \mu(t)z(t) > 0$ for large t.

That is, if $x(t)$ is a solution of (3) that is of one sign for all large $t \in [a, \infty)_{\mathbb{T}}$, then $z(t) := \frac{r(t)x^\Delta(t)}{x(t)}$ satisfies (4) with $r(t) + \mu(t)z(t) > 0$ for large t. Conversely, if $z(t)$ solves (4) with $r(t)+\mu(t)z(t) > 0$ for large t, then (3) has a solution $x(t)$ which is of one sign for all large t.

We will use this lemma to show that a nonoscillatory solution of (1) leads to a solution of the Riccati dynamic inequality (4). In order to do this, we introduce the auxiliary functions $H(t, t_1)$ and $\eta_i(t, t_1)$ defined by

$$H(t, t_1) := \int_{t_1}^{t} \frac{1}{r(s)}\, \Delta s, \quad \text{and} \quad \eta_i(t, t_1) := \frac{H(\tau_i(t), t_1)}{H(\sigma(t), t_1)}, \quad l \leq i \leq n.$$

We may then establish the following result.

Lemma 2. *Let $x(t)$ be a solution of* (1) *which satisfies*

$$x(t) > 0, \quad x^\Delta(t) > 0, \quad (r(t)x^\Delta(t))^\Delta \leq 0$$

for all $t \geq \tau_i(t) \geq T \geq a$. Then for each $1 \leq i \leq n$ we have

$$x(\tau_i(t)) > \eta_i(t, T)x^\sigma(t), \quad t \geq \tau_i(t) > T.$$

Proof. For $t \geq \tau_i(t) > T \geq a$ we have

$$\begin{aligned} x(\sigma(t)) - x(\tau_i(t)) &= \int_{\tau_i(t)}^{\sigma(t)} x^\Delta(s)\Delta s \\ &= \int_{\tau_i(t)}^{\sigma(t)} \frac{1}{r(s)} r(s)x^\Delta(s)\Delta s \\ &\leq r(\tau_i(t))x^\Delta(\tau_i(t)) \int_{\tau_i(t)}^{\sigma(t)} \frac{\Delta s}{r(s)} \end{aligned}$$

which yields

$$x^\sigma(t) \leq x(\tau_i(t)) + r(\tau_i(t))x^\Delta(\tau_i(t))H(\sigma(t), \tau_i(t)).$$

Dividing both sides of this inequalilty by $x(\tau_i(t))$ we get

$$\frac{x^\sigma(t)}{x(\tau_i(t))} \leq 1 + \frac{r(\tau_i(t))x^\Delta(\tau_i(t))}{x(\tau_i(t))} H(\sigma(t), \tau_i(t)). \tag{5}$$

Also, we have

$$\begin{aligned} x(\tau_i(t)) - x(T) &= \int_T^{\tau_i(t)} x^\Delta(s)\Delta s \\ &\geq r(\tau_i(t))x^\Delta(\tau_i(t)) \int_T^{\tau_i(t)} \frac{\Delta s}{r(s)} \end{aligned}$$

and so

$$\begin{aligned} x(\tau_i(t)) &\geq x(T) + r(\tau_i(t))x^\Delta(\tau_i(t))H(\tau_i(t), T) \\ &> r(\tau_i(t))x^\Delta(\tau_i(t))H(\tau_i(t), T). \end{aligned}$$

Therefore, we have

$$\frac{r(\tau_i(t))x^\Delta(\tau_i(t))}{x(\tau_i(t))} < \frac{1}{H(\tau_i(t), T)}. \tag{6}$$

Hence, from (5) and (6) we have

$$\begin{aligned} \frac{x^\sigma(t)}{x(\tau_i(t))} &< 1 + \frac{H(\sigma(t), \tau_i(t))}{H(\tau_i(t), T)} \\ &= \frac{H(\sigma(t), T)}{H(\tau_i(t), T)} = \frac{1}{\eta_i(t, T)}. \end{aligned}$$

This gives us the desired result

$$x(\tau_i(t)) > x^\sigma(t)\eta_i(t, T).$$

Q.E.D.

Lemma 3. *Assume $q_i(t) \geq 0$, $1 \leq i \leq n$, and $\sum_{i=1}^{n} q_i(t) \not\equiv 0$ for large t. Let x be a solution of (1) with $x(t) > 0$, $t \in [t_0, \infty)_{\mathbb{T}}$ and assume further that*

$$\int_{t_0}^{\infty} \frac{\Delta t}{r(t)} = \infty.$$

Then there exists a $T \in [t_0, \infty)_{\mathbb{T}}$ such that

$$x(t) > 0, \quad x^{\Delta}(t) > 0, \quad \text{and} \quad (r(t)x^{\Delta}(t))^{\Delta} \leq 0$$

for $t \in [T, \infty)_{\mathbb{T}}$.

Proof. We can suppose that $t_1 \geq t_0$ is such that $x(t) > 0$, $x(\tau_i(t)) > 0$, $t \geq t_1$, for all $1 \leq i \leq n$. Then we have

$$(r(t)x^{\Delta}(t))^{\Delta} = -\sum_{i=1}^{n} q_i(t)x(\tau_i(t)) \leq 0, \quad t \in [t_1, \infty)_{\mathbb{T}},$$

and so $r(t)x^{\Delta}(t)$ is decreasing for $t \in [t_1, \infty)_{\mathbb{T}}$. Therefore, if $x^{\Delta}(t_2) \leq 0$ for some $t_2 \in [t_1, \infty)_{\mathbb{T}}$, then it follows that

$$r(t)x^{\Delta}(t) \leq 0, \quad t \in [t_2, \infty)_{\mathbb{T}}.$$

If $x^{\Delta}(t_3) < 0$ for some $t_3 \geq t_2$, then an integration gives

$$\begin{aligned} \int_{t_3}^{t} x^{\Delta}(s)\Delta s &= x(t) - x(t_3) \\ &\leq r(t_3)x^{\Delta}(t_3) \int_{t_3}^{t} \frac{\Delta s}{r(s)} \\ &\to -\infty, \quad \text{as} \quad t \to \infty, \end{aligned}$$

which gives us a contradiction. Hence, $x^{\Delta}(t) \equiv 0$ for $t \in [t_2, \infty)_{\mathbb{T}}$ and this means $x(t) \equiv$ *constant* for $t \in [t_2, \infty)_{\mathbb{T}}$. But, then

$$(r(t)x^{\Delta}(t))^{\Delta} \equiv 0 \equiv -\sum_{i=1}^{n} q_i(t)x(\tau_i(t)) \not\equiv 0,$$

which is a contradiction. Hence, it follows that

$$x^{\Delta}(t) > 0, \quad t \in [t_1, \infty)_{\mathbb{T}}, \quad \text{and} \quad (r(t)x^{\Delta}(t))^{\Delta} \leq 0, \quad t \in [t_1, \infty)_{\mathbb{T}}.$$

Q.E.D.

§2. Main results

We may now apply the previous lemmas to obtain our first oscillation result.

Theorem 4. *Assume $r(t) > 0$ with $\int_a^\infty 1/r(t)\,\Delta t = \infty$ and assume that $q_i(t) \geq 0$, $1 \leq i \leq n$, and $\sum_{i=1}^n q_i(t) \not\equiv 0$, for all sufficiently large t. If*

$$(r(t)x^\Delta)^\Delta + Q(t,T)x^\sigma = 0, \tag{7}$$

where, for $t \in (T,\infty)_{\mathbb{T}}$,

$$Q(t,T) := \sum_{i=1}^n \eta_i(t,T)q_i(t),$$

is oscillatory on $(T,\infty)_{\mathbb{T}}$ for all sufficiently large T, then all solutions of (1) are oscillatory.

Proof. If not, assume that $x(t)$ is a solution of (1) of one sign for $t \geq t_1 \geq a$ and without loss of generality let us suppose that $x(t) > 0$, $t \in [t_1,\infty)_{\mathbb{T}}$. Then by Lemma 2 and Lemma 3, there exists a $T \in [t_1,\infty)_{\mathbb{T}}$, sufficiently large, such that

$$x(t) > 0, \quad x^\Delta(t) > 0, \quad t \in [T,\infty)_{\mathbb{T}},$$

$$x(\tau_i(t)) \geq \eta_i(t,T)x^\sigma(t), \quad \tau_i(t) \in (T,\infty)_{\mathbb{T}}, \quad \text{for all} \quad 1 \leq i \leq n,$$

and (7) is oscillatory on $[a,\infty)_{\mathbb{T}}$. Consequently, we have that $x(t) > 0$ satisfies $x^\Delta(t) > 0$ and $\quad (r(t)x^\Delta)^\Delta + Q(t,T)x^\sigma \leq 0, \quad t \in (T,\infty)_{\mathbb{T}}$.

If we set $z(t) := \frac{r(t)x^\Delta(t)}{x(t)}$, then $z(t) > 0$ and

$$\begin{aligned} z^\Delta(t) &= \frac{x(t)(r(t)x^\Delta(t))^\Delta - r(t)(x^\Delta(t))^2}{x(t)x^\sigma(t)} \\ &\leq -Q(t,T) - \frac{1}{r(t)}z^2(t)\frac{x(t)}{x^\sigma(t)} \\ &= -Q(t,T) - \frac{z^2(t)}{r(t)}\frac{x(t)}{x(t)+\mu(t)x^\Delta(t)} \\ &= -Q(t,T) - \frac{z^2(t)}{r(t)+\mu(t)z(t)}. \end{aligned}$$

Therefore, since $r(t) + \mu(t)z(t) > 0$ and z is a solution of

$$z^\Delta + Q(t,T) + S[z](t) \leq 0$$

for large t, we get by Lemma 1 that the linear equation

$$(r(t)x^{\Delta})^{\Delta} + Q(t,T)x^{\sigma} = 0 \tag{8}$$

is nonoscillatory on $(T,\infty)_{\mathbb{T}}$. This contradiction proves the result.

Q.E.D.

We may establish a number of corollaries by using Theorem 4 and known criteria for linear second order dynamic equations (cf. [2], [5-8], [10] and [12-15]). For example, we have the following result.

Corollary 5. *Assume*

$$\int_T^\infty \frac{\Delta s}{r(s)} = \infty = \int_T^\infty Q(s,T)\Delta s.$$

Then all solutions of (1) *are oscillatory.*

Proof. Corollary 5 follows from the Fite–Wintner–Leighton criterion which says that all solutions of (3) are oscillatory (cf. [2]) if

$$\int_a^\infty \frac{1}{r(t)}\Delta t = \infty = \int_a^\infty q(t)\Delta t.$$

Q.E.D.

For the case $r(t) \equiv 1$ and a single delay, (1) becomes

$$x^{\Delta\Delta} + q(t)x(\tau(t)) = 0. \tag{9}$$

In this case, $\eta(t,T) = \frac{\tau(t)-T}{\sigma(t)-T}$, so that

$$Q(t,T) = \frac{\tau(t)-T}{\sigma(t)-T}q(t) \sim \frac{\tau(t)}{\sigma(t)}q(t) \quad \text{as} \quad t \to \infty.$$

Therefore, if

$$\int_a^\infty \frac{\tau(t)}{\sigma(t)}q(t)\Delta t = \infty,$$

then all solutions of (9) are oscillatory.

We next consider the dynamic equation

$$x^{\Delta\Delta} + \frac{\gamma}{t\tau(t)}x(\tau(t)) = 0. \tag{10}$$

We have the following.

Corollary 6. *All solutions of* (10) *are oscillatory if* $\gamma > \frac{1}{4}$ *and* $\lim_{t\to\infty} \frac{\mu(t)}{t} = 0$.

Proof. We use the fact that all solutions of

$$x^{\Delta\Delta} + \frac{\gamma}{t\sigma(t)} x^{\sigma} = 0 \tag{11}$$

are oscillatory if $\gamma > \frac{1}{4}$ and $\lim_{t\to\infty} \frac{\mu(t)}{t} = 0$ (cf. [12]) along with Theorem 4. Q.E.D.

§3. Examples

In this section we give examples of our main results.

Example 7. *If $\mathbb{T}$ is any time scale with $\lim_{t\to\infty} \frac{\mu(t)}{t} = 0$ (e.g., $\mathbb{T} = \mathbb{R}$ or $\mathbb{T} = \mathbb{Z}$), then all solutions of* (11)*, or more generally*

$$x^{\Delta\Delta} + \sum_{i=1}^{n} \frac{\gamma_i}{t\tau_i(t)} x(\tau_i(t)) = 0$$

are oscillatory provided

$$\bar{\gamma} := \sum_{i=1}^{n} \gamma_i > \frac{1}{4}.$$

To see this, we observe that in this case

$$\begin{aligned} Q(t,T) &= \sum_{i=1}^{n} \eta_i(t,T) q_i(t) \\ &= \sum_{i=1}^{n} \frac{\gamma_i}{t\tau_i(t)} \left(\frac{\tau_i(t) - T}{\sigma(t) - T} \right) \end{aligned}$$

and

$$\sum_{i=1}^{n} \frac{\gamma_i}{t\tau_i(t)} \left(\frac{\tau_i(t) - T}{\sigma(t) - T} \right) \sim \frac{\bar{\gamma}}{t\sigma(t)} \quad \text{as} \quad t \to \infty.$$

Therefore, since (11) *with γ replaced by $\bar{\gamma}$ is oscillatory, the result now follows from Theorem 4.*

Example 8. *If $\mathbb{T} = q^{\mathbb{N}_0}$, $q > 1$, then the q-difference equation*

$$x^{\Delta\Delta} + \frac{\gamma}{t\sigma(t)} x^{\sigma} = 0$$

is oscillatory iff $\gamma > \frac{1}{(\sqrt{q}+1)^2}$ (cf. [5], [12]*). Therefore, the delay q-difference equation (where $\tau_i : \mathbb{T} = q^{\mathbb{N}_0} \to \mathbb{T}$)*

$$x^{\Delta\Delta} + \sum_{i=1}^{n} \frac{\gamma_i}{t\tau_i(t)} x(\tau_i(t)) = 0$$

is oscillatory provided

$$\bar{\gamma} = \sum_{i=1}^{n} \gamma_i > \frac{1}{(\sqrt{q}+1)^2}.$$

Additional examples may be readily given. We leave this to the interested reader.

References

[1] D. Anderson, R. Krueger and A. Peterson, Delay dynamic equations with stability, Adv. Difference Equ., **2006** (2006), 1–19.

[2] M. Bohner, L. Erbe and A. Peterson, Oscillation for nonlinear second order dynamic equations on a time scale, J. Math. Anal. Appl., **301** (2005), 491–507.

[3] M. Bohner and A. Peterson, Dynamic Equations on Time Scales: An Introduction with Applications, Birkhäuser, Boston, 2001.

[4] M. Bohner and A. Peterson, Advances in Dynamic Equations on Time Scales, Editors, Birkhäuser, Boston, 2003.

[5] M. Bohner and M. Ünal, Kneser's theorem in q-calculus, J. Phys. A, **38** (2005), 6729–6739.

[6] L. Erbe, Oscillation criteria for second order linear equations on a time scale, Canad. Appl. Math. Quart., **9** (2001), 1–31.

[7] L. Erbe and A. Peterson, Comparison theorems of Hille–Wintner type for dynamic equations on time scales, Proc. Amer. Math. Soc., **133** (2005), 3243–3253.

[8] L. Erbe, A. Peterson and P. Řehák, Comparison theorems for linear dynamic equations on time scales, J. Math. Anal. Appl., **275** (2002), 418–438.

[9] S. Hilger, Analysis on measure chains - A unified approach to continuous and discrete calculus, Results Math., **18** (1990), 18–56.

[10] E. Hille, Nonoscillation theorems, Trans. Amer. Math. Soc., **64** (1948), 234–252.

[11] B. Kaymakçalan, V. Lakshmikantham and S. Sivasundaram, Dynamical Systems on Measure Chains, Kluwer Academic Publishers, Boston, 1996.

[12] P. Řehák, Half-linear dynamic equations on time scales: IVP and oscillatory properties, Nonlinear Funct. Anal. Appl., **7** (2002), 361–404.

[13] P. Řehák, Function sequence technique for half-linear dynamic equations on time scales, Panamer. Math. J., **16** (2006), 31–56.

[14] C.-T. Taam, Nonoscillatory differential equations, Duke Math. J., **19** (1952), 493–497.

[15] A. Wintner, On the comparison theorem of Kneser–Hille, Math. Scand., **5** (1957), 255–260.

Department of Mathematics
University of Nebraska-Lincoln
Lincoln, NE 68588-0130
USA

E-mail address: lerbe@math.unl.edu
apeterso@math.unl.edu

Advanced Studies in Pure Mathematics 53, 2009
Advances in Discrete Dynamical Systems
pp. 225–235

Analysis of an evolutionary variational inequality arising in elasticity quasi-static contact problems

Nicolae Pop

Abstract.

In this paper, we present the strong and the variational form of a dynamic contact problem with friction. We derive the result and obtain an incremental formulation by space discretization with finite element method and by time discretization with finite difference method of the dynamic contact problem. As well we describe the solution strategies for spatially discrete system and the condition for stability of the solution. Dificulties caused by the discontinuity of the Coulomb's friction law (passage from sliding to adhesion), are tried to be passed over using the Newton-Raphson iterative techniques, and their success depends also on the small value of the parameter , from the regularized friction law.

§1. Introduction

The present paper is a continuation of the analysis presented in [10], which consists in a numerical analysis of a dynamic contact problem with friction presented in [11]. The problem is intended to model the physical situation of two elastically deforming bodies that come into contact with friction and are obeying the normal compliance law, where the normal stress is depending on statistical parameters of the contact surface's profile.

First we give a classical and a variational formulation of the continuous dynamic contact problem. Afterwards, we derive the result and

---

Received December 26, 2006.
Revised October 12, 2007.
2000 *Mathematics Subject Classification.* 74S05, 74M10, 74M15, 74H15, 74H30, 39A11.
Key words and phrases. Unilateral contact problem, dry friction laws, finite element method, finite difference method, Newmark algorithm, Newton-Raphson method.

obtain an incremental formulation by space discretization with finite element method and by time discretization with finite difference method of the dynamic contact problem and we present the solution strategies for spatially discrete system and the condition for stability of the solution.

In the present paper we compare the quasi-static contact problem with the dynamic contact problem. In the case of the quasi-static contact problem the inertial forces are neglected, but for the dynamic contact problem case, the inertial terms cannot be neglected.

§2. Formulation of the problem

Let us consider two linear elastic bodies that at a given time $t = 0$ occupy domains Ω^1 and $\Omega^2 \subset R^d$, respectively, where $d = 2$ or $d = 3$. The boundary of each body, is divided into three subregions such that

$$\partial\Omega^1 = \Gamma^1 = \overline{\Gamma}^1_U \cup \overline{\Gamma}^1_N \cup \overline{\Gamma}^1_C \text{ and } \partial\Omega^2 = \Gamma^2 = \overline{\Gamma}^2_U \cup \overline{\Gamma}^2_N \cup \overline{\Gamma}^2_C,$$

which are topologically open, and disjoint, only Γ^1_C and Γ^2_C being accepted to have common points, and $mes(\Gamma^i_N) > 0$, $i = 1, 2$.

The displacement $\overline{u}\,(t, x)$ will be prescribed on $\Gamma_U = \overline{\Gamma}^1_U \cup \overline{\Gamma}^2_U$ and traction $\overline{h}\,(t, x)$ is to be given on $\Gamma_N = \Gamma^1_N \cup \Gamma^2_N$. For the beginning, the boundary $\Gamma_N = \Gamma^1_N \cup \Gamma^2_N$ is considered without tensions. At the same time the stress vector $\sigma^{(n)}(u)$ is defined, oriented towards the exterior of the boundary $\partial\Omega = \partial\Omega^1 \cup \partial\Omega^2$. The initial displacement $u\,(0, x) = u_0\,(x)$, the initial velocity $\dot{u}\,(0, x) = u_1\,(x)$ and the density of body force f are also given.

So long as the two bodies do not touch each other, the field of the displacements will be the solution of a boundary value problem of the differential equations of elastodynamics. If the two bodies touch one another, then in the contact boundary there are forces strong enough to prevent the interaction (penetration) of the two bodies. The condition that needs to be expressed in order to describe this process is called *the contact condition.* Beside these forces there may appear in the contact area friction forces as well, which are described by a law of friction. The contact problem in a time interval $[0, t_E]$ with $t_E > 0$, has the following form:

The elastodynamic equation on $\Omega = \Omega_1 \cup \Omega_2$

$$\rho\ddot{u}\,(t, x) - \sigma_{ij,j}\,(u\,(t, x)) = f\,(t, x) \text{ on } [0, t_E] \times \Omega. \tag{2.1}$$

The boundary conditions

$$u\,(t, x) = \bar{u}\,(t, x) \text{ on } [0, t_E] \times \Gamma_U \tag{2.2}$$

$$\sigma^{(n)}(u)(t,x) = \bar{h}(t,x) \text{ on } [0,t_E] \times \Gamma_N \tag{2.3}$$

and the initial conditions are $u(0,x) = u_0(x)$, the initial velocity $\dot{u}(0,x) = u_1(x)$. The condition of non-penetrating bodies will be understood as a geometric *contact condition.* It is approximated by the inequation $u_N^R(x,t) \leq g(x)$, where $g(x)$ is the initial gap between the bodies and u_N^R is the relative displacement between the contact boundaries. We shall use the following notations for the normal and tangential components of the displacements and of the stress vector:

$u_N = u \cdot n = u_i n_i^1$, $u_T = u - u_N n^1$, $\sigma_N = \sigma_{ij} n_i^1 n_j^1$, $1 \leq i,j \leq d$ and n^1 is the outward normal unit vector on Γ^1.

In the normal interface response case, we have

$$\sigma_N(u) = -c_N(u_N^R - g)_+^{m_N}, \text{ on} \Gamma_N \times (0,t_E) \tag{2.4}$$

if

$$\dot{u}_N^R \leq g \Rightarrow \quad |\sigma_T(u)| \leq c_T(u_N^R - g)_+^{m_T}$$

else

$$\sigma_T(u) = c_T(u_N^R - g)_+^{m_T}$$

Where c_N, m_N, c_T, m_T are material constants depending on the interface properties and b_+={b,0}. The following steps, similar with those of Oden and Martins [9], the nonlinear elastodynamics problem can be formally equivalent to the following variational problem:

Problem P1. *Find the function u:[0,t]→ V s.t.*

$$\begin{aligned} &\langle \ddot{u}(t), v - \dot{u}(t)\rangle + a(u(t), v - \dot{u}(t)) + \langle P(u(t)), v - \dot{u}(t)\rangle + \\ &+ j(u(t), v) - j(u(t), \dot{u}(t)) \geq \langle f(t), v - \dot{u}(t)\rangle, \ \forall v \in V \end{aligned} \tag{2.5}$$

with the initial conditions:

$$u(0,x) = u_0(x) \text{ and} \dot{u}(0,x) = u_1(x). \tag{2.6}$$

We have assumed here, for simplicity, that $\rho \equiv 1$. The following notations and definitions were also used:

$$V = \left\{ v \in \left[H^1(\Omega)\right]^d \mid v = 0 \text{ a.e. on } \Gamma_C \right\}, \tag{2.7}$$

the space of admissible displacements (velocities);

$$a : V \times V \to \mathrm{R}, \quad a(u,v) = \int_\Omega C_{ijkl}\varepsilon_{ij}(u)\varepsilon_{kl}(v)dx, \tag{2.8}$$

the virtual work produced by the action of the stress $\sigma_{ij}(u)$ on the strains $\varepsilon_{ij}(v)$;

$$(2.9) \qquad P : V \to V', \quad \langle P(u), v\rangle = \int_{\Gamma_C} c_N (u_N^R - g)_+^{m_N} v_N ds,$$

the virtual work produced by the normal pressure on the displacement (velocity) v;

$$(2.10) \qquad j : V \times V \to \mathrm{R}, \quad j(u,v) = \int_{\Gamma_C} c_T \left(u_N^R - g\right)_+^{m_T} |v_T^R| ds,$$

the virtual power produced by the frictional force on the velocity v ;

$$(2.11) \quad f(t) \in V', \ \langle L, v\rangle \equiv \langle f(t), v\rangle = \int_{\Omega} f(t) v dx + \int_{\Gamma_C} \overline{h}\gamma(v) ds,$$

the virtual work produced by the external forces.

Here $\langle\cdot,\cdot\rangle$ denotes duality pairing on $V \times V'$ where V' is the topological dual of of V; γ is the trace operator mapping from $\left[H^1(\Omega)\right]^d$ onto $\left[H^{\frac{1}{2}}(\Gamma)\right]^d$ which may be decomposed into normal component $\gamma_N(v)$ and tangential component $\gamma_T(v)$. For simplicity of notation, the later are denoted as v_N and v_T, respectively. We also observe that the boundary integrals on Γ_C are well defined for $1 \le m_N,\ m_T \le 3$ if d=3 and for $1 \le m_N,\ m_T < \infty$ if d=2. These conditions on m_N and m_T basically com from the embedding theorem which states that for $v \in \left[H^1(\Omega)\right]^d$,$\gamma(v) \in [L^q(\Gamma_C)]^d$ with $2 \le q \le \infty$ for d=2, and with $2 \le q \le 4$ for d=3.

One of the directions for the approximation of the Problem P1 is a family of regularized problems which lead insteads to of a variational equality. For this, it approximates the functional j:$V \times V \to \mathrm{R}$ which is nondifferentiable in the second argument (velocity) by a family of functionals j_ε convex and differentiable on the second argument:

$$(2.12) \quad j_\varepsilon : V \times V \to \mathrm{R}, \quad j_\varepsilon(u,v) = \int_{\Gamma_C} c_T \left(u_N^R - g\right)_+^{m_T} \Psi_\varepsilon\left(v_T^R\right) ds,$$

where the function Ψ_ε:$[L^q(\Gamma_C)]^d \to L^q(\Gamma_C)$, is an approximation of the function $|\cdot|$: $[L^q(\Gamma_C)]^d \to L^q(\Gamma_C)$ and is defined for $\varepsilon > 0, \quad \xi \in [L^q(\Gamma_C)]^d$ and a.e. $x \in \Gamma_C$, according to

$$(2.13) \qquad \Psi_\varepsilon(\xi) = \begin{cases} \varepsilon\left|\frac{\xi}{\varepsilon}\right|^2 \left(1 - \frac{1}{3}\left|\frac{\xi}{\varepsilon}\right|\right) & if \ \ |\xi(x)| \le \varepsilon \\ \varepsilon\left(\left|\frac{\xi}{\varepsilon}\right| - \frac{1}{3}\right) & if \ \ |\xi(x)| > \varepsilon \end{cases}$$

The regularized form of Problem P1, is:

Problem P1ε. *Find the functions* $u_\varepsilon : [0, t_E] \rightarrow V s.t.$

$$(2.14)\ \langle \ddot{u}_\varepsilon (t), v \rangle + a(u_\varepsilon (t), v) + \langle P(u_\varepsilon (t)), v \rangle + \langle j_\varepsilon (u_\varepsilon (t), \dot{u}_\varepsilon (t)), v \rangle = \langle f(t), v \rangle, \ \forall v \in V$$

with the initial conditions:

$$u_\varepsilon (0, x) = u_0 (x) \ \ and \ \ \dot{u}_\varepsilon (0, x) = u_1 (x) . \tag{2.15}$$

We observe that now we have a variational equation instead of a variational inequality. However, the regularized friction conditions are now the form:

$$\sigma_T = -c_T \left(u_N^R - g \right)_+^{m_T} F_\varepsilon \left(\dot{u}_T^R \right) \tag{2.16}$$

with

$$F_\varepsilon \left(\dot{u}_T^R \right) = \begin{cases} \left(2 - \left| \frac{\dot{u}_T^R}{\varepsilon} \right| \right) \frac{\dot{u}_T^R}{\varepsilon} & if\ \dot{u}_T^R \leq \varepsilon \\ \frac{\dot{u}_T^R}{\left| \dot{u}_T^R \right|} & if\ \dot{u}_T^R > \varepsilon \end{cases} .$$

where F_ε is the friction coefficient. The choice of ε will be dictated only by the desired proximity of the solutions of the Problems P1 and P1ε and the corresponding computational costs associated.

§3. Spatially semi-discrete approximations of the evolutionary variational equality

Two types of semi-discrete approximation scheme can be developed. In the first type, when we replace the infinite-dimensional space V by a finite-dimensional subspace V_h (the element finite spaces), leading to a finite-dimensional system of ordinary differential equations (3.3). Such approximation schemes can be termed as spatially semi-discrete scheme. In the second type, we replace the time derivatives by finite differences that lead to elliptic variational equations over infinite-dimensional space V at each time step. Such approximation schemes can be termed at temporally semi-discrete schemes. In this paper, we consider the first type, spatially semi-discrete schemes, because after a spatially semi-discrete discretization, we can employ some ODE solvers for mathematical solution, see [8].

Using standard finite element procedures, approximate version of Problem P1ε can be constructed in finite-dimensional subspaces V_h ($\subset V$

$\subset V'$). For certain (h) the approximate displacements, velocities and accelerations at each time t are elements of

$$V_h, \quad r^h(t), \quad \dot{r}^h(t), \quad \ddot{r}^h(t) \in V_h.$$

Within each element Ω_h^e $(e = 1, \dots, N_h)$ the components of the displacements, velocities and accelerations are expressed in the form:

$$r_k^h(t,x) = \sum_{I=1}^{N_e} r_k^I(t) N_I(x), \quad \dot{r}_k^h(t,x) = \sum_{I=1}^{N_e} \dot{r}_k^I(t) N_I(x),$$
$$\ddot{r}_k^h(t,x) = \sum_{I=1}^{N_e} \ddot{r}_k^I(t) N_I(x), \tag{3.1}$$

where $k = 2$ *or* 3, N_e =number of node of the element, $r_k^I(t)$, $\dot{r}_k^I(t)$, $\ddot{r}_k^I(t)$ are the nodal values of the displacements, velocities and accelerations, at the time t and N_I is the element shape function associated with the nodal point. The finite element version of the Problem P21ε with N_h^Ω representing the number of the nodes of finite element mesh of Ω, is equivalent to the following matrix problem:

Problem MP31ε^h. *Find the function* $r : [0, t_E] \to \mathrm{R}^{d \times N_h^\Omega}$, *s.t.*

$$M\ddot{r}(t) + Kr(t) - P(r(t)) + J(r(t), \dot{r}(t)) = F(t), \tag{3.2}$$

with the initial conditions $r(0) = p_0, \quad \dot{r}(0) = p_1$.

Here we have introduced the following matrix notations:

$r(t), \dot{r}(t), \ddot{r}(t)$: the column vectors of nodal displacements, velocities and accelerations, respectively;

M : standard mass matrix, K: standard stiffness matrix, $F(t)$: consistent nodal exterior forces vector, $P(r(t))$: vector of consistent nodal forces on Γ_C $J(r(t), \dot{r}(t))$: vector of consistent nodal friction forces on Γ_C, p_0, p_1: initial nodal displacement, velocity. The components of the element vector ${}^{(e)}P$ are of the form: ${}^{(e)}P = -\int_{{}^{(e)}\Gamma_C^2} \sigma_N \cdot n \cdot N_I ds$ and the components of the element vector ${}^{(e)}J$ are of the form: ${}^{(e)}J = -\int_{{}^{(e)}\Gamma_C^2} \sigma_T \cdot N_I ds$.

In order to obtain the components of the element vector P and J is used a contact finite element, see [10], [11].

The algorithms that we shall use for solving the discrete dynamical system involve variants of standard schemes used in nonlinear structural dynamics calculations: the Newmark-type algorithm or the central-difference scheme.

Using the Newton-Raphson method to solve the variational equation obtained at time t_k introduced, into the variational equation P31ε^h, the relations which define the Newmark-type algorithm or the central-difference scheme the following system of algebraic linear equations is obtained to solve at each iteration.

Let us consider a partition of the time interval $I = \bigcup_{k=1}^{N} [t_{k-1}, t_k]$ with $0 = t_0 < t_1 < \cdots < t_N = t_E$, we denote $\Delta t = t_k - t_{k-1}$ for the length of the sub-interval $[t_{k-1}, t_k]$. Two general classes are generally of interest. In the quasistatic case the inertial loads are taken to be insignificant in comparison with the applied loads and internal forces. In the dynamic case, the inertial terms are restored and issued associated with temporal accuracy and stability that come to the fore (must analysis).

If we denote in ordinary differential equation (3.2), $K_N \equiv K - P + J$ (where the matrices P and J are nonlinear), $d_k \equiv r(t_k)$, $v_k \equiv \dot{r}(t_k)$, $a_k \equiv \ddot{r}(t_k)$, then (3.2) becomes:

$$Ma(t) + K_N(d(t)) = F(t) \tag{3.3}$$

For the analysis of this problem, must begin with *quasistatic case*:

$$K_N(d(t)) = F(t) \tag{3.4}$$

subject to only one initial condition $d(0) = d_0$.

Note that the time variable t may correspond to real time, but doesn't need to have physical meaning for rate independent behavior. For example, it is common for t to be taken as a generic parameterization for the applied loading on the system. With the above notation, the incremental load approach attempts (in each increment Δt) to find d_{k+1}, given the solution d_k at time level t_k, such that

$$K_N(d_{k+1}) = F(t_{k+1}). \tag{3.5}$$

By introducing the concept of a residual vector $R(d_{k+1})$:

$$R(d_{k+1}) := F(t_{k+1}) - K_N(d_{d+1}) \tag{3.6}$$

Solution of (3.5) therefore amounts to the finding of the root of the equation

$$R(d_{k+1}) = 0. \tag{3.7}$$

The physical meaning that is starting with an initial equilibrium state at t_k, so that $R(d_k) = 0$, we introduce a prescribed load increment $\Delta F = F(t_{k+1}) - F(t_k)$, and attempt to find that displacement increment $d_{k+1} - d_k$ that will restore equilibrium (eq. (3.5)).

In *dynamic case* inertial terms can not be neglected, and we note that it is possible in contrast to the quasi-static problem, the variable t in the dynamic case has the interpretation of real time.

Problem P32. Find approximations d_{k+1}, v_{k+1}, a_{k+1} at time t_{k+1} from equation (3.5), with given displacement vector d_k, velocity v_k and acceleration a_k at time t_k.

$$Ma_{k+1} + K_N (d_{k+\alpha}) = F (t_{k+\alpha}) \tag{3.8}$$

$$d_{k+\alpha} = \alpha d_{k+1} + (1 - \alpha) d_k,$$

$$d_{k+1} = d_k + \Delta t v_k + \frac{\Delta t^2}{2} [(1 - 2\beta) a_k + 2\beta a_{k+1}],$$

$$v_{k+1} = v_k + \Delta t [(1 - \gamma) a_k + \gamma a_{k+1}],$$

where α, β and γ are algorithmic parameters that define the stability and accuracy characteristics of the method. In particular, when $\alpha = 1$, the algorithm reduces to the classical Newmark algorithm.

A wide range of algorithms exists corresponding to the different available choices of these parameters, we illustrate two methods: *implicit methods*, which is simply member of the Newmark family obtained by setting $\alpha = 1$, $\beta = 1/4$ and $\gamma = 1/2$ in (3.8). In this case, the integrator is second order accurate and unconditionally stable for linear problems, meaning that the spectral radii of the integrator remains less than 1 in modulus, for any time step Δt. For *explicit methods*, let us take $\alpha = 1$, $\beta = 0$ and $\gamma = 1/2$, substitute into (3.8) and examining the central difference algorithm. This integrator, for explicit method, is second order accurate and only conditionally stable, meaning that linearized stability is only retained when Δt is less than some critical limit. This limit, sometimes called the *Courant stability limit*, can be shown to be as follows $\Delta t \leq {}^2\!/_{\omega_{\max}}$, where $\omega_{\max}$ is the highest modal natural frequency in the mesh. This frequency can be estimated: $\omega_{max} \approx 2 \left({}^c\!/_h\right)_{\max}$ where c and h are the sound speed and characteristic mesh size, respectively. Finally, we find that $\Delta t \leq \left({}^h\!/_c\right)_{\min}$.

This meaning that the time step may by no larger than the amount of time required for a sound wave to traverse the element in the mesh having the smallest transit time. This fact tells us that explicit methods are appropriate only for those problems featuring very high frequency response, for problems featuring low frequency response, the implicit methods are is highly desirable, albeit at the cost of explicit updates in each increment. The implicit and explicit methods are valid only for linear or linearized problems. In this section we give a general framework for solving the nonlinear discrete equations associated with computation of an unknown state at step t_{k+1}, in either context of a quasi-static (3.7) and dynamic contact problem formulation as in (3.8). In either case, the equation to be solved takes the form $R(d_{k+1}) = 0$, where the R is considered to be a nonlinear function of the solution vector d_{k+1}.

The general concept of a *Newton-Raphson iterative solution technique* consists in the linearization of a nonlinear equation. This concept is applied to equation (3.8) and is defined in iteration j by $R\left(d_{k+1}^j\right) + \left[{}^{\partial R}/_{\partial d}\right]_{d_{k+1}^j} \Delta d_k^j = 0$, followed by the update $d_{k+1}^{j+1} = d_{k+1}^j + \Delta d_k^j$.

Iteration on j typically continue until the Euclidian norm $\left\|\Delta d_k^j\right\|$ is smaller than some tolerance.

The residual at iteration j , from (3.8) is of the form

$$(3.9) \qquad \begin{aligned} R\left(d_{k+1}^j\right) \quad &:= \quad F(t_{k+1}) - K_N\left(d_{k+1}^j\right) - \frac{4}{\Delta t^2} M d_{k+1}^j \\ &+ \quad M\left(a_k + \Delta t v_k + \frac{4}{\Delta t^2} d_k^j\right) = 0 \end{aligned}$$

which take the form $\left[({}^4/_{\Delta t})M + K_L\left(d_{k+1}^j\right)\right] \Delta d_k^j = R\left(d_{k+1}^j\right)$, where the stiffness matrix $K_L\left(d_{k+1}^j\right)$ is given as

$$K_L\left(d_{k+1}^j\right) = \left({}^{\partial K_N}/_{\partial d}\right)_{d_{k+1}^j}.$$

We note that a variety of iterative procedures exist as alternatives to the Newton-Raphson nonlinear solution procedure (quasi-Newton, secant methods etc.).

The scheme of solving the linearized dynamic contact problems is the following:

(i) initialization the set of the iterative count $t_k = 0,\ \Delta t = 0,\ k = 0,\ j = 0,\ d_k^j = 0$;

(ii) compute the mass matrix M, the standard stiffness matrix K and a dynamic residual R;

(iii) compute the contact nodal forces P and the contact friction forces J;

(1) compute the normal gap g_N^j;
(2) check for contact finite element status:
IF g_N^j >TOL then out of contact
ELSE in contact. Check for frictional stick or slip contact status
ENDIF

(3) compute total matrix K_L and residual R, this involves to compute

the vectors $\left\{a^j_{k+1}, v^j_{k+1}, d^j_{k+1}\right\}$, from k=0 to N, (k=N=>$t_{k+1}=t_E$);
(IV) check for convergence:
IF$\left|\Delta d^j_{k+1} - \Delta d^{j+1}_{k+1}\right|$ <TOL1 then converge and exit.
ELSE go to step (VI)
(V) update the displacement field $d^{j+1}_{k+1} = d^j_{k+1} + \Delta d^j_k$;
(VI) set j=j+1 and go back to (ii);

§4. Conclusions

1) It is known that the matrix K_N is ill conditioned, if we split the normal and tangential stress from contact boundary, in blocks diagonal matrices, these block matrices contain coefficients of the same size order, and with this procedure we obtain a better conditioned matrix.

2) The discontinuity of the Coulomb's friction law at zero sliding velocity is a major source of computational difficulties in friction problems. Even though, in the algorithms described in this and the previous sections, a regularized form of that law is used, those difficulties cannot be completely avoided. The situation which may arise when using the methods described here with a constant step is the following: in unloading situations (passage from sliding to adhesion) the Newton-Raphson iterative techniques may fail to converge if ε is very small and the step too large. For small values of ε the radius of converge of the iterative scheme used is very small due to the step change in Ψ_ε on the interval $[-\varepsilon, \varepsilon]$.

3) The critical situations arise in transitions from sliding to adhesion, because these are the most important (dramatic) changes. In order to avoid these difficulties that appear in the iterative solutions sequence, we decrease the time step until two successive solutions are not too far apart.

References

[1] L. E. Anderson, A quasistatic frictional problem with normal compliance, Nonlinear Anal., **16** (1991), 347–369.

[2] M. Cocu, Unilateral contact problems with friction for an elastoviscoplastic material with internal state variable, In: Proc. Contact Mechanics Int. Symp., (ed. A. Curnier), PPUR, 1992, pp. 207–216.

[3] S. Drabla and M. Sofonea, Analysis of a Signorini problem with friction, IMA J. Appl. Math., **63** (1999), 113–130.

[4] G. Duvaut, Loi de frottement non locale, J. Méc. The. Appl., spcial issue, 1982, 73–78.

[5] J. Jarušek and C. Eck, Dynamic contact problems with friction in linear viscoelasticity, C. R. Acad. Sci. Paris Sér. I Math., **322** (1996), 497–502.

[6] N. Kikuchi and J. T. Oden, Contact Problems in Elasticity: A Study of Variational Inequalities and Finite Element Methods, SIAM Stud. Appl. Math., **8**, SIAM Philadelphia, 1988.

[7] A. Klarbring, A. Mikelić and M. Shillor, Frictional contact problems with normal compliance, Internat. J. Engrg. Sci., **26** (1988), 811–832.

[8] T. A. Laursen, Computational contact impact mechanics. Fundamentals of modeling interfacial phenomena in nonlinear finite element analysis, Springer-Verlag, 2003.

[9] J. T. Oden and J. A. C. Martins, Models and computational methods for dynamic friction phenomena, Comput. Methods Appl. Mech. Engrg., **52** (1986), 527–634.

[10] N. Pop, On the existence of the solution for the equations modeling contact problems, In: Proceedings of the Third International Palestinian Conference, Mathematics & Mathematics Education, (eds. S. Elaydi, E. S. Titi, S. K. Saleh and R. Abu-Saris), World Scientific, 2000, pp. 196–207.

[11] N. Pop, A numerical formulation for quasistatic frictional contact problem, Advance in Difference Equations, Proceedings of the Second International Conference on Difference Equations, 1995, (eds. S. Elaydi, I. Gyori, G. Ladas), Gordon and Breach, 1997, pp. 517–524.

[12] A. Signorini, Sopra alcune questioni di elastostatica, Atti. Soc. Ital. Progr. Sci., 1933.

[13] D. Tabor, Friction – The present state of our understanding, J. Lubr. Technol., **103** (1981), 169–170.

[14] K. L. Woo and T. R. Thomas, Contact of rough surface: A review of experimental work, Wear, **58** (1980), 331–340.

North University of Baia Mare
Romania

E-mail address: nicpop@gmail.com

Advanced Studies in Pure Mathematics 53, 2009
Advances in Discrete Dynamical Systems
pp. 237–259

Dissipative delay endomorphisms and asymptotic equivalence

Christian Pötzsche[1]

Abstract.

Using an invariant manifold theorem we demonstrate that the dynamics of nonautonomous dissipative delayed difference equations (with delay M) is asymptotically equivalent to the long-term behavior of an N-dimensional first order difference equation (with $N \leq M$) – assumed the nonlinearity is small Lipschitzian on the absorbing set. As consequence we obtain a result of Kirchgraber that multi-step methods for the numerical solution of ordinary differential equations are essentially one-step methods, and generalize it to varying step-sizes.

§1. Motivation and introduction

Scalar real delay (or higher order) difference equations appear in a variety of applications, ranging from biosciences (Mackey-Glass, Wazewska-Czyzewska and Lasota equation, etc.) to numerical analysis (multi-step methods for the numerical solution of ordinary, or discretizations of delay differential equations). Provided their delay is M (or equivalently, their order is $M + 1$), they can be formulated as first order equations in $\mathbb{R}^{M+1}$ and become accessible to the theory of discrete dynamical systems. It is well-known that the possible complexity of their long-time behavior depends on the dimension of the state space, i.e., the size of M. Ideally, one wants to keep M as small as possible, but frequently the delay is dictated from the given model.

In this article the geometry of delay difference equations is studied using invariant manifold theory. The problems under consideration are

Received October 30, 2006.

2000 *Mathematics Subject Classification.* 39A11, 37D10, 65L06.

Key words and phrases. Delay difference equation, pullback attractor, attractive invariant manifold.

[1]Research supported by the Deutsche Forschungsgemeinschaft.Present address: Munich University of Technology, Centre for Mathematical Sciences, Boltzmannstraße 3, D-85748 Garching, Germany

assumed to be dissipative with contractive linear part and a locally Lipschitzian nonlinearity. Provided the Lipschitz constants on the absorbing set are small, we can associate an asymptotically equivalent lower dimensional equation to the original problem. Hence, we are able to reduce the dimension of the state space for a delay difference equation without loosing information on its long-time behavior.

Such a global reduction principle is well-known in the area of dissipative evolutionary partial differential equations. Here the existence of a so-called *inertial manifold* ensures that the corresponding infinite-dimensional semiflow asymptotically behaves as the solutions of a finite-dimensional ordinary differential equation (the *inertial form*; cf., e.g., [SY02, pp. 569ff, Chapter 8]). In order to study the persistence of inertial manifolds under (numerical) discretizations one needs flexible invariant manifold theorems for mappings. Such results have been derived for instance in [Pöt07a, Pöt07b] and can be simplified to Theorem C.1 of this paper, allowing an application to delay difference equations. Attractive invariant manifold theorems for mappings already date back to [Har64] and were also considered by [KS78] with generalizations in [NS92]. Differing from center-unstable manifolds they provide a reduction principle on the whole absorbing set and not only in a small neighborhood.

Our basic contribution is to discuss nonautonomous difference equations instead of discrete dynamical systems (mappings). Dissipativity is understood in the sense of pullback convergence (cf., e.g., [Klo00]) and invariant manifolds generalize to invariant fiber bundles.

After some necessary terminology is presented in Section 2, we provide a simple criterion for dissipativity in Section 3 and demonstrate its applicability to nonautonomous versions of various well-studied delay difference equations. The following Section 4 equips us with sufficient conditions that delay difference equations (with delay M) feature asymptotically the same dynamical behavior as first order difference equations in $\mathbb{R}^N$ with $N \leq M$. As an application we show in the final Section 5 that strongly stable multi-step methods for the numerical integration of ordinary differential equations are conjugated to one-step methods. In the autonomous constant step-size setting this is a result due to [Kir86] with generalizations to general linear methods by [Sto93]. Our approach is intended to demonstrate how such results can be lifted to schemes with varying step-sizes and nonautonomous ordinary differential equations. Due to the lack of space, our analysis is somehow crude neglecting convergence issues; for that we refer the interested reader to the nice and comprehensive discussion in [Sto93]. Finally, for the reader's convenience, the Appendix contains some notions important for our nonautonomous perspective, namely nonautonomous sets, exponential dichotomies and invariant fiber bundles.

Concerning terminology, $\mathbb{Z}_\kappa^+ := [\kappa, \infty) \cap \mathbb{Z}$ for integers $\kappa \in \mathbb{Z}$. We index real N-tupels $x \in \mathbb{R}^N$ according to $x = (x_{-N+1}, \ldots, x_0)$ and to provide largely explicit assumptions we use the norm $\|x\| := \max_{j=-N+1}^{0} |x_j|$ throughout. For the open ball in $\mathbb{R}^N$ centered in 0 with radius $r > 0$ we simply write B_r. Given a matrix $A \in \mathbb{R}^{N \times M}$ we denote its transpose by $A^T A \in \mathbb{R}^{M \times N}$, the ith row, jth column element by $A_{i,j} \in \mathbb{R}$, and obtain the induced norm

$$\|A\| = \max_{i=-N+1}^{0} \sum_{j=-M+1}^{0} |A_{i,j}| .$$

Moreover, $\operatorname{im} A$ denotes the *range* (*image*) and $\ker A$ the *kernel* (*nullspace*) of A. The identity matrix on $\mathbb{R}^N$ is I_N and O_N the zero matrix.

§2. Delay difference equations versus delay endomorphisms

This paper deals with scalar nonautonomous $M+1$-th order difference equations

$$\boxed{x(k+1) = f(k, x(k-M), \ldots, x(k-1), x(k)),} \tag{2.1}$$

where $M \in \mathbb{Z}_0^+$ is interpreted as *delay*, $f : \mathbb{Z} \times I^{M+1} \to I$ is the *right hand side* and $I \subseteq \mathbb{R}$ an interval. Given an initial time $\kappa \in \mathbb{Z}$, a *solution* of the *delay difference equation* is a sequence $\phi : \mathbb{Z}_{\kappa-M}^+ \to I$ satisfying the identity (2.1) and initial value problems are well-posed, if beyond $\phi(\kappa) \in I$ also supplementary values $\phi(\kappa - M), \cdots, \phi(\kappa - 1) \in I$ are known. The unique solution $\varphi(\cdot; \kappa, \xi_{-M}, \ldots, \xi_0) : \mathbb{Z}_{\kappa-M}^+ \to I$ starting at time $\kappa \in \mathbb{Z}$ with initial values $x(\kappa + j) = \xi_j$ for $-M \leq j \leq 0$ is the *general solution* of (2.1). It satisfies the *cocycle property*

$$\varphi(k; l, \varphi(l-M; \kappa, \xi), \ldots, \varphi(l; \kappa, \xi)) = \varphi(k; \kappa, \xi)$$
$$\text{for all } k \geq l - M,\ l \geq \kappa - M,$$

where we have abbreviated $\xi = (\xi_{-M}, \ldots, \xi_0) \in I^{M+1}$.

From the above it is clear that the natural state space for (2.1) is $M+1$-dimensional. In order to make this more precise, we introduce the mapping $\hat{f} : \mathbb{Z} \times I^{M+1} \to I^{M+1}$,

$$\hat{f}(k, x_{-M}, \dots, x_0) := \begin{pmatrix} x_{-M+1} \\ \vdots \\ x_0 \\ f(k, x_{-M}, \dots, x_{-1}, x_0) \end{pmatrix},$$

denoted as *delay endomorphism* associated with (2.1). Having this notion at hand, the delay difference equation (2.1) is equivalent to the first order difference equation

$$\boxed{x(k+1) = \hat{f}(k, x(k))} \tag{2.2}$$

in the higher dimensional set $I^{M+1} \subseteq \mathbb{R}^{M+1}$ in the following sense: The solutions $\phi : \mathbb{Z}^+_{\kappa - M} \to I$ of (2.1) and $\hat{\phi} : \mathbb{Z}^+_\kappa \to I^{M+1}$ of (2.2) are related by the identities

$$\phi_j(k) = \hat{\phi}_0(k+j) \quad \text{for all } -M \le j < 0, \qquad \phi_0(k) = \hat{\phi}_0(k) \tag{2.3}$$

and the general solution $\hat{\varphi}$ of (2.2) can be interpreted as discrete (2-parameter) semiflow of the delay difference equation (2.1).

In case the mapping f is linear, we can write (2.1) as

$$\boxed{x(k+1) = \sum_{j=-M}^{0} \ell_j(k) x(k+j)} \tag{2.4}$$

with sequences $\ell_{-M}, \dots, \ell_0 : \mathbb{Z} \to \mathbb{R}$, and borrowing terminology from [HNW93, pp. 402ff] we denote the delay endomorphism associated with the linear equation (2.4) as *companion matrix*

(2.5)

$$L(k) := \begin{pmatrix} 0 & 1 & & & \\ & 0 & 1 & & \\ & & \ddots & \ddots & \\ & & & 0 & 1 \\ \ell_{-M}(k) & \ell_{-M+1}(k) & \dots & \ell_{-1}(k) & \ell_0(k) \end{pmatrix} \quad \text{for all } k \in \mathbb{Z}.$$

The equivalence between (2.1) and (2.2) enables us to apply many results from the theory of nonautonomous dynamical systems to delay difference equations (see Appendix A for some terminology needed in the following). Here, due to the special algebraic structure of the delay endomorphism $\hat{f}$, certain assumptions can be simplified.

For the purpose of studying the dynamical behavior of (2.1) we restrict to a class of delay difference equations, whose solutions eventually enter a bounded set.

Definition 2.1. A delay difference equation (2.1) is said to be *pullback dissipative*, if I has a bounded subset $A \subseteq I$ such that for all bounded $B \subseteq I$ there exists a $N = N(B) \in \mathbb{Z}_0^+$ such that

$$\varphi(k; k-n, \xi) \in A \quad \text{for all } k \in \mathbb{Z},\ n \geq N,\ \xi \in B^{M+1}.$$

One denotes A as *absorbing set* of (2.1).

Remark 2.2. (1) The existence of an absorbing set $A \subseteq I$ has far reaching consequences. For a continuous right hand side $f : \mathbb{Z} \times I^{M+1} \to I$ one can prove the existence of a *global pullback attractor* $\mathcal{A}$ for (2.1) (cf. [Klo00, Theorem 3.6]). This is a uniquely determined nonautonomous set $\mathcal{A} \subseteq \mathbb{Z} \times A^{M+1}$ with the following properties:

(a) $\mathcal{A}(k)$ is compact,
(b) $\mathcal{A}(k) = \hat{\varphi}(k; \kappa, \mathcal{A}(\kappa))$ for all $\kappa \leq k$ (*invariance*), and
(c) $\lim_{n\to\infty} \operatorname{dist}(\hat{\varphi}(\kappa, \kappa-n, \xi), \mathcal{A}(\kappa-n)) = 0$ for all $\kappa \in \mathbb{Z}$, $\xi \in I^{M+1}$ (*attractivity*).

Dynamically the pullback attractor $\mathcal{A}$ consists of all pairs $(\kappa, \xi_{-M}, \ldots, \xi_0) \in \mathcal{A}$ such that there exists a bounded solution $\phi : \mathbb{Z} \to I$ of (2.1) with $\phi(\kappa - M) = \xi_{-M}, \ldots, \phi(\kappa) = \xi_0$. In particular, $\mathcal{A}$ contains stationary, periodic, homoclinic and heteroclinic solutions of (2.1).

(2) For bounded intervals I the delay difference equation (2.1) is trivially pullback dissipative with absorbing set I.

§3. Dissipative delay difference equations

For the specific scalar delay difference equation

$$\boxed{x(k+1) = \lambda x(k) + g(k, x(k-M), \ldots, x(k-1), x(k)),} \tag{3.1}$$

we can deduce dissipativity criteria under the assumption $\lambda \in (0, 1)$ and $g : \mathbb{Z} \times I^{M+1} \to I$. They are based on a simple lemma guaranteeing forward boundedness of solutions.

Lemma 3.1. *For each $\kappa \in \mathbb{Z}$ the following holds:*

(a) *If there exists a $K^+ \geq 0$ such that $g(k, x_{-M}, \ldots, x_0) \leq K^+$ for all $k \in \mathbb{Z}_\kappa^+$ and $x_{-M}, \ldots, x_0 \in I$, then the general solution φ of (3.1) satisfies*

$$\varphi(k; \kappa, \xi_{-M}, \ldots, \xi_0) \leq \lambda^{k-\kappa}\xi_0 + \frac{K^+}{1-\lambda} \quad \textit{for all } k \in \mathbb{Z}_\kappa^+,\ \xi_{-M}, \ldots, \xi_0 \in I.$$

(b) *If* $-I \subseteq I$ *and there exists a* $K^- \geq 0$ *such that* $-K^- \leq g(k, x_{-M}, \dots, x_0)$ *for all* $k \in \mathbb{Z}_\kappa^+$ *and* $x_{-M}, \dots, x_0 \in I$, *then the general solution* φ *of* (3.1) *satisfies*

$$-\lambda^{k-\kappa}\xi_0 - \frac{K^-}{1-\lambda} \leq \varphi(k; \kappa, \xi_{-M}, \dots, \xi_0)$$

for all $k \in \mathbb{Z}_\kappa^+$, $\xi_0, \dots, \xi_{-M} \in I$.

Proof. Let $\kappa \in \mathbb{Z}$ and abbreviate the tupel $\xi = (\xi_{-M}, \dots, \xi_0) \in I^{M+1}$.

(a) Define $\psi(k) := \lambda^{\kappa-k}\varphi(k, \kappa, \xi)$. Then one has

$$\begin{aligned} \psi(j+1) - \psi(j) &= \lambda^{\kappa-j-1}\left(\varphi(j+1; \kappa, \xi) - \lambda\varphi(j; \kappa, \xi)\right) \\ &\overset{(3.1)}{=} \lambda^{\kappa-j-1} g(j, \varphi(j; \kappa, \xi), \dots, \varphi(j-M; \kappa, \xi)) \\ &\leq K^+\lambda^{\kappa-j-1} \end{aligned}$$

for all $j \in \mathbb{Z}_\kappa^+$ and summation yields

$$\begin{aligned} \psi(k) - \psi(\kappa) &= \sum_{j=\kappa}^{k-1}(\psi(j+1) - \psi(j)) \\ &\leq \frac{K^+}{\lambda}\sum_{j=0}^{k-\kappa-1}\lambda^{-j} = \frac{K^+}{\lambda - 1}\left(1 - \lambda^{\kappa-k}\right) \end{aligned}$$

for all $k \in \mathbb{Z}_\kappa^+$. The definition of ψ implies our claimed estimate.

(b) If $\varphi(\cdot; \kappa, \xi)$ is the general solution of (3.1), then $-\varphi(\cdot; \kappa, -\xi)$ is the general solution of the transformed delay difference equation

$$x(k+1) = \lambda x(k) - g(k, -x(k), -x(k-1), \dots, -x(k-M)).$$

Utilizing this observation we obtain from (a) that $-\varphi(k; \kappa, -\xi) \leq -\lambda^{k-\kappa}\xi_0 + \frac{K^-}{1-\lambda}$ for all $k \in \mathbb{Z}_\kappa^+$ and this establishes the estimate in (b). Q.E.D.

Proposition 3.2 (pullback dissipativity). (a) *If* $I \subseteq [0, \infty)$ *and if there exists a* $K^+ \geq 0$ *such that* $0 \leq g(k, x_{-M}, \dots, x_0) \leq K^+$ *for all* $k \in \mathbb{Z}$ *and* $x_{-M}, \dots, x_0 \in I$, *then* (3.1) *is pullback dissipative with an absorbing set* $A = [0, R^+]$, *where* $R^+ > \frac{K^+}{1-\lambda}$.

(b) *If* $-I \subseteq I$ *and if there exist* $K^\pm \geq 0$ *such that* $-K^- \leq g(k, x_{-M}, \dots, x_0) \leq K^+$ *for all* $k \in \mathbb{Z}$ *and* $x_{-M}, \dots, x_0 \in I$, *then* (3.1) *is pullback dissipative with an absorbing set* $A = [R^-, R^+]$, *where* $R^+ > \frac{K^+}{1-\lambda}$ *and* $R^- < \frac{-K^-}{1-\lambda}$.

Remark 3.3. (1) In both cases the delay difference equation (3.1) possesses a global pullback attractor, provided $g : \mathbb{Z} \times I^{M+1} \to I$ is continuous (cf. Remark 2.2).

(2) A further dissipativity criterion for (3.1) can be obtained by an application of [CM04, Theorem 5.2] to the associated delay endomorphism. In addition, this enables us to weaken the boundedness assumption on g.

Proof. Let $B \subseteq I$ be bounded, i.e., $B \subseteq [b_-, b_+]$ with reals $b_- < 0 < b_+$, and choose the initial state $\xi = (\xi_{-M}, \dots, \xi_0) \in B^{M+1}$.

(a) The above Lemma 3.1(a) implies

$$0 \leq \varphi(k; k-n, \xi) \leq \lambda^n \xi_0 + \frac{K^+}{1-\lambda} \leq \lambda^n b_+ + \frac{K^+}{1-\lambda}$$

and due to $\lambda \in (0,1)$ there exists a $N = N(B) \in \mathbb{Z}_0^+$ such that $\lambda^n b_+ \in \left[0, R^+ - \frac{K^+}{1-\lambda}\right]$ for $n \geq N$, thus $\varphi(k; k-n, B) \subseteq [0, R^+]$ for all $k \in \mathbb{Z}$, $n \geq N$.

(b) This can be shown analogously to (a) using Lemma 3.1(a) and (b). Q.E.D.

Now we present some dissipative delay difference equations. The *Mackey-Glass equation* models dynamics of *haematopoiesis*, i.e., white blood cell production in the human body (cf. [MG77]).

Example 3.4 (Mackey-Glass equation). Let $I = [0, \infty)$ and $(\beta_k)_{k\in\mathbb{Z}}$ be a bounded sequence in I. The discrete nonautonomous *Mackey-Glass equation* is given by

$$x(k+1) = \lambda x(k) + \frac{\beta_k}{1 + x(k-M)^p},$$

where $g : \mathbb{Z} \times I \to \mathbb{R}$, $g(k,y) := \frac{\beta_k}{1+y^p}$ and the parameter $p > 0$. From Proposition 3.2(a) we obtain that $[0, R]$ is an absorbing set for $R > \frac{1}{1-\lambda} \sup_{k\in\mathbb{Z}} \beta_k$. A further delay difference equation suggested by Mackey and Glass reads as

$$x(k+1) = \lambda x(k) + \frac{\beta_k x(k-M)}{1 + x(k-M)^p},$$

where $g : \mathbb{Z} \times I \to \mathbb{R}$, $g(k,y) := \frac{\beta_k y}{1+y^p}$ and parameters $p > 1$. Since the function $y \mapsto \frac{y}{1+y^p}$ has the maximal value $\frac{1}{p}(p-1)^{1-1/p}$ for $y = (p-1)^{-1/p}$ we derive from Proposition 3.2(a) that the interval $[0, R]$ is an absorbing set for $R > \frac{1}{p(1-\lambda)}(p-1)^{1-1/p} \sup_{k\in\mathbb{Z}} \beta_k$.

Another important model fitting into our approach is the discrete *Wazewska-Czyzewska and Lasota equation* describing the erythropoietic (red blood-cell) system (cf. [LWC76]), i.e., the survival of red blood-cells in an animal.

Example 3.5 (Wazewska-Lasota equation). Let $I = [0,\infty)$ and $(\beta_k)_{k\in\mathbb{Z}}$, $(\gamma_k)_{k\in\mathbb{Z}}$ be two sequences in I, where $(\beta_k)_{k\in\mathbb{Z}}$ is assumed to be bounded. We investigate the equation

$$x(k+1) = \lambda x(k) + \beta_k e^{-\gamma_k x(k-M)},$$

where $g : \mathbb{Z} \times I \to \mathbb{R}$, $g(k,y) := \beta_k e^{-\gamma_k y}$. From Proposition 3.2(a) we deduce that $[0,R]$ is an absorbing set, if one has $R > \frac{1}{1-\lambda}\sup_{k\in\mathbb{Z}} \beta_k$.

Our final example will be a delay difference equation defined on the reals. It is a discrete model for the behavior of a single, self-excitatory neuron with graded delayed response (cf. [Her94]).

Example 3.6. Let $I = \mathbb{R}$ and $(\beta_k)_{k\in\mathbb{Z}}$, $(\gamma_k)_{k\in\mathbb{Z}}$ be real sequences with bounded $(\beta_k)_{k\in\mathbb{Z}}$. Consider the discrete delay difference equation

$$x(k+1) = \lambda x(k) + \beta_k \tanh(\gamma_k x(k-M)),$$

where $g : \mathbb{Z} \times \mathbb{R} \to \mathbb{R}$, $g(k,y) := \beta_k \tanh(\gamma_k y)$. Using Proposition 3.2(b) we see that the compact interval $[-R,R]$ is an absorbing set, if $R > \frac{1}{1-\lambda}\sup_{k\in\mathbb{Z}} |\beta_k|$.

Remark 3.7. For the autonomous version of the delay difference equations introduced in Example 3.4–3.6 one can use a criterion of [Iva94, Invariance property] to show that the respective minimal absorbing intervals are invariant.

§4. Asymptotic equivalence for delay difference equations

In this section we consider delay difference equations

$$\boxed{x(k+1) = \sum_{j=-M}^{0} \ell_j(k)x(k+j) + g(k, x(k-M), \dots, x(k))} \tag{4.1}$$

with $\ell_{-M}, \dots, \ell_0 : \mathbb{Z} \to \mathbb{R}$ and a function $g : \mathbb{Z} \times \mathbb{R}^{M+1} \to \mathbb{R}$. The linear part (2.4) is said to have an *exponential dichotomy*, if the associated companion matrix L (see (2.5)) satisfies the conditions given in Definition B.1. In particular for an autonomous linear part (2.4), an exponential dichotomy is determined by the roots of the characteristic polynomial

$$z^{M+1} - \sum_{j=-M}^{0} \ell_j z^{M+j} = 0$$

(cf. Appendix B), which are the eigenvalues of the constant companion matrix L.

Now we are in a position to state our main result. Its assumptions guarantee the existence of an exponentially attractive positively invariant submanifold of the state space for (4.1), which dominates the dynamical behavior. At first glance Theorem 4.1 seems technical, but an interpretation will be given below before we prove it.

Theorem 4.1 (asymptotic equivalence). *Assume the following holds for* (4.1)*:*

(i) *The linear delay difference equation* (2.4) *admits an exponential dichotomy with growth rates* $0 < \Lambda < \lambda \leq 1$*, constants* $K^{\pm} \geq 1$ *and projectors* $P_{\pm}$ *with* $N_{\pm} \equiv \dim \operatorname{im} P_{\pm}(k)$.

(ii) *One has* $\sup_{k\in\mathbb{Z}} |g(k,0,\ldots,0)| < \infty$ *and with functions* $l^{\pm} : [0,\infty) \to [0,\infty)$ *the following local Lipschitz conditions hold for all* $k \in \mathbb{Z}$, $r \geq 0$*:*

$$\max_{j=-M}^{0} |P_{\pm}(k+1)_{j,0}|\, |g(k,x) - g(k,\bar{x})| \leq l^{\pm}(r)\, \|x - \bar{x}\| \quad \textit{for all } x, \bar{x} \in B_r. \tag{4.2}$$

(iii) *The nonlinear delay equation* (4.1) *is pullback absorbing with absorbing set* $A \subseteq \mathbb{R}$.

Choose $\rho > 0$ *so large that* $\bar{A}^{M+1} \subseteq B_\rho$ *and suppose the following* spectral gap condition*:*

$$K^- l^-(\rho) + K^+ l^+(\rho) + \max\left\{K^- l^-(\rho), K^+ l^+(\rho)\right\} < \frac{\lambda - \Lambda}{4}, \tag{4.3}$$

$$4K^- l^-(\rho) < \lambda. \tag{4.4}$$

Then there exists a Lyapunov transformation $T : \mathbb{Z} \to \mathbb{R}^{(M+1)\times(M+1)}$*, a nonempty open nonautonomous set* $\mathcal{O} \subseteq \mathcal{P}_-$ *and a continuous function* $w : \mathcal{O} \to \mathbb{R}^{M+1}$ *with* $w(k,x) \in \mathcal{P}_+(k)$, $\operatorname{Lip}_2 w < 1$ *such that the following holds:*

(a) *For every solution* $\phi : \mathbb{Z}^+_{\kappa-M} \to I$ *of the delay difference equation* (4.1) *there exists a constant* $C \geq 0$ *and a further solution* $\phi^* : \mathbb{Z}^+_{\kappa^*-M} \to I$ *of* (4.1) *satisfying*

$$\|\phi(k) - \phi^*(k)\| \leq C\gamma^{k-\kappa} \quad \textit{for all } k \in \mathbb{Z}^+_{\kappa^*}, \tag{4.5}$$

where $\gamma \in (\Lambda, 1)$, $\kappa \leq \kappa^*$ *and* $(\phi^*(\kappa^* - M), \ldots, \phi^*(\kappa^*))^T = \eta + w(\kappa, \eta)$ *with* $\eta \in \mathcal{O}(\kappa^*)$.

(b) *The solution* ϕ^* *allows the representation*

$$\begin{pmatrix} \phi^*(k-M) \\ \vdots \\ \phi^*(k)) \end{pmatrix} = T(k)^{-1}\psi(k;\kappa^*, T(\kappa^*)\eta) + w(k, T(k)^{-1}\psi(k;\kappa^*, T(\kappa^*)\eta))$$

for all $k \geq \kappa^*$, *where* ψ *is the general solution of the reduced equation*

$$\begin{pmatrix} y_{-M}(k+1) \\ \vdots \\ y_{-N_-}(k+1) \\ y_{1-N_-}(k+1) \\ \vdots \\ y_0(k+1) \end{pmatrix} = T(k+1)L(k) \begin{pmatrix} 0 \\ \vdots \\ 0 \\ y_{1-N_-}(k) \\ \vdots \\ y_0(k) \end{pmatrix} + G(k, y_{1-N_-}(k), \ldots, y_0(k)) \tag{4.6}$$

and using $S(k) := T(k)^{-1}$ *we have defined the nonlinearity*

$$G(k, y_{1-N_-}, \ldots, y_0) := g\left(k, S(k)\begin{pmatrix} 0 \\ \vdots \\ 0 \\ y_{1-N_-} \\ \vdots \\ y_0 \end{pmatrix} + w\left(k, S(k)\begin{pmatrix} 0 \\ \vdots \\ 0 \\ y_{1-N_-} \\ \vdots \\ y_0 \end{pmatrix}\right)\right)\begin{pmatrix} 0 \\ \vdots \\ 0 \\ S_-(k+1)_{1-N_-,0} \\ \vdots \\ S_-(k+1)_{0,0} \end{pmatrix}.$$

In order to provide an interpretation of Theorem 4.1 we remark that the canonical state space for the reduced equation (4.7) is $\mathbb{R}^{N_-}$, since its solutions are uniquely determined by N_- initial values. The relation (4.5) guarantees that every solution ϕ of (4.1) is asymptotically equivalent to another solution ϕ^* of (4.1), which is uniquely driven by the first order difference equation (4.7). Consequently, the long-term behavior of the delay difference equation (4.1) with state space $\mathbb{R}^{M+1}$ is ultimately determined by the N_--dimensional problem (4.7).

Remark 4.2. While it is a problem of linear algebra to construct the Lyapunov transformation $T : \mathbb{Z} \to \mathbb{R}^{(M+1)\times(M+1)}$ from the dichotomy data for L (see [Pöt98, p. 166, Lemma A.6.1]), it is practically impossible to derive an explicit expression for the mapping w, which satisfies the *invariance equation* (cf. (C.5)) for $\kappa \in \mathbb{Z}$, $\eta = (\eta_{-M}, \dots, \eta_0) \in \mathcal{P}_-(\kappa)$ given by

$$
\begin{aligned}
w\Bigg(\kappa + 1, &\eta_{-M+1}, \dots, \eta_0, \sum_{j=-M}^{0} f_j(\kappa)\eta_j + g(\kappa, \eta + w(\kappa, \eta))\Bigg)_i \\
&= w(\kappa, \eta)_{i+1} \quad \text{for all } - M \le i < 0, \\
w\Bigg(\kappa + 1, &\eta_{-M+1}, \dots, \eta_0, \sum_{j=-M}^{0} f_j(\kappa)\eta_j + g(\kappa, \eta + w(\kappa, \eta))\Bigg)_0 \\
&= \sum_{j=-M}^{0} \ell_j(\kappa) w(\kappa, \eta)_j + g(\kappa, \eta + w(\kappa, \eta)) \sum_{j=-M}^{0} P_+(\kappa + 1)_{j,0}.
\end{aligned}
$$

Nonetheless, under differentiability assumptions on the nonlinearity g one can use this functional equation to compute Taylor approximations to w. It is worth to point out that is a dynamical and not an algebraic problem (see [PR05b, PR05b]). Such approximations seem of little use in the present more global setting. Therefore, Theorem 4.1 is a primarily theoretical result.

Proof of Theorem 4.1. We successively establish both assertions.

(*a*) Our main technical tool in the proof will be Theorem C.1. Thereto, in this first step we are going to show that the delay endomorphism associated with (4.1) satisfies its assumptions. We consider the first order difference equation (C.1), where $L(k)$ is the companion matrix of (2.4) and nonlinearity $F(k, x) := (0, \dots, 0, g(k, x_{-M}, \dots, x_0))^T$.

- ad $(H)_1$: This is just the present assumption (i). Note that we have $\lambda \le 1$.
- ad $(H)_2$: Obviously, (C.2) is satisfied and, thanks to the particular algebraic structure of the mapping F, the local Lipschitz estimate (C.3) reduces to (4.2).
- ad $(H)_3$: From (iii) it is easy to see that the nonautonomous set $\mathbb{Z} \times A^{M+1}$ is uniformly pullback absorbing w.r.t. (C.1).

Consequently, Theorem C.1 guarantees the existence of an attractive nonautonomous set $\mathcal{W} \subseteq \mathbb{Z} \times \mathbb{R}^{M+1}$ being graph of a function w with properties stated in Theorem C.1(a). Let $\phi : \mathbb{Z}^+_{\kappa-M} \to I$ be a solution of (4.1) and $\hat{\phi}$ be the associated solution of (C.1) by virtue of the identities (2.3). Referring to Theorem C.1(b) there exist $\kappa^* \in \mathbb{Z}^+_\kappa$, $\eta^* \in \mathcal{P}_-(\kappa^*)$ such that (C.6) holds with $\xi^* = \eta^* + w(\kappa^*, \eta^*)$. Then the last component $\phi^* := \hat{\varphi}(\cdot; \kappa^*, \xi^*)_0$ is a solution of (4.1) and the asymptotic equivalence (4.5) holds true. Thus, we have shown assertion (a).

(*b*) Reflecting our above construction, the solution ϕ^* allows the representation (cf. (2.3))

$$(\phi^*(k-M), \ldots, \phi^*(k))^T = \tilde{\varphi}(k; \kappa^*, \eta^*) + w(k, \tilde{\varphi}(k; \kappa^*, \eta^*)) \quad \text{for all } k \geq \kappa^*,$$

where $\tilde{\varphi}$ is the general solution of the reduced equation (C.7) in the pseudo-unstable bundle $\mathcal{P}_-$. Referring to, for instance [Pöt98, p. 33, Satz 1.5.7], we know from the exponential dichotomy assumption (i) that the companion matrix (2.5) is kinematically similar to a block-diagonal matrix. In particular, there exists a bounded sequence $T : \mathbb{Z} \to \mathbb{R}^{(M+1)\times(M+1)}$ of invertible matrices, such that also the sequence of inverses $T(\cdot)^{-1} : \mathbb{Z} \to \mathbb{R}^{(M+1)\times(M+1)}$ is bounded, and one has

$$T(k)P_-(k)T(k)^{-1} \equiv \begin{pmatrix} 0_{N_+} & \\ & I_{N_-} \end{pmatrix} \quad \text{on } \mathbb{Z}.$$

Thus, the transformation $y = T(k)x$ brings (C.7) into the form (4.7) and the general solution ψ of this N_--dimensional equation satisfies $\psi(k; \kappa^*, T(\kappa)\eta^*) = T(k)\tilde{\varphi}(k; \kappa^*, \eta^*)$ for all $k \geq \kappa^*$. Q.E.D.

As further illustration of Theorem 4.1 we discuss a class of examples sufficiently large to include the Examples 3.4–3.6. On the other hand, it is simple enough to illuminate our lengthy, but quantitative assumptions, in particular on the exponential dichotomy.

Example 4.3. Let $\lambda \in (0, 1)$ and $g_0 : \mathbb{Z} \times \mathbb{R} \to \mathbb{R}$ be a function such that the following holds:

- g_0 is bounded, i.e., $|g_0(k, x)| \leq K$ for all $k \in \mathbb{Z}$, $x \in \mathbb{R}$,
- the partial derivative $D_2 g_0$ exists and with some $R > \frac{K}{1-\lambda}$ one has

$$l := \sup_{k\in\mathbb{Z}} \sup_{|x|\leq R} |D_2 g_0(k, x)| < \infty.$$

These assumptions and Proposition 3.2(b) guarantee that the single delay difference equation

$$x(k+1) = \lambda x(k) + g_0(k, x(k-M)) \tag{4.7}$$

is pullback dissipative with absorbing set $[-R, R]$. The associated constant companion matrix[1]

$$L = \begin{pmatrix} 0 & 1 & & \\ & \ddots & \ddots & \\ & & & 1 \\ & & & \lambda \end{pmatrix} \in \mathbb{R}^{(M+1)\times(M+1)}$$

has the spectrum $\sigma(L) = \{0, \lambda\}$, which implies that L possesses a (Λ, λ)-decomposition for any real $\Lambda \in (0, \lambda)$. Moreover, L admits an exponential dichotomy with growth rates Λ, λ, constants $K^{\pm} = 2(1+\lambda^{-M})\Lambda^{-M}$ and constant projectors

$$P_+ = \begin{pmatrix} 1 & 0 & & \dots & -\lambda^{-M} \\ & 1 & 0 & \dots & -\lambda^{-M+1} \\ & & \ddots & \ddots & \vdots \\ & & & 1 & -\lambda^{-1} \\ & & & & 0 \end{pmatrix}, \quad P_- = \begin{pmatrix} 0 & 0 & & \dots & \lambda^{-M} \\ & 0 & 0 & \dots & \lambda^{-M+1} \\ & & \ddots & \ddots & \vdots \\ & & & 0 & \lambda^{-1} \\ & & & & 1 \end{pmatrix}$$

with $\dim \operatorname{im} P_- = 1$. Having this information available, it is not difficult to verify that our spectral gap conditions (4.3) and (4.4) reduce to

$$48(1+\lambda^{-M})(\lambda\Lambda)^{-M} l < \lambda - \Lambda, \qquad 8(1+\lambda^{-M})(\lambda\Lambda)^{-M} l < \lambda,$$

respectively. Provided these conditions hold, Theorem 4.1 ensures that the dynamical behavior of the delay difference equation (4.7) is asymptotically equivalent to a scalar first order difference equation (note dim im $P_- = 1$). However, this example clearly demonstrates that the assumptions (4.3)–(4.4) are rather restrictive and indicates that large delays require very weak nonlinearities, i.e., small local Lipschitz constants for g_0.

§5. Strongly stable multi-step methods

Our second application of Theorem C.1 comes from the numerical integration of ordinary differential equations (see, for instance, [HNW93]). Thanks to a theoretically interesting result of Kirchgraber (cf. [Kir86],

---

[1]Note that this particular splitting of (4.7) into linear part and nonlinearity is quite arbitrary and not required from Theorem 4.1. Another choice would be writing (4.7) as $x(k+1) = \lambda_0 x(k) + g(k, x(k-M), x(k))$ with nonlinearity $g(k, x_{-M}, x_0) = (\lambda - \lambda_0)x_0 + g_0(k, x_{-M})$, where $\lambda_0 \in (0, 1)$. Then we obtain a different gap condition.

see also [Sto93]) we know that strongly stable multi-step methods are asymptotically equivalent to one-step methods, hence essentially one-step methods. Here we will indicate how to generalize this to multi-step methods with varying step-sizes.

To keep our presentation accessible and simple we restrict to scalar, yet nonautonomous ordinary differential equations

$$\boxed{\dot{x} = f(t,x),} \tag{5.1}$$

equipped with an initial condition $x(\tau) = \xi$ with $\tau, \xi \in \mathbb{R}$. For further simplicity let us suppose $f : \mathbb{R} \times \mathbb{R} \to \mathbb{R}$ is continuous with

$$\operatorname{Lip}_2 f < \infty, \qquad \sup_{t \in \mathbb{R}} |f(t,0)| < \infty;$$

then the general solution $\chi(\cdot;\tau,\xi) : \mathbb{R} \to \mathbb{R}$ of (5.1) is well-defined (cf., e.g., [Har64]).

In order to discretize the problem (5.1) we prescribe a discrete set of time steps $\{t_k\}_{k\in\mathbb{Z}}$ satisfying $t_0 = \tau$, $t_k < t_{k+1}$ for all $k \in \mathbb{Z}$ and define the ratios $\omega_k := \frac{t_{k+1}-t_k}{t_k - t_{k-1}}$. Note that for constant step-sizes one has the identity $\omega_k \equiv 1$ on $\mathbb{Z}$. Following [HNW93, p. 396, Section III.5], a *linear variable step-size multi-step method* to solve (5.1) is a difference equation

$$\begin{aligned} x(k+1) = &- \sum_{j=-M}^{0} \alpha_{M+j}(\omega_k, \dots, \omega_{k-M}) x(k+j) \\ &+ (t_{k+1} - t_k) \sum_{j=-M}^{0} \beta_{M+j}(\omega_k, \dots, \omega_{k-M}) f(t_{k+j}, x(k+j)), \end{aligned} \tag{5.2}$$

which obviously fits in the framework of our delay difference equation (4.1) with

$$\ell_j(k) = -\,\alpha_{M+j}(\omega_k, \dots, \omega_{k-M}) \quad \text{for all } -M \le j \le 0,\ k \in \mathbb{Z},$$

$$g(k, x_{-M}, \dots, x_0) = (t_{k+1} - t_k) \sum_{j=-M}^{0} \beta_{M+j}(\omega_k, \dots, \omega_{k-M}) f(t_{k+j}, x_j).$$

Then the values $\varphi(k; 0, \xi_{-M}, \cdot, \xi_{-1}, \xi)$ obtained from (5.2) approximate the solution $\chi(\cdot;\tau,\xi)$ of the ordinary differential equation (5.1) at times $t = t_k$.

For explicit examples of variable step-size multi-step methods we again refer to [HNW93, p. 396, Section III.5] and remark that the following Hypothesis is typically satisfied.

Hypothesis 5.1. *Suppose the multi-step method* (5.2) *satisfies:*

(o) *There exists a* $H > 0$ *and a neighborhood* $\Omega \subseteq (0, \infty)$ *of* 1 *such that*

$$0 < t_{k+1} - t_k \leq H, \qquad \omega_k \in \Omega \quad \text{for all } k \in \mathbb{Z},$$

(i) $\alpha_0, \dots, \alpha_M : \Omega^{M+1} \to \mathbb{R}$ *are continuous and* $\beta_0, \dots, \beta_M : \Omega^{M+1} \to \mathbb{R}$ *are bounded,*

(ii) *the method* (5.2) *is of* order $p \geq 0$, *i.e.,*

$$1 + \sum_{j=-M}^{0} \alpha_{M+j}(\omega_k, \dots, \omega_{k-M}) \equiv 0 \quad \text{on } \mathbb{Z},$$

(iii) *for* $\omega_k \equiv 1$ *the method* (5.2) *is* strongly stable, *i.e., all roots* $z_{-M}, \dots, z_0 \in \mathbb{C}$ *of*

$$z^{M+1} + \sum_{j=-M}^{0} \alpha_{M+j}(1, \dots, 1) z^{M+j} = 0$$

lie inside the open unit disc $B_1 \subseteq \mathbb{C}$ *except the simple root* $z_0 = 1$.

Remark 5.2. Let $L_0 \in \mathbb{R}^{(M+1)\times(M+1)}$ be the companion matrix of the constant step-size formula (where $\omega_k \equiv 1$). By Hypothesis 5.1(iii) we can transform L_0 into block diagonal form

$$T^{-1} L_0 T = \begin{pmatrix} \tilde{L}_0 & \\ & 1 \end{pmatrix}, \quad \tilde{L}_0 = \begin{pmatrix} z_{-M} & \varepsilon_{-M} & & \\ & \ddots & \ddots & \\ & & \ddots & \varepsilon_{-2} \\ & & & z_{-1} \end{pmatrix} \in \mathbb{R}^{M\times M}$$

with an invertible matrix $T \in \mathbb{R}^{(M+1)\times(M+1)}$, where $|\varepsilon_j| < 1 - |z_j|$ for all $-M \leq j < 0$ and

$$\|\tilde{L}_0\| < 1. \tag{5.3}$$

This is an easy consequence of the Jordan canonical form (yielding a transformation T_0 and values $\varepsilon_j \in \{0, 1\}$) and an appropriate multiplication of the columns of T_0 (yielding that (5.3) holds).

The following perturbation result ensures that the linear part of (5.2) admits an exponential dichotomy for weakly varying time steps.

Lemma 5.2. *Assume Hypothesis 5.1 holds and choose* $\Lambda \in \left(\|\tilde{L}_0\|, 1\right)$. *Then there exists a neighborhood* $\Omega_0 \subseteq \Omega$ *of* 1 *such that* $\omega_k \in \Omega_0$ *for all* $k \in \mathbb{Z}$ *implies that (2.4) admits an exponential dichotomy with growth rates* $\Lambda, 1$, *constants* $K^{\pm} = \|T\| \|T^{-1}\|$ *and constant projectors*

$$P_+ = T \begin{pmatrix} I_M & \\ & 0 \end{pmatrix} T^{-1}, \qquad P_- = T \begin{pmatrix} 0_M & \\ & 1 \end{pmatrix} T^{-1}$$

with $\dim \operatorname{im} P_- = 1$, *where* $T \in \mathbb{R}^{(M+1)\times(M+1)}$ *is the transformation matrix from Remark 5.2.*

Proof. Note that the last column of T, the eigenvector of L_0 corresponding to the eigenvalue 1, is given by $e_0 = (1, \dots, 1)^T$. In addition, by Hypothesis 5.1(ii) this vector $e_0 \in \mathbb{R}^{M+1}$ is also an eigenvector of each companion matrix $L(k)$, $k \in \mathbb{Z}$. Therefore, T transforms every $L(k)$ into block diagonal form

$$T^{-1} L(k) T = \begin{pmatrix} \tilde{L}(k) & \\ & 1 \end{pmatrix} \quad \text{for all } k \in \mathbb{Z},$$

where the matrix sequence $\tilde{L} : \mathbb{Z} \to \mathbb{R}^{M\times M}$ satisfies

$$\begin{aligned} \|\tilde{L}(k)\| &\le \|\tilde{L}_0\| + \|\tilde{L}(k) - \tilde{L}_0\| \le \|\tilde{L}_0\| + \|T^{-1}[L(k) - L_0]T\| \\ &\le \|\tilde{L}_0\| + \|T\| \|T^{-1}\| \sum_{j=-M}^{0} |\alpha_j(\omega_k, \dots, \omega_{k-M}) - \alpha_j(1, \dots, 1)| \\ &\qquad \text{for all } k \in \mathbb{Z}. \end{aligned}$$

Thus, by continuity of the functions $\alpha_j : \Omega^{M+1} \to \mathbb{R}$ (cf. Hypothesis 5.1(i)) and (5.3) we know that for each $\Lambda \in \left(\|\tilde{L}_0\|, 1\right)$ there exists a neighborhood $\Omega_0 \subseteq \Omega$ of 1 such that $\|\tilde{L}(k)\| \le \Lambda$, provided $\omega_k \in \Omega_0$ holds for all $k \in \mathbb{Z}$. Now it is straight forward to verify that the sequence L satisfies the dichotomy estimates (B.1). Q.E.D.

For the initialization of a multi-step method (5.2), one needs a *starting procedure* S. This is a mapping $S : \mathbb{Z} \times \mathbb{R} \to \mathbb{R}^{M+1}$ providing starting values $\xi_{-M}, \dots, \xi_0 \in \mathbb{R}$ for (5.2) according to

$$S(\kappa, \xi) = (\xi_{-M}, \dots, \xi_0);$$

i.e., for each incomplete initial condition $x(\kappa) = \xi$ for (5.2) the starting procedure S delivers a full set of initial values $\xi_{-M}, \dots, \xi_0$. Typically the value of a starting procedure will be the iterates of a one-step scheme to integrate (5.1).

Hypothesis 5.3. *Suppose the starting procedure $S : \mathbb{Z}\times\mathbb{R} \to \mathbb{R}^{M+1}$ satisfies for all $\kappa \in \mathbb{Z}$ that $P_-S(\kappa,\cdot) : \mathbb{R} \to \operatorname{im} P_-$ is a homeomorphism.*

Theorem 5.4. *Assume that Hypotheses 5.1, 5.4 hold, choose $\Lambda \in (\|\tilde{L}_0\|, 1)$ and that the discrete time steps are small and weakly varying in the sense of*

(5.4)

$$\sup_{(\theta_0,\dots,\theta_M)\in\Omega_0^{M+1}} \sum_{j=-M}^{0} |\beta_j(\theta_0,\dots,\theta_M)| \operatorname{Lip}_2 f\, H < \frac{1-\Lambda}{6K^+K^-}, \quad \omega_k \in \Omega_0$$

with constants $K^\pm \geq 1$ and the neighborhood $\Omega_0 \subseteq \Omega$ of 1 from Lemma 5.3. Then there exists a continuous function $w : \mathbb{Z} \times \operatorname{im} P_- \to \operatorname{im} P^+$ with $\operatorname{Lip}_2 w < 1$ and graph

$$\mathcal{W} := \left\{(\kappa, \eta + w(\kappa,\eta)) \in \mathbb{Z} \times \mathbb{R}^{M+1} : \eta \in \operatorname{im} P_-\right\},$$

as well as a scalar nonautonomous difference equation

$$\boxed{x(k+1) = G(k, x(k))} \tag{5.5}$$

with continuous right hand side $G : \mathbb{Z} \times \mathbb{R} \to \mathbb{R}$ such that the following holds for all $\kappa \in \mathbb{Z}$:

(a) *The multi-step method (5.2) (with starting procedure S) and the scalar difference equation (5.5) are conjugated on $\mathcal{W}$, i.e., their general solutions φ and ψ, respectively, satisfy*

$$S_k \circ \psi(k;\kappa,\cdot) = \begin{pmatrix} \varphi(k-M;\kappa,\cdot)\circ S_\kappa \\ \vdots \\ \varphi(k;\kappa,\cdot)\circ S_\kappa \end{pmatrix} \quad \text{for all } k \in \mathbb{Z}_\kappa^+$$

with a homeomorphism $S_k : \mathbb{R} \to \mathcal{W}(k)$,

(b) *there is a $\gamma \in (\Lambda, 1)$ such that for all $\xi_{-M},\dots,\xi_0 \in \mathbb{R}$ there exist a constant $C \geq 0$ and an initial value $\eta \in \mathbb{R}$ with*

$$\left\| \begin{pmatrix} \varphi(k-M;\kappa,\xi_{-M},\dots,\xi_0) \\ \vdots \\ \varphi(k;\kappa,\xi_{-M},\dots,\xi_0) \end{pmatrix} - S_k\psi(k;\kappa,\eta) \right\| \leq C\gamma^{k-\kappa} \quad \text{for all } k \in \mathbb{Z}_\kappa^+.$$

Proof. In this proof we can use the global version of Theorem C.1 described in Remark C.2. Thereto, let L be the companion matrix associated with the multi-step method (5.2), define the mapping $F : \mathbb{Z} \times \mathbb{R}^{M+1} \to \mathbb{R}^{M+1}$ by

$$F(k,x) := \left(0,\dots,0,(t_{k+1}-t_k)\sum_{j=-M}^{0}\beta_{M+j}(\omega_k,\dots,\omega_{k-M})f(t_{k+j},x_j)\right)^T .$$

and $\hat{\varphi}$ is the general solution of (C.1). The assumptions of Theorem C.1 (or Remark C.2.) hold:

- ad $(H)_1$: By Lemma 5.3 the companion matrix L admits an exponential dichotomy with $\Lambda, 1$ and $\dim \operatorname{im} P_- = 1$.
- ad $(H)_2$: An easy estimate shows that (5.4) implies (C.2), while the other assumptions are not necessary in the present global setting.

Thus, there exists a function w with graph $\mathcal{W}$ as claimed above. Let $\kappa \in \mathbb{Z}$ be given.

(a) With the starting procedure S from Hypothesis 5.4 we define the mapping $S_\kappa : \mathbb{R} \to \mathcal{W}(\kappa)$ by $S_\kappa(x) := P_-S(\kappa,x) + w(\kappa, P_-S(\kappa,x))$. Thanks to Hypothesis 5.4 and $\operatorname{Lip}_2 w < 1$ we know that S_κ is a homeomorphism. Thus, let us define the right hand side $G : \mathbb{Z} \times \mathbb{R} \to \mathbb{R}$ of (5.5) as $G(\kappa,x) := S_{\kappa+1}^{-1}(L(\kappa)S_\kappa(x) + F(\kappa, S_\kappa(x)))$, which is obviously continuous. Mathematical induction yields $S_k \circ \psi(k;\kappa,\cdot) = \hat{\varphi}(k;\kappa,\cdot) \circ S_\kappa$ for $k \in \mathbb{Z}_\kappa^+$ and using (2.3) we get assertion (a).

(b) Let $\xi_{-M},\dots,\xi_0 \in \mathbb{R}$ and set $\xi = (\xi_{-M},\dots,\xi_0)$. Referring to Theorem C.1(b) there exists a unique initial value $\xi^* \in \mathcal{W}(\kappa)$ such that

$$\|\hat{\varphi}(k;\kappa,\xi) - \hat{\varphi}(k;\kappa,\xi^*)\| \le C\gamma^{k-\kappa} \quad \text{for all } k \in \mathbb{Z}_\kappa^+$$

(due to our global assumptions we have $\kappa^* = \kappa$). Then (b) follows with $\eta := S_\kappa^{-1}(\xi^*)$. Q.E.D.

§Appendix A. Nonautonomous dynamics

Let $D \subseteq \mathbb{R}^N$ be nonempty. We consider a first order nonautonomous difference equation

$$\text{(A.1)} \qquad \boxed{x(k+1) = f(k, x(k))}$$

with *right hand side* $f : \mathbb{Z} \times D \to D$. A sequence $\phi : \mathbb{Z}_\kappa^+ \to D$, $\kappa \in \mathbb{Z}$, satisfying the identity $\phi(k+1) \equiv f(k,\phi(k))$ on $\mathbb{Z}_\kappa^+$ is called *solution* of (A.1). We define the *general solution* $\varphi(\cdot;\kappa,\xi)$, $\xi \in D$, of (A.1) as unique solution satisfying the initial condition $x(\kappa) = \xi$. We remark that in general $\varphi(k;\kappa,\cdot)$ does not exist for $k < \kappa$.

A set $\mathcal{A} \subseteq \mathbb{Z} \times D$ is called *nonautonomous set* with *k-fiber* $\mathcal{A}(k) := \{x \in D : (k,x) \in \mathcal{A}\}$ for $k \in \mathbb{Z}$. Such a set $\mathcal{A}$ is said to be *positively invariant* w.r.t. (A.1), if $f(k, \mathcal{A}(k)) \subseteq \mathcal{A}(k+1)$ holds, and it is called *invariant*, if one has equality $f(k, \mathcal{A}(k)) = \mathcal{A}(k+1)$ for $\kappa \leq k$. Moreover, we denote (A.1) as difference equation *in* $\mathcal{A}$, if $\mathcal{A}$ is positively invariant.

§Appendix B. Discrete dichotomies

We deal with matrix-valued sequences $L : \mathbb{Z} \to \mathbb{R}^{N \times N}$ with *evolution operator*

$$\Phi(k,\kappa) := \begin{cases} I_N & \text{for } k = \kappa \\ L(k-1) \cdots L(\kappa) & \text{for } k > \kappa \end{cases};$$

if the linear mapping $L(k) : \mathbb{R}^N \to \mathbb{R}^N$, $k \in \mathbb{Z}$, is invertible (possibly between appropriate subspaces of $\mathbb{R}^N$), then $\Phi(k,\kappa) := L(k)^{-1} \cdots L(\kappa - 1)^{-1}$ for $k < \kappa$.

Definition B.1 (exponential dichotomy). Let $0 < \Lambda < \lambda$ and $K^{\pm} \geq 1$ be given. Then a matrix sequence $L : \mathbb{Z} \to \mathbb{R}^{N \times N}$ is said to possess an *exponential dichotomy*, if there exist complementary projections $P_{\pm} : \mathbb{Z} \to \mathbb{R}^{N \times N}$ with $P_-(k+1)L(k) = L(k)P_-(k)$, the mappings

$$L(k)|_{\operatorname{im} P_-(k)} : \operatorname{im} P_-(k) \to \operatorname{im} P_-(k+1)$$

are invertible with associate evolution operator $\bar{\Phi}(k,\kappa)$, and the dichotomy estimates hold:

(B.1)
$$\|\Phi(k,l)P_+(l)\| \leq K^+ \Lambda^{k-l}, \quad \left\|\bar{\Phi}(l,k)P_-(k)\right\| \leq K^- \lambda^{l-k} \quad \text{for all } l \leq k.$$

Of particular importance are the two nonautonomous sets

$$\mathcal{P}_- := \left\{(k,x) \in \mathbb{R}^N : x \in \operatorname{im} P_-(k)\right\},$$
$$\mathcal{P}_+ := \left\{(k,x) \in \mathbb{R}^N : x \in \operatorname{im} P_+(k)\right\},$$

denoted as *pseudo-unstable* and *pseudo-stable bundle*, respectively.

In general it is difficult to determine wether a given matrix sequence admits an exponential dichotomy. Nonetheless, for the special cases of constant and periodic matrices a dichotomy is fully determined by spectral properties. Thereto, we need certain preliminaries from linear algebra (cf. [HS74, pp. 109–133]).

For given $0 < \Lambda < \lambda$ we say a matrix $T \in \mathbb{R}^{N \times N}$ possesses a *(Λ, λ)-decomposition*, if the disjoint sets $\sigma^+ := \{\nu \in \sigma(T) : |\nu| \leq \Lambda\}$, $\sigma^- := \{\nu \in \sigma(T) : \lambda \leq |\nu|\}$ are nonempty with $\sigma(T) = \sigma^+ \cup \sigma^-$, i.e.,

$\sigma(T)$ can be separated by an annulus with center 0 and radii $\Lambda < \lambda$. This at hand, we introduce the direct sums

$$V_T^{\pm} := \bigoplus_{\substack{\nu\in\sigma^{\pm},\\ \Im\nu=0}} \ker\left(T-\nu I_N\right)^N \oplus \bigoplus_{\substack{\nu\in\sigma^{\pm},\\ \Im\nu>0}} \ker\left(T^2-2\Re\nu T+|\nu|^2 I_N\right)^N$$

and integers $n_{\pm} := \dim V_T^{\pm}$. Let $\{x_1^{\pm}, \dots, x_{n_{\pm}}^{\pm}\}$ be a basis of $V_T^{\pm}$. Using the invertible matrix $C := (x_1^+, \dots, x_{n_+}^+, x_1^-, \dots, x_{n_-}^-) \in \mathbb{R}^{N\times N}$ we introduce the complementary projections

$$Q_T^+ := C\begin{pmatrix} I_{n_+} & \\ & 0_{n_-}\end{pmatrix}C^{-1}, \qquad Q_T^- := C\begin{pmatrix} 0_{n_+} & \\ & I_{n_-}\end{pmatrix}C^{-1},$$

fulfilling $\ker Q_T^{\pm} = V_T^{\mp}$ and $\operatorname{im} Q_T^{\pm} = V_T^{\pm}$.

Constant matrix sequences: Suppose $L(k)$ is independent of $k \in \mathbb{Z}$, i.e., $L(k) \equiv L$. An eigenvalue ν of $L \in \mathbb{R}^{N\times N}$ is said to be *semisimple*, if its algebraic and geometric multiplicities coincide. If the matrix L possesses a (Λ, λ)-decomposition and eigenvalues of L with modulus Λ and λ are semisimple, then L possesses an exponential dichotomy with growth rates Λ, λ and constant invariant projectors $P_{\pm} = Q_A^{\pm}$ (cf. [PR05a, Proposition 2.1]).

Periodic matrix sequences: Suppose $L(k)$ is ω-periodic, i.e., $L(k) \equiv L(k+\omega)$ on $\mathbb{Z}$. Note that the matrix $M_\omega(k) := \Phi(k+\omega, k)$, $k \in \mathbb{Z}$, has the same eigenvalues as the so-called *monodromy matrix* $M_\omega(0)$. They are the *Floquet multipliers* of L. If the monodromy matrix $M_\omega(0)$ possesses an $(\Lambda^\omega, \lambda^\omega)$-decomposition and the Floquet multipliers with modulus Λ^ω and λ^ω are semisimple, then L possesses an exponential dichotomy with growth rates Λ, λ and ω-periodic invariant projectors $P_{\pm}(k) := Q_{M_\omega(k)}^{\pm}$ for all $k \in \mathbb{Z}$ (cf. [PR05a, Proposition 2.2]).

§Appendix C. Attractive fiber bundles

Now we turn to nonlinear first order difference equations

$$\text{(C.1)} \qquad \boxed{x(k+1) = L(k)x(k) + F(k, x(k)),}$$

where $L : \mathbb{Z} \to \mathbb{R}^{N\times N}$ and $F : \mathbb{Z}\times\mathbb{R}^N \to \mathbb{R}^N$ specifies the nonlinearity. For the general solution of (C.1) we write $\hat{\varphi}$. Then our invariant manifold theorem reads as follows:

Theorem C.1 (attractive fiber bundles). *Assume the following holds for* (C.1)*:*

$(H)_1$ *The sequence $L : \mathbb{Z} \to \mathbb{R}^{N\times N}$ has an exponential dichotomy with projectors $P_\pm$, growth rates $0 < \Lambda < \lambda$ and constants $K^\pm$.*

$(H)_2$ *One has the growth condition*

$$\sup_{k<\kappa} \|F(k,0)\| \, \lambda^{\kappa-k} < \infty \quad \textit{for all } \kappa \in \mathbb{Z} \tag{C.2}$$

and there exist functions $l^\pm : [0,\infty) \to [0,\infty)$ such that for all $r > 0$, $k \in \mathbb{Z}$ we have

(C.3)
$$\|P_\pm(k+1)\,[F(k,x) - F(k,\bar{x})]\| \le l^\pm(r)\,\|x - \bar{x}\| \quad \textit{for all } x, \bar{x} \in B_r.$$

$(H)_3$ *The difference equation* (C.1) *possesses a* uniformly pullback absorbing set $\mathcal{A} \subseteq \mathbb{Z} \times \mathbb{R}^N$, *i.e., $\mathcal{A}$ is bounded and for every nonempty bounded subset $\mathcal{B} \subseteq \mathbb{Z}\times\mathbb{R}^N$ there exists $N = N(\mathcal{B}) \in \mathbb{Z}_0^+$ such that*

$$\hat{\varphi}(k; k-n, \mathcal{B}(k-n)) \subseteq \mathcal{A}(k) \quad \textit{for all } k \in \mathbb{Z},\ n \ge N.$$

Choose $\rho > 0$ so large that $\bar{\mathcal{A}} \subseteq \mathbb{Z}\times B_\rho$ and suppose the following spectral gap condition*:*

$$K^- l^-(\rho) + K^+ l^+(\rho) + \max\left\{K^- l^-(\rho), K^+ l^+(\rho)\right\} < \tfrac{\lambda-\Lambda}{4},$$
$$4K^- l^-(\rho) < \lambda. \tag{C.4}$$

Then there exists a nonautonomous set $\mathcal{W} \subseteq \mathbb{Z} \times \mathbb{R}^N$ (denoted as attractive invariant fiber bundle*), which is positively invariant w.r.t.* (C.1), *and possesses the following properties:*

(a) *$\mathcal{W}$ is graph of a function $w : \mathcal{O} \to \mathbb{R}^N$ over a nonempty open nonautonomous set $\mathcal{O} \subseteq \mathcal{P}_-$, i.e., $\mathcal{W} = \{(\kappa, \eta + w(\kappa,\eta)) : (\kappa,\eta) \in \mathcal{O}\}$, the functions $w(\kappa,\cdot) : \mathcal{O}(\kappa) \to \mathcal{P}_+(\kappa)$ are well-defined and satisfy:*

  (a_1) *They are globally Lipschitzian with* $\operatorname{Lip}_2 w < 1$,

  (a_2) *one has the functional equation* (invariance equation)

$$w(\kappa+1, \eta_1) = L(\kappa) w(\kappa,\eta) + P_+(\kappa+1) F(\kappa, \eta + w(\kappa,\eta)), \tag{C.5}$$

for all $(\kappa,\eta) \in \mathcal{O}$ such that $\eta_1 := L(\kappa)\eta + F(\kappa, \eta + w(\kappa,\eta)) \in \mathcal{O}(\kappa+1)$,

(b) *$\mathcal{W}$ is* asymptotically complete, *i.e., for every pair $(\kappa,\xi) \in \mathbb{Z} \times \mathbb{R}^N$ there exists a point $(\kappa^*, \xi^*) \in \mathcal{W}$ with $\kappa \le \kappa^*$ such that*

$$\|\hat{\varphi}(k;\kappa,\xi) - \hat{\varphi}(k;\kappa^*,\xi^*)\| \le C\gamma^{k-\kappa} \quad \textit{for all } k \in \mathbb{Z}_{\kappa^*}^+, \tag{C.6}$$

and a $\gamma \in (\Lambda, \lambda)$, where the real $C \ge 0$ depends boundedly on κ, ξ.

(c) *The nonautonomous difference equation*

$$x(k+1) = L(k)x(k) + P_-(k+1)F\big(k, x(k) + w(k, x(k))\big) \tag{C.7}$$

in the pseudo-unstable bundle $\mathcal{P}_-$ *is denoted as* reduced equation *of* (C.1) *and the general solution* $\tilde{\varphi}$ *of* (4.7) *is related to* $\hat{\varphi}$ *by*

$$\hat{\varphi}(k;\kappa,\eta + w(\kappa,\eta)) = \tilde{\varphi}(k;\kappa,\eta) + w(k,\tilde{\varphi}(k;\kappa,\eta)) \quad \textit{for all } \eta \in \mathcal{O}(\kappa).$$

Remark C.2. If the Lipschitz constants $l^{\pm} : [0,\infty) \to [0,\infty)$ are bounded, it is possible to disclaim assumptions $(H)_3$, (C.4) and to deduce a stronger global version of Theorem C.1: Then the function w is defined on $\mathcal{P}_-$ and in Theorem C.1(b) the point $(\kappa^*, \xi^*) \in \mathcal{W}$ is uniquely determined with $\kappa^* = \kappa$ (cf. [Pöt07a, Proposition 2.1 and Theorem 2.5]).

Proof. The above theorem is a special case of [Pöt07b, Theorem 4.1]. Q.E.D.

References

[CM04] D. N. Cheban and C. Mammana, Invariant manifolds, global attractors and almost periodic solutions of nonautonomous difference equations, Nonlinear Anal., **56** (2004), 465–484.

[Har64] P. Hartman, Ordinary Differential Equations, John Wiley & Sons, New York, 1964.

[Her94] A. V. M. Herz, Global analysis of recurrent neural networks, In: Models of Neural Networks, Vol. 3, (eds. E. Domany, J.L. van Hemman and K. Schulten), Springer, New York, 1994.

[HNW93] E. Hairer, S. P. Nørsett and G. Wanner, Solving Ordinary Differential Equations I — Nonstiff Problems, 2nd revised ed., Springer, Berlin-Heidelberg-New York, 1993.

[HS74] M. W. Hirsch and S. Smale, Differential Equations, Dynamical Systems, and Linear Algebra, Academic Press, Boston-New York-San Diego, 1974.

[Iva94] A. F. Ivanov, On global stability in a nonlinear discrete model, Nonlinear Anal., **23** (1994), 1383–1389.

[Kir86] U. Kirchgraber, Multi-step methods are essentially one-step methods, Numer Math., **48** (1986), 85–90.

[Klo00] P. E. Kloeden, Pullback attractors in nonautonomous difference equations, J. Difference Equations Appl., **6** (2000), 33–52.

[KS78] U. Kirchgraber and E. Stiefel, Methoden der analytischen Störungsrechnung und ihre Anwendungen (in German), B. G. Teubner, Stuttgart, 1978.

[LWC76] A. Lasota and M. Wazewska-Czyzewska, Matematyczne problemy dynamiki ukladu krwinek czerwonych (in Polish), Mat. Stosowana, **6** (1976), 23–40.

[MG77] M. C. Mackey and L. Glass, Oscillation and chaos in physiological control systems, Science, **197** (1977), 287–289.

[NS92] K. Nipp and D. Stoffer, Attractive invariant manifolds for maps: Existence, smoothness and continuous dependence on the map, Research Report 92-11, Seminar für Angewandte Mathematik, ETH Zürich, 1992.

[Pöt98] C. Pötzsche, Nichtautonome Differenzengleichungen mit invarianten und stationären Mannigfaltigkeiten (in German), thesis, Univ. Augsburg, 1998.

[Pöt07a] ———, Attractive invariant fiber bundles, Appl. Anal., **86** (2007), 687–722.

[Pöt07b] ———, Discrete inertial manifolds, submitted (2007).

[PR05a] C. Pötzsche and M. Rasmussen, Local approximation of invariant fiber bundles: An algorithmic approach, In: Difference Equations and Discrete Dynamical Systems, (eds. L. J. S. Allen, B. Aulbach, S. Elaydi and R. J. Sacker), World Scientific, New Jersey, London, 2005, pp. 155–170.

[PR05b] ———, Taylor approximation of invariant fiber bundles, Nonlinear Anal., **60** (2005), 1303–1330.

[Sto93] D. Stoffer, General linear methods: connection to one step methods and invariant curves, Numer. Math., **64** (1993), 395–407.

[SY02] G. R. Sell and Y. You, Dynamics of evolutionary equations, Appl. Math. Sci., **143**, Springer, Berlin-New York-Heidelberg, 2002.

Christian Pötzsche
School of Mathematics
University of Minnesota
Minneapolis MN, 55455
USA

E-mail address: `christian.poetzsche@ma.tum.de`

Advanced Studies in Pure Mathematics 53, 2009
Advances in Discrete Dynamical Systems
pp. 261–269

A Willett type criterion with the best possible constant for linear dynamic equations

Pavel Řehák

Abstract.

We establish oscillation criteria for the linear dynamic equation $(r(t)y^{\Delta})^{\Delta} + p(t)y^{\sigma} = 0$. These criteria can be understood as an extension of the classical Willett criterion. What is special on these new results is that the constant involved in the criteria, which is equal to the "magic" 1/4 in the differential equations case, is in fact no more constant. In general case, it depends on the asymptotic behavior of the coefficients p, r, and primarily on the asymptotic behavior of graininess. In addition, we prove that the value of this new "constant" is the best possible.

§1. Introduction

Consider the linear dynamic equation

$$(r(t)y^{\Delta})^{\Delta} + p(t)y^{\sigma} = 0, \tag{1}$$

where $r(t) > 0$ and $p(t)$ are rd-continuous functions defined on a time scale interval $[a, \infty)$, $a \in \mathbb{T}$, and a time scale $\mathbb{T}$ is assumed to be unbounded from above.

If $\mathbb{T} = \mathbb{R}$, then (1) reduces to the Sturm-Liouville differential equation

$$(r(t)y')' + p(t)y = 0. \tag{2}$$

Received October 12, 2007.
2000 *Mathematics Subject Classification.* 34C10, 39A11, 39A12, 39A13.
Supported by the Grants KJB1019407 of the Grant Agency of ASCR and 201/04/0580 of the Czech Grant Agency, and by the Institutional Research Plan AV0Z010190503.

It is known, see e.g. [14], that under the assumptions $\int_a^\infty 1/r(s)\,ds = \infty$ and $\int_t^\infty p(s)\,ds \geq 0$ ($\not\equiv 0$) for large t, equation (2) is oscillatory provided

$$\liminf_{t\to\infty} \left(\int_t^\infty p(s)\,ds\right)^{-1} \int_t^\infty \frac{1}{r(s)} \left(\int_s^\infty p(\tau)\,d\tau\right)^2 ds > \frac{1}{4}.$$

The constant 1/4 on the right-hand side is the best possible. For historical reasons, we call this criterion as of Willett type.

The aim of this paper is to show how this criterion can be extended to linear dynamic equation (1). In particular, we will see that the constant 1/4 is no more constant in general setting. In fact, in our criterion for (1), the new "constant" depends on the asymptotic behavior of the coefficients p, r and primarily on the asymptotic behavior of graininess. In addition, we will prove that this new constant is the best possible under quite mild assumptions. Thus a situation may happen (with $\mathbb{T} \neq \mathbb{R}$) where the constant is strictly greater than 1/4, but still sharp.

Oscillatory properties of (1) or of closely related objects have been studied e.g. in [1, 2, 4, 5, 6, 7, 8, 9, 11, 12, 13]. Recall that (1) covers many of well-studied equations, like linear differential equations, linear difference equations, linear h-difference equations, and linear q-difference equations. Comparing our results with the existing ones, they are new even in the discrete case ($\mathbb{T} = \mathbb{Z}$) as well as in many other cases. There is a result in [12], which is very close to our new one, namely a Willett type criterion for half-linear dynamic equation. However, only the case which corresponds to the constant 1/4 is discussed there. The newness of the presented result lies particularly in considering general case, where graininess becomes to play very important role, and this makes the problem quite complex. Thanks to this general setting, much wider class of equations can be examined.

The paper is organized as follows. In the next section we recall some important concepts and state preliminary results that are crucial to prove the main results. Generalized Willett type theorems are presented in the third section. Both cases are examined, $\int_a^\infty 1/r(s)\,\Delta s = \infty$ and $\int_a^\infty 1/r(s)\,\Delta s < \infty$.

§2. Basic concepts and preliminary results

We assume that the reader is familiar with the notion of time scales. Thus note just that $\mathbb{T}$, σ, f^σ, μ, f^Δ, and $\int_a^b f^\Delta(s)\,\Delta s$ stand for time scale, forward jump operator, $f \circ \sigma$, graininess, delta derivative of f, and delta integral of f from a to b, respectively. See [10], which is the initiating paper of the time scale theory written by Hilger, and the monographs [2, 3] by Bohner and Peterson containing a lot of information on time scale calculus.

We will proceed with some essentials of oscillation theory of (1). First note that we are interested only in nontrivial solutions of (1). We say that a solution y of (1) has a *generalized zero* at t in case $y(t) = 0$. If $\mu(t) > 0$, then we say that y has a *generalized zero* in $(t, \sigma(t))$ in case $y(t)y^\sigma(t) < 0$. A nontrivial *solution* y of (1) is called *oscillatory* if it has infinitely many generalized zeros; note that the uniqueness of IVP excludes the existence of a cluster point which is less than ∞. Otherwise it is said to be *nonoscillatory.* In view of the fact that the Sturm type separation theorem extends to (1) (see e.g. [11]), we have the following equivalence: One solution of (1) is oscillatory if and only if every solution of (1) is oscillatory. Hence we may speak about *oscillation* or *nonoscillation of equation* (1). Recall that the principal statements, like the Sturmian theory (Reid type roundabout theorem, Sturm type separation and comparison theorems) for (1), can be established under the mere assumption $r(t) \neq 0$ and the basic concepts, especially generalized zero, have to be adjusted, see e.g. [1] or [11].

The next lemma, called the function sequence technique, plays a crucial role in proving the main results. Its proof is based on the equivalence between nonoscillation of (1) and solvability of the Riccati type integral inequality

$$w(t) \geq \int_t^\infty p(s)\,\Delta s + \int_t^\infty w^2(s)/(r(s) + \mu(s)w(s))\,\Delta s.$$

Lemma 1 ([12]). *Assume that* $\int_a^\infty 1/r(s)\,\Delta s = \infty$, $\int_t^\infty p(s)\,\Delta s \geq 0$ *and nontrivial for large* t. *Define the function sequence* $\{\psi_k(t)\}$ *by*

$$\psi_0(t) = \int_t^\infty p(s)\,\Delta s, \quad \psi_1(t) = \int_t^\infty \frac{\psi_0^2(s)}{r(s) + \mu(s)\psi_0(s)}\,\Delta s,$$

$$\psi_{k+1}(t) = \int_t^\infty \frac{(\psi_0(s) + \psi_k(s))^2}{r(s) + \mu(s)(\psi_0(s) + \psi_k(s))}\,\Delta s, \quad k = 1, 2, \ldots.$$

Then equation (1) *is nonoscillatory if and only if there exists* $t_0 \in [a, \infty)$ *such that* $\lim_{k\to\infty} \psi_k(t) = \psi(t)$ *for* $t \geq t_0$, *i.e., the sequence* $\{\psi_k(t)\}$ *is well defined and pointwise convergent.*

The following lemma, which exploits the transformation of dependent variable, will be useful in the case when $\int_a^\infty 1/r(s)\,\Delta s$ converges.

Lemma 2 ([9]). *Assume that* h *is an rd-continuously delta differentiable function with* $h(t) \neq 0$. *Then* $y = hu$ *transforms equation* (1) *into the equation* $(\tilde{r}(t)u^\Delta)^\Delta + \tilde{p}(t)u^\sigma = 0$ *with* $\tilde{r} = rhh^\sigma$ *and*

$\tilde{p} = h^\sigma[(rh^\Delta)^\Delta + ph^\sigma]$. *This transformation preserves oscillatory properties.*

§3. Main results

Now we are ready to prove the main result, an extension of the Willett criterion.

Theorem 1. *Let*

$$\int_a^\infty \frac{1}{r(s)}\,\Delta s = \infty. \tag{3}$$

Assume that

$$\int_t^\infty p(s)\,\Delta s \ge 0 \quad \textit{and nontrivial for large } t. \tag{4}$$

Denote

$$N^* := \limsup_{t\to\infty} \frac{\mu(t)\int_t^\infty p(s)\,\Delta s}{r(t)} \tag{5}$$

and let $N^* < \infty$. *If*

(6)

$$\liminf_{t\to\infty} \left(\int_t^\infty p(s)\,\Delta s\right)^{-1} \int_t^\infty \frac{1}{r(s)} \left(\int_s^\infty p(\tau)\,\Delta\tau\right)^2 \Delta s > \frac{(N^*+1)^2}{4},$$

then equation (1) *is oscillatory.*

Moreover, if there exists the limit

$$N := \lim_{t\to\infty} \frac{\mu(t)\int_t^\infty p(s)\,\Delta s}{r(t)} < \infty, \tag{7}$$

then the constant $(N+1)^2/4$ *in* (6) *is the best possible.*

Proof. We will apply Lemma 1 and use its notation. Condition (6) can be rewritten as

$$\int_t^\infty \psi_0^2(s)/r(s)\,\Delta s \ge \delta\psi_0(t)$$

for large t, say $t \ge t_0 \ge a$, where $\delta > (N^*+1)^2/4$. Then

$$\begin{aligned}
\psi_1(t) &= \int_t^\infty \frac{\psi_0^2(s)}{r(s)} \cdot \frac{r(s)}{r(s)+\mu(s)\psi_0(s)}\,\Delta s \\
&\ge \inf_{s\ge t_0} \frac{r(s)}{r(s)+\mu(s)\psi_0(s)} \int_t^\infty \frac{\psi_0^2(s)}{r(s)}\,\Delta s \\
&\ge \Gamma(t_0,1)\delta\psi_0(t) = \delta_1\psi_0(t),
\end{aligned}$$

where $\delta_1 = \Gamma(t_0, 1)\delta$ and

$$\Gamma(t_0, x) = \inf_{t \geq t_0} \frac{1}{1 + x\mu(t)\psi_0(t)/r(t)}.$$

Further, since $x \mapsto x^2/(y + zx)$ is increasing for $x > 0, y > 0, z > 0$, similarly as above we have

$$\begin{aligned}\psi_2(t) &\geq (\delta_1 + 1)^2 \int_t^\infty \frac{\psi_0^2(s)}{r(s) + (\delta_1 + 1)\mu(s)\psi_0(s)} \Delta s \\ &\geq (\delta_1 + 1)^2 \Gamma(t_0, \delta_1 + 1)\delta\psi_0(t) = \delta_2\psi_0(t),\end{aligned}$$

where $\delta_2 = (\delta_1 + 1)^2 \Gamma(t_0, \delta_1 + 1)\delta$. Similarly, by induction,

$$\psi_k(t) \geq \delta_k \psi_0(t), \quad k = 1, 2, \ldots, \tag{8}$$

where $\delta_1 = \Gamma(t_0, 1)\delta$ and

$$\delta_{k+1} = (\delta_k + 1)^2 \Gamma(t_0, \delta_k + 1)\delta, \quad k = 1, 2, \ldots. \tag{9}$$

First assume that there exists the limit $N = \lim_{t\to\infty} \mu(t) \int_t^\infty p(s)\Delta s / r(t)$. Then taking the limit as $t_0 \to \infty$, (9) yields

$$\delta_{k+1} = \frac{(\delta_k + 1)^2}{1 + (\delta_k + 1)N}\delta. \tag{10}$$

Since $x \mapsto (x+1)^2/[1 + (x+1)N]$ is increasing, $\delta_k \leq \delta_{k+1}$ $[\delta_k \geq \delta_{k+1}]$ implies $\delta_{k+1} \leq \delta_{k+2}$ $[\delta_{k+1} \geq \delta_{k+2}]$. Hence, $\{\delta_k\}$ is monotone. We claim that $\delta_k \to \infty$ as $k \to \infty$. If not, let $\delta_k \to L < \infty$ as $k \to \infty$. Clearly, we cannot have $L = 0$, otherwise, from (10), $0 = 1/(1+N) > 0$. Thus $L > 0$. From (10) we obtain

$$L = \frac{(L + 1)^2}{1 + (L + 1)N}\delta. \tag{11}$$

Next we show that (11) has no real solution. Indeed, (11) can be rewritten as

$$(N - \delta)L^2 + (1 + N - 2\delta)L - \delta = 0. \tag{12}$$

Note that $N \neq \delta$, otherwise, in view of $\delta > (N+1)^2/4$, we get $(N-1)^2 < 0$, contradiction. The discriminant of (12) is equal to $(N+1)^2 - 4\delta$, and so it is negative since we have assumed $\delta > (N+1)^2/4$. This proves that $\delta_k \to \infty$ as $k \to \infty$, which implies $\psi_k(t) \to \infty$ as $k \to \infty$ for $t \geq t_0$, by (8). Consequently, (1) is oscillatory, in view of Lemma 1. Next we examine the case when

$$\liminf_{t\to\infty} \frac{\mu(t)\int_t^\infty p(s)\Delta s}{r(t)} < \limsup_{t\to\infty} \frac{\mu(t)\int_t^\infty p(s)\Delta s}{r(t)} = N^*.$$

Then we have

$$\lim_{t_0\to\infty} \Gamma(t_0, \delta_k + 1) = \frac{1}{1 + (\delta_k + 1)N^*}$$

and hence, using the same arguments as above, we come to the conclusion that $\delta_k \to \infty$ as $k \to \infty$. This implies again oscillation of (1).

In the second part we prove that the constant $(N+1)^2/4$ in (6) is the best possible provided (7) holds. To show this, consider the Euler type dynamic equation

$$y^{\Delta\Delta} + \frac{\lambda}{t\sigma(t)} y^\sigma = 0, \tag{13}$$

where λ is a positive real parameter, i.e., equation (1) where $r(t) \equiv 1$ and $p(t) = \lambda/(t\sigma(t))$. Assume that there exists the limit $M := \lim_{t\to\infty} \mu(t)/t < \infty$. Then, for N defined by (7), we have $N = \lambda M$. Further,

$$\begin{aligned}
&\left(\int_t^\infty p(s)\,\Delta s\right)^{-1} \int_t^\infty \frac{1}{r(s)} \left(\int_s^\infty p(\tau)\,\Delta\tau\right)^2 \Delta s \\
&= \frac{t}{\lambda} \int_t^\infty \frac{\lambda^2}{s^2}\,\Delta s \\
&= \lambda t \int_t^\infty \frac{1}{s\sigma(s)} \cdot \frac{\sigma(s)}{s}\,\Delta s \\
&\geq \lambda \inf_{s\geq t} \frac{\sigma(s)}{s}.
\end{aligned}$$

In order to be (6) fulfilled, it suffices to take λ such that

$$\lambda \liminf_{t\to\infty} \frac{\sigma(t)}{t} > \frac{(N+1)^2}{4}. \tag{14}$$

Since $\sigma(t)/t = 1 + \mu(t)/t \to 1 + M$ as $t \to \infty$ and $N = \lambda M$, (14) is equivalent to

$$\lambda(M+1) > \frac{(\lambda M + 1)^2}{4}. \tag{15}$$

If $M=0$, then clearly $\lambda>1/4$ implies oscillation of (13) by the first part of Theorem 1. Solving quadratic inequality (15) when $M>0$ and using the first part of Theorem 1 combined with the Sturm type comparison theorem, we find that (13) is oscillatory provided

$$\lambda>\frac{M+2-2\sqrt{M+1}}{M^2}=\frac{\left(\sqrt{M+1}-1\right)^2}{M^2}=\frac{1}{\left(\sqrt{M+1}+1\right)^2}.$$

Altogether, for any $M\in[0,\infty)$ we have that $\lambda>\left(\sqrt{M+1}+1\right)^{-2}$ implies oscillation of (13). From [13] we know that $\lambda_0=\left(\sqrt{M+1}+1\right)^{-2}$ is the oscillation constant for (13), i.e., (13) is oscillatory for all $\lambda>\lambda_0$ and nonoscillatory for all $\lambda<\lambda_0$. Now, since $N=\lambda M$, we see that N cannot be lowered, and this proves that $(N+1)^2/4$ in (14) and so in (6) is the best possible. Q.E.D.

Using the transformation of dependent variable and Theorem 1 we can easily treat the complementary case to (3), namely $\int_a^\infty 1/r(s)\,\Delta s$ converges.

Theorem 2. *Let*

$$\int_a^\infty \frac{1}{r(s)}\,\Delta s<\infty.$$

Assume that

$$\int_t^\infty\left(\int_{\sigma(s)}^\infty\frac{1}{r(\tau)}\,\Delta\tau\right)^2 p(s)\,\Delta s\geq 0 \quad \textit{and nontrivial for large } t.$$

Denote $R(t):=\int_t^\infty 1/r(s)\,\Delta s$. *Let*

$$\tilde{N}^*:=\limsup_{t\to\infty}\frac{\mu(t)\int_t^\infty(R^\sigma(s))^2p(s)\,\Delta s}{r(t)R(t)R^\sigma(t)}<\infty.$$

If

$$\text{(16)}\quad \liminf_{t\to\infty}\left(\int_t^\infty(R^\sigma(s))^2p(s)\,\Delta s\right)^{-1}\int_t^\infty\frac{\left(\int_s^\infty(R^\sigma(\tau))^2p(\tau)\,\Delta\tau\right)^2}{r(s)R(s)R^\sigma(s)}\,\Delta s$$
$$>\frac{\tilde{N}^*+1}{4},$$

then equation (1) *is oscillatory.*

Moreover, if there exists the limit

$$\tilde{N}:=\lim_{t\to\infty}\frac{\mu(t)\int_t^\infty(R^\sigma(s))^2p(s)\,\Delta s}{r(t)R(t)R^\sigma(t)}<\infty,$$

then the constant $(\tilde{N}+1)^2/4$ *in* (16) *is the best possible.*

Proof. Denote $R(t) := \int_t^\infty 1/r(s)\,\Delta s$. First note that by Lemma 2, the transformation $y = hu$ with $h(t) = R(t)$ transforms (1) into the equation $(\tilde{r}(t)u^\Delta)^\Delta + \tilde{p}(t)u^\sigma = 0$, where $\tilde{r}(t) = R(t)R^\sigma(t)r(t)$ and $\tilde{p}(t) = (R^\sigma(t))^2 p(t)$. Since $(1/R(t))^\Delta = 1/\tilde{r}(t)$, we get that $\int_a^\infty 1/\tilde{r}(s)\,\Delta s = \infty$. Applying now Theorem 1 to the transformed equation and using the fact that oscillatory properties are preserved after the transformation, we get the statement. Q.E.D.

References

[1] R. P. Agarwal and M. Bohner, Quadratic functionals for second order matrix equations on time scales, Nonlinear Anal., **33** (1998), 675–692.

[2] M. Bohner and A. C. Peterson, Dynamic Equations on Time Scales: An Introduction with Applications, Birkhäuser, Boston, 2001.

[3] M. Bohner and A. Peterson, Advances in Dynamic Equations on Time Scales, Birkhäuser, Boston, 2003.

[4] M. Bohner and S. H. Saker, Oscillation of second order nonlinear dynamic equations on time scales, Rocky Mountain J. Math., **34** (2004), 1239–1254.

[5] M. Bohner and M. Ünal, Kneser's theorem in q-calculus, J. Phys. A, **38** (2005), 6729–6739.

[6] O. Došlý and S. Hilger, A necessary and sufficient condition for oscillation of the Sturm-Liouville dynamic equation on time scales, J. Comput. Appl. Math., **141** (2002), 147–158.

[7] L. H. Erbe and S. Hilger, Sturmian theory on measure chains, Differential Equations Dynam. Systems, **1** (1993), 223–244.

[8] L. H. Erbe and A. C. Peterson, Some recent results in linear and nonlinear oscillation, Dynam. Systems Appl., **13** (2004), 381–395.

[9] L. H. Erbe, A. C. Peterson and P. Řehák, Integral comparison theorems for second order linear dynamic equations, In: Difference Equations, Special Functions and Orthogonal Polynomials, Proceedings of the International Conference, Munich, Germany, 25-30 July 2005, (eds. S. Elaydi, J. Cushing, R. Lasser, A. Ruffing, V. Papageorgiou and W. Van Assche), World Scientific, 2007, pp. 497–506.

[10] S. Hilger, Analysis on measure chains — a unified approach to continuous and discrete calculus, Results Math., **18** (1990), 18–56.

[11] P. Řehák, Half-linear dynamic equations on time scales: IVP and oscillatory properties, Nonlinear Funct. Anal. Appl., **7** (2002), 361–404.

[12] P. Řehák, Function sequence technique for half-linear dynamic equations on time scales, Panamer. Math. J., **16** (2006), 31–56.

[13] P. Řehák, How the constants in Hille–Nehari theorems depend on time scales, Adv. Difference Equ., **2006** (2006), 1–15.

[14] D. Willett, On the oscillatory behavior of the solutions of second order linear differential equations, Ann. Polon. Math., **21** (1969), 175–194.

Pavel Řehák
Mathematical Institute
Academy of Sciences of the Czech Republic
Žižkova 22, CZ-61662 Brno,
Czech Republic

E-mail address: rehak@math.muni.cz

Advanced Studies in Pure Mathematics 53, 2009
Advances in Discrete Dynamical Systems
pp. 271–282

On asymptotic stability of linear stochastic Volterra difference equations with respect to a fading perturbation

John A. D. Appleby[1], Markus Riedle and Alexandra Rodkina

Abstract.

The paper concerns necessary and sufficient conditions on the fading intensity of a state–independent stochastic perturbation for the asymptotic stability of a linear stochastic Volterra difference equation. In broad terms, it is shown here that the results obtained in the deterministic case are robust to fading stochastic perturbations which are independent of the state, once it is known that these perturbations fade more rapidly than an identifiable critical rate.

§1. Introduction

This note can be considered as an extension of the continuous–time approaches and results of the paper [1] to the discrete case, in which necessary and sufficient conditions for the stability of linear Itô–Volterra equations with deterministically fading noise intensity is obtained. Similar conditions on the noise intensity are established in the continuous case in [2]. Almost sure asymptotic stability results for autonomous scalar non-linear difference equations without delays were obtained in [3] and in [4].

The outline of the note is as follows. Section 2 lists and comments upon the general results obtained. Section 3 shows how the general

---

Received October 12, 2007.

Key words and phrases. Volterra difference equation, stochastic Volterra difference equation, resolvent, almost sure asymptotic stability.

[1]The first author was partially supported by an Albert College Fellowship awarded by Dublin City University's Research Advisory Panel. The third author was partially supported by Programme Committee of the 11th International Conference of Difference Equations and by a New Initiative Fellowship awarded by University if the West Indies.

results can be applied to specific problems, and shows the connections between conditions required for the almost sure asymptotic stability of linear Volterra difference equations and corresponding continuous time Itô–Volterra equations driven by Brownian motion. Finally, Section 4 contains proofs of the results.

§2. Results

The following standard notation is employed. $\mathbb{N}$ denotes the natural numbers, $\mathbb{Q}^+$ the positive rational numbers, $\mathbb{R}$ the real numbers, and $\mathbb{R}^r$ r–dimensional real space, for $r \in \mathbb{N}$. If E is a subset of $\mathbb{R}$ then the characteristic function $\chi_E : \mathbb{R} \to \mathbb{R}$ is defined by

$$\chi_E(x) = \begin{cases} 1, & \text{if } x \in E, \\ 0, & \text{if } x \notin E. \end{cases}$$

The standard basis vectors in $\mathbb{R}^r$ are denoted by $\mathbf{e}_1, \dots, \mathbf{e}_r$. The standard innerproduct of x and y in $\mathbb{R}^r$ is denoted by $\langle x, y \rangle$. The $d \times d$ real valued identity matrix is denoted by I_d. We say that a real sequence $a = \{a(n) : n \in \mathbb{N}\}$ obeys $a \in \ell^1(\mathbb{N}; \mathbb{R})$ if $\sum_{n \in \mathbb{N}} |a(n)| < \infty$. We say that the $d \times r$ matrix–valued sequence $a = \{a(n) = (a(n))_{ij} : n \in \mathbb{N}\}$ is in $\ell^1(\mathbb{N}; \mathbb{R}^d \times \mathbb{R}^r)$ if each entry $a_{ij} \in \ell^1(\mathbb{N}; \mathbb{R})$, $i = 1, \dots, d$, $j = 1, \dots, r$. The convolution of two sequences $f = \{f(n) : n \in \mathbb{N}\}$ and $g = \{g(n) : n \in \mathbb{N}\}$, $f * g$, is a sequence defined by

$$(f * g)(n) = \sum_{k=0}^{n} f(n-k) g(k), \quad n \in \mathbb{N}.$$

The $*$ notation is also used here for the convolution of distribution functions; however, the type of convolution being used will be clear from the context.

We consider the stochastic Volterra difference equation

(2.1)

$$X(n+1) = X(n) + \sum_{k=0}^{n} A(n-k) X(k) + \sigma(n) \xi(n+1), \quad n \in \mathbb{N}; \quad X(0) = \zeta.$$

Here $A \in \ell^1(\mathbb{N}; \mathbb{R}^d \times \mathbb{R}^d)$, $\sigma(n) \in \mathbb{R}^d \times \mathbb{R}^r$ for all $n \in \mathbb{N}$. Each $\xi(n)$ is a random vector in $\mathbb{R}^r$. We make the following standing assumption about ξ through the paper:

(2.2)
$\xi = \{\xi(n) : n \in \mathbb{N}\}$ is a sequence of independent $\mathbb{R}^r$–valued random vectors;

(2.3)
$\xi_j(n) := \langle \xi(n), \mathbf{e}_j \rangle$ have distribution function F, $\forall n \in \mathbb{N}, \forall j = 1, \dots, r$;

(2.4) $\quad \mathbb{E}[\xi_j(n)] = 0 \quad \mathbb{E}[\xi_j(n)^2] = 1, \quad \forall n \in \mathbb{N}, \forall j = 1, \dots, r$;

(2.5) $\quad$ for fixed n, $\{\xi_j(n)\}_{j=1}^r$ are independent random variables.

Let R be the resolvent defined by

$$R(n+1) = R(n) + \sum_{k=0}^{n} A(n-k)R(k), \quad n \in \mathbb{N}; \quad R(0) = I_d, \tag{2.6}$$

so that each $R(n) \in \mathbb{R}^d \times \mathbb{R}^d$. The significance of the resolvent in this context is that it allows X to be written purely in terms of the perturbation, according to the variation of parameters formula

$$X(n) = R(n)\zeta + \sum_{k=1}^{n} R(n-k)\sigma(k-1)\xi(k), \quad n = 1, 2, \dots \tag{2.7}$$

Everywhere in the paper we suppose that for A and R (defined by (2.6)) the following condition holds true:

$$A \in \ell^1(\mathbb{N}; \mathbb{R}^d \times \mathbb{R}^r), \quad R \in \ell^1(\mathbb{N}; \mathbb{R}^d \times \mathbb{R}^d). \tag{2.8}$$

The importance of the process Σ defined by $\Sigma(n+1) := \sigma(n)\xi(n+1)$ is that its almost sure asymptotic stability is equivalent to the almost sure asymptotic stability of X, provided the underlying deterministic resolvent R is summable. This result is proven in Proposition 1 below. Consequently, it is unnecessary to study the correlated *unbounded* delay equation for X directly; instead it is enough to determine the conditions under which an independent white noise process with *zero* lag tends to zero. Moreover, an explicit formula is not generally available for X, because it depends on R, which is in general not known in closed form. On the other hand, however, the formula for Σ could hardly be simpler.

Proposition 1. *Suppose that the sequence of random variables* $\xi = \{\xi(n) : n \in \mathbb{N}\}$ *obeys* (2.2)–(2.5). *Let* A *and* R *obey (2.8) and let* X *be defined by* (2.1). *Then the following are equivalent:*

(a) $\lim_{n\to\infty} \sigma(n)\xi(n+1) = 0$, *a.s.*

(b) $\lim_{n\to\infty} X(n) = 0$, *a.s.*

There is a simple *sufficient* condition which ensures $\sigma(n)\xi(n+1) \to 0$ as $n \to \infty$ a.s.

Proposition 2. *Suppose that $\xi = \{\xi(n) : n \in \mathbb{N}\}$ is a $\mathbb{R}^r$–valued sequence of random vectors which obeys (2.4), and $\sigma = \{\sigma(n) : n \in \mathbb{N}\}$ is a $\mathbb{R}^d \times \mathbb{R}^r$–valued sequence. If $\sum_{n=1}^{\infty} |\sigma(n)|^2 < \infty$, then $\sigma(n)\xi(n+1) \to 0$ as $n \to \infty$, a.s.*

The merit of this summability condition on σ is that it does not require independence of the stochastic sequence; however, it is not clear how close it comes to being optimal in ensuring almost sure stability of X. In the rest of the paper, we concern ourselves with establishing necessary and sufficient conditions on σ and the random sequence ξ which give stability.

Next, we need to connect the condition on the stochastic equation (2.1) to the conditions on the data. This is achieved by the following result.

Lemma 3. *Let $\sigma = \{\sigma(n) : n \in \mathbb{N}\}$ be a $\mathbb{R}^d \times \mathbb{R}^r$–valued sequence. Suppose that the sequence of random variables $\xi = \{\xi(n) : n \in \mathbb{N}\}$ obeys (2.2)–(2.5). Define the distribution function $F_{n,i,j}$ by*

$$F_{n,i,j}(x) = \begin{cases} F(x/|\sigma_{ij}(n)|), & \sigma_{ij}(n) > 0, \\ 1 - F(-x/|\sigma_{ij}(n)|), & \sigma_{ij}(n) < 0, \\ \chi_{[0,\infty)}(x), & \sigma_{ij}(n) = 0, \end{cases} \tag{2.9}$$

and $F_{n,i}$ as the convolution of the distributions $F_{n,i,j}$ for $j = 1, \ldots, r$. Then

$$\sum_{i=1}^{d}\sum_{n=1}^{\infty}[1 - F_{n,i}(\varepsilon) + F_{n,i}(-\varepsilon)] < \infty \quad \text{for all } \varepsilon \in \mathbb{Q}^+ \tag{2.10}$$

is equivalent to $\sigma(n)\xi(n+1) \to 0$ as $n \to \infty$ a.s.

In the scalar case ($d = r = 1$), this simplifies to give

Lemma 4. *Let $\sigma = \{\sigma(n) : n \in \mathbb{N}\}$ be a real sequence. Suppose that $\xi = \{\xi(n) : n \in \mathbb{N}\}$ is a sequence of independent random variables with the same distribution function F with $\mathbb{E}[\xi(n)] = 0$ and $\mathbb{E}[\xi(n)^2] = 1$.*

(a) *The following statements are equivalent:*

$$\sum_{n=0}^{\infty} 1 - F(\varepsilon/|\sigma(n)|) + F(-\varepsilon/|\sigma(n)|) < \infty, \quad \text{for all } \varepsilon \in \mathbb{Q}^+; \tag{2.11}$$

$$\lim_{n\to\infty} \sigma(n)\xi(n+1) = 0, \quad a.s. \tag{2.12}$$

(b) *The following statements are equivalent:*

$$\sum_{n=0}^{\infty} 1 - F(\varepsilon/|\sigma(n)|) + F(-\varepsilon/|\sigma(n)|) = \infty, \quad \textit{for some } \varepsilon \in \mathbb{Q}^+; \tag{2.13}$$

$$\limsup_{n\to\infty} |\sigma(n)\xi(n+1)| > \varepsilon, \quad \textit{a.s.} \tag{2.14}$$

In the general case, by combining Lemma 3 and Proposition 1, we get the following necessary and sufficient conditions on F and $\sigma(n)$ such that $X(n) \to 0$ a.s. as $n \to \infty$.

Theorem 5. *Let $\sigma = \{\sigma(n) : n \in \mathbb{N}\}$ be a $\mathbb{R}^d \times \mathbb{R}^r$–valued sequence. Suppose the sequence of random variables $\xi = \{\xi(n) \in \mathbb{R}^r : n \in \mathbb{N}\}$ obeys* (2.2)– (2.5). *Let A and R obey (2.8) and let X be defined by* (2.1). *Then the following are equivalent:*

(a) (2.10) *holds, where $F_{n,i}$ and $F_{n,i,j}$ are as defined in Lemma 3;*
(b) $\lim_{n\to\infty} X(n) = 0$, *a.s.*

In the scalar case, by combining Lemma 4 and Proposition 1, we may establish a more easily–expressed result which characterises asymptotic stability.

Theorem 6. *Let $\sigma = \{\sigma(n) : n \in \mathbb{N}\}$ be a real sequence. Suppose $\xi = \{\xi(n) : n \geq 1\}$ is a sequence of independent and identically distributed random variables with the same distribution function F such that $\mathbb{E}[\xi(n)] = 0$, $\mathbb{E}[\xi(n)^2] = 1$. Let A and R obey (2.8) and let X be defined by* (2.1).

(a) *If F and σ obey* (2.11), *then $\mathbb{P}[\lim_{n\to\infty} X(n) = 0] = 1$.*
(b) *If F and σ do not obey* (2.11), *then $\mathbb{P}[\lim_{n\to\infty} X(n) = 0] = 0$.*

§3. Applications and extensions of main results

If we additionally assume some monotonicity in $n \mapsto \sigma^2(n)$, we can recover the critical rate of convergence in the case when the white noise process is Gaussian. We first state the result in the scalar case and then indicate how the result can be generalized to the finite dimensional case.

Corollary 7. *Let $\sigma = \{\sigma(n) : n \in \mathbb{N}\}$ be a real sequence. Let $\xi = \{\xi(n) : n \in \mathbb{N}\}$ be a sequence of independently and identically distributed standard normal random variables. Let A and R obey (2.8) and let X be defined by* (2.1).

(a) *If*

$$\sum_{n=1}^{\infty} \sigma(n) \exp\left(-\frac{\varepsilon^2}{2\sigma^2(n)}\right) < \infty, \quad \textit{for all } \varepsilon \in \mathbb{Q}^+, \tag{3.1}$$

then $\mathbb{P}[\lim_{n\to\infty} X(n) = 0] = 1$.

(b) *If* (3.1) *does not hold, then* $\mathbb{P}[\lim_{n\to\infty} X(n) = 0] = 0$.

If, moreover, $n \mapsto \sigma^2(n)$ *is non–increasing, then the following are equivalent:*

(A) $\lim_{n\to\infty} \sigma^2(n) \log n = 0$;
(B) $\mathbb{P}[\lim_{n\to\infty} X(n) = 0] > 0$;
(C) $\mathbb{P}[\lim_{n\to\infty} X(n) = 0] = 1$.

A proof of the following finite–dimensional analogue of Corollary 7 is not given, but left to the interested reader as an exercise. The key observations in proving this result are that (i) the i–th component of $\sigma(n)\xi(n+1)$, which is $Y^{(i)}(n) = \sum_{j=1}^{r} \sigma_{ij}(n)\xi_j(n+1)$ is normally distributed $\mathcal{N}\left(0, \sum_{j=1}^{r} \sigma_{ij}^2(n)\right)$, and (ii) for each $i = 1,\dots,d$, $Y^{(i)} = \{Y^{(i)}(n) : n \in \mathbb{N}\}$ is a sequence of independent random variables.

Corollary 8. *Let* $\sigma = \{\sigma(n) : n \in \mathbb{N}\}$ *be a* $\mathbb{R}^d \times \mathbb{R}^r$*–valued sequence. Let* $\xi = \{\xi(n) \in \mathbb{R}^r : n \in \mathbb{N}\}$ *be a sequence of random vectors obeying* (2.2)–(2.5) *such that* $\xi_j(n) = \langle \xi(n), \mathbf{e}_j \rangle$ *are standard normal random variables,* $j = 1,\dots,r$. *Define* $\sigma_i(n) \geq 0$ *by* $\sigma_i^2(n) = \sum_{j=1}^{r} \sigma_{ij}^2(n)$, $i = 1,\dots,d$. *Let* A *and* R *obey (2.8) and let* X *be defined by* (2.1).

(a) *If*

$$\sum_{i=1}^{d}\sum_{n=1}^{\infty} \sigma_i(n) \exp\left(-\frac{\varepsilon^2}{2\sigma_i^2(n)}\right) < \infty, \quad \text{for all } \varepsilon \in \mathbb{Q}^+, \tag{3.2}$$

then $\mathbb{P}[\lim_{n\to\infty} X(n) = 0] = 1$.

(b) *If* (3.2) *does not hold, then* $\mathbb{P}[\lim_{n\to\infty} X(n) = 0] = 0$.

If, moreover, each of the maps $n \mapsto \sigma_i^2(n)$, $i = 1,\dots,d$, *is non–increasing, then the following are equivalent:*

(A) $\lim_{n\to\infty} \|\sigma(n)\|^2 \log n = 0$;
(B) $\mathbb{P}[\lim_{n\to\infty} X(n) = 0] > 0$;
(C) $\mathbb{P}[\lim_{n\to\infty} X(n) = 0] = 1$.

These two results are analogous to those found in [1], which studies the asymptotic behaviour of linear continuous–time stochastic Volterra equations of Itô–type where the driving semimartingale is Brownian motion. The necessary and sufficient conditions on the rate of decay of the noise intensity σ given in (A) is the same as that required in [1]. The rate of decay $\sigma^2(n) \log n \to 0$ has also been shown to be necessary and sufficient for almost sure asymptotic stability in a class of nonlinear stochastic delay–differential equations studied in [2]. This logarithmic decay condition was shown to be necessary for a class of non–delay gradient dynamical system in Chan and Williams [5].

Now give a result for a scalar equation in which the distribution of ξ has power–law decay in the tails, with the same rate of decay being present in each tail.

Corollary 9. *Let $\xi = \{\xi(n) : n \in \mathbb{N}\}$ be a sequence of independently and identically distributed random variables such that $\mathbb{E}[\xi(n)] = 0$, $\mathbb{E}[\xi(n)^2] = 1$ and whose common distribution function F obeys*

$$\lim_{x\to-\infty} F(x)|x|^\beta = c_1, \quad \lim_{x\to\infty}[1 - F(x)]x^\beta = c_2,$$

for some $\beta > 2$, and positive c_1 and c_2. Let A and R obey (2.8) and let X be defined by (2.1).

(a) *If $\sum_{n=1}^\infty |\sigma(n)|^\beta < \infty$, then $\mathbb{P}[\lim_{n\to\infty} X(n) = 0] = 1$.*
(b) *If $\sum_{n=1}^\infty |\sigma(n)|^\beta = \infty$, then $\mathbb{P}[\lim_{n\to\infty} X(n) = 0] = 0$.*

To prove the results in this section, the following auxiliary results are needed.

Lemma 10. *Let $\xi = \{\xi(n) : n \in \mathbb{N}\}$ be a sequence of independently and identically distributed random variables, with common distribution function F which does not have compact support. If* (2.12) *holds, then* $\lim_{n\to\infty}\sigma(n) = 0$.

Lemma 11. *Suppose that $(a(n))_{n\geq 0}$ is a non–negative and non–increasing sequence such that $\sum_{n=0}^\infty a(n) < \infty$. Then $na(n) \to 0$ as $n \to \infty$.*

§4. Proofs

4.1. Proof of Proposition 1

We show first that (a) implies (b). If $\Sigma(n+1) := \sigma(n)\xi(n+1) \to 0$ as $n \to \infty$, a.s., it follows immediately from the fact that $R \in \ell^1(\mathbb{N}; \mathbb{R}^d \times \mathbb{R}^d)$ that $(R * \Sigma)(n) \to 0$ as $n \to \infty$, a.s. Moreover, because $R \in \ell^1(\mathbb{N}; \mathbb{R}^d \times \mathbb{R}^d)$, we have $R(n) \to 0$ as $n \to \infty$. Therefore by (2.7) we have that $X(n) \to 0$ as $n \to \infty$, a.s.

To show (b) implies (a), suppose $X(n) \to 0$ as $n \to \infty$ a.s. Rewriting (2.1) gives $\sigma(n)\xi(n+1) = X(n+1) - X(n) - \sum_{k=0}^n A(n-k)X(k)$. Since $A \in \ell^1(\mathbb{R}^{d\times d})$, we have $(A * X)(n) \to 0$ as $n \to \infty$, a.s. Therefore, $\sigma(n)\xi(n+1) \to 0$ as $n \to \infty$, a.s.

4.2. Proof of Proposition 2

Consider $Y^{(i)}(n) = \sum_{j=1}^{r} \sigma_{ij}(n)\xi_j(n+1)$, $i = 1, \dots, d$. Then $\sigma(n)\xi(n+1) \to 0$ as $n \to \infty$, a.s. once $Y^{(i)}(n) \to 0$ a.s. as $n \to \infty$. Evidently $\mathbb{E}[Y^{(i)}(n)] = 0$, while

$$\mathbb{E}[Y^{(i)}(n)^2] \le \sum_{j=1}^{r}\sum_{l=1}^{r} |\sigma_{ij}(n)||\sigma_{il}(n)| = \left(\sum_{j=1}^{r} |\sigma_{ij}(n)|\right)^2.$$

This estimate, Chebyshev's inequality, and $|\sigma(n)\xi(n+1)|_1 = \sum_{i=1}^{d} |Y^{(i)}(n)|$ give

$$\sum_{n=1}^{\infty} \mathbb{P}[|\sigma(n)\xi(n+1)|_1 > \varepsilon] \le \frac{1}{\varepsilon^2}\sum_{n=1}^{\infty}\sum_{i=1}^{d}\left(\sum_{j=1}^{r} |\sigma_{ij}(n)|\right)^2, \quad \text{for any } \varepsilon > 0.$$

The righthand side can be bounded by an n–independent multiple of $\sum_{n=1}^{\infty} |\sigma(n)|_F^2$, which is finite by hypothesis. The result now follows from the Borel–Cantelli lemma.

4.3. Proof of Lemma 3

Fix $\varepsilon \in \mathbb{Q}^+$. Define $Y^{(i)}(n) = \sum_{j=1}^{r} \sigma_{ij}(n)\xi_j(n+1)$, $i = 1, \dots, d$. Proving $\sigma(n)\xi(n+1) \to 0$ as $n \to \infty$ a.s. is equivalent to proving that $Y^{(i)}(n) \to 0$ as $n \to \infty$ a.s. for each $i = 1, \dots, d$. Define $A_{n,\varepsilon}^{(i)} = \{\omega : |Y^{(i)}(n)| > \varepsilon\}$. Then $\mathbb{P}[\bar{A}_{n,\varepsilon}^{(i)} \text{ ev.}] = 1$ for every $\varepsilon \in \mathbb{Q}^+$ is equivalent to $\lim_{n\to\infty} Y^{(i)}(n) = 0$, a.s. Indeed, since for each fixed i the sequence $Y^{(i)} = \{Y^{(i)}(n) : n \in \mathbb{N}\}$ is an independent sequence, by the Borel–Cantelli lemma $\mathbb{P}[\bar{A}_{n,\varepsilon}^{(i)} \text{ ev.}] = 1$ for all $\varepsilon \in \mathbb{Q}^+$ is equivalent to $\sum_{n=1}^{\infty} \mathbb{P}[A_{n,\varepsilon}^{(i)}] < \infty$, for all $\varepsilon \in \mathbb{Q}^+$. Hence $\sigma(n)\xi(n+1) \to 0$ as $n \to \infty$ a.s. is equivalent to $\sum_{n=1}^{\infty} \mathbb{P}[|Y^{(i)}(n)| > \varepsilon] < \infty$ for all $\varepsilon \in \mathbb{Q}^+$. Let $F_{n,i}$ be the distribution function of $Y^{(i)}(n)$, F the distribution function of ξ_i, and $F_{n,i,j}$ the distribution function of $\sigma_{ij}(n)\xi_j(n+1)$. Then

$$F_{n,i,j}(x) = \mathbb{P}[\sigma_{ij}(n)\xi_j(n+1) \le x] = \begin{cases} F(x/|\sigma_{ij}(n)|), & \sigma_{ij}(n) > 0, \\ 1 - F(-x/|\sigma_{ij}(n)|), & \sigma_{ij}(n) < 0. \end{cases}$$

If $\sigma_{ij}(n) = 0$, $F_{n,i,j}(x) = 1$ if $x \ge 0$ and $F_{n,i,j}(x) = 0$ if $x < 0$. Thus $F_{n,i,j}(x) = \chi_{[0,\infty)}(x)$. Since for each fixed n, $(\xi_j(n))_{j=1}^{r}$ are independent random variables, $F_{n,i} = F_{n,i,1} * \dots * F_{n,i,r}$, so $\mathbb{P}[|Y^{(i)}(n)| > \varepsilon] = 1 - F_{n,i}(\varepsilon) + F_{n,i}(-\varepsilon)$, completing the proof.

4.4. Proof of Lemma 4

Define $Y(n) = \sigma(n)\xi(n+1)$ and $A_n^\varepsilon = \{\omega \in \Omega : |Y(n,\omega)| > \varepsilon\}$. Thus, if $\sigma(n) \neq 0$, we have $\mathbb{P}[A_n^\varepsilon] = 1 - F(\varepsilon/|\sigma(n)|) + F(-\varepsilon/|\sigma(n)|)$, and $\mathbb{P}[A_n^\varepsilon] = 0$ if $\sigma(n) = 0$. To prove part (a), by the Borel–Cantelli Lemma and independence of the A_n^ε, $\mathbb{P}[\bar{A}_n^\varepsilon \text{ ev.}] = 1$ is equivalent to $\sum_{n=1}^\infty \mathbb{P}[A_n^\varepsilon] < \infty$. Therefore

$$\limsup_{n\to\infty} |Y(n)| \le \varepsilon \text{ on } \Omega_\varepsilon, \text{ with } \mathbb{P}[\Omega_\varepsilon] = 1 \text{ is equivalent to } \sum_{n=1}^\infty \mathbb{P}[A_n^\varepsilon] < \infty,$$

so by considering this equivalence for all $\varepsilon \in \mathbb{Q}^+$, we conclude that (2.11) is equivalent to (2.12). To prove part (b), the independence of the A_n^ε and the Borel–Cantelli lemma imply that $\mathbb{P}[\bar{A}_n^{\varepsilon_0} \text{ ev.}] = 0$ for some $\varepsilon > 0$ is equivalent to $\sum_{n=1}^\infty \mathbb{P}[A_n^{\varepsilon_0}] = \infty$ for some $\varepsilon_0 > 0$. Thus $\mathbb{P}[A_n^{\varepsilon_0} \text{ i.o.}] = 1$ for some $\varepsilon_0 > 0$ is equivalent to $\sum_{n=1}^\infty \mathbb{P}[A_n^{\varepsilon_0}] = \infty$ for some $\varepsilon_0 > 0$. Therefore we conclude that (2.13) is equivalent to (2.14).

4.5. Proof of Corollary 7

Let the common distribution function be F. Then

$$\lim_{x\to-\infty} F(x)|x|e^{\frac{1}{2}x^2} = (2\pi)^{-1/2}, \lim_{x\to-\infty} [1 - F(x)]|x|e^{\frac{1}{2}x^2} = (2\pi)^{-1/2}.$$

To prove part (a), notice by Lemma 10 that we must have $\sigma(n) \to 0$ as $n \to \infty$. Now, let ε be a fixed positive rational number. Therefore, the estimates

$$\lim_{n\to\infty} \frac{F(-\varepsilon/|\sigma(n)|)}{\frac{|\sigma(n)|}{\varepsilon} \cdot e^{-\frac{1}{2}\frac{\varepsilon^2}{\sigma^2(n)}}} = \frac{1}{\sqrt{2\pi}}, \quad \lim_{n\to\infty} \frac{1 - F(\varepsilon/|\sigma(n)|)}{\frac{|\sigma(n)|}{\varepsilon} \cdot e^{-\frac{1}{2}\frac{\varepsilon^2}{\sigma^2(n)}}} = \frac{1}{\sqrt{2\pi}},$$

hold and so

$$\lim_{n\to\infty} [1 - F(\varepsilon/|\sigma(n)|) + F(-\varepsilon/|\sigma(n))]\varepsilon|\sigma(n)|^{-1} e^{\frac{1}{2}\frac{\varepsilon^2}{\sigma^2(n)}} = 2(2\pi)^{-1}. \tag{4.1}$$

Thus

$$\sum_{n=1}^\infty |\sigma(n)|\varepsilon^{-1} \exp\{-2^{-1}\varepsilon^2\sigma^{-2}(n)\} < \infty \tag{4.2}$$

implies $\mathbb{P}[X(n) \to 0 \text{ as } n \to \infty] = 1$. To prove part (b), if the sum in (4.2) is infinite for some $\varepsilon \in \mathbb{Q}^+$, then $\limsup_{n\to\infty} |\sigma(n)\xi(n+1)| > 0$ a.s., and so a brief perusal of the proof of Proposition 1 reveals that $\mathbb{P}[\lim_{n\to\infty} X(n) = 0] = 0$. It is easily seen that the implication holds under the hypothesis (a) above.

Now suppose that $n \mapsto \sigma^2(n)$ is non–increasing. First, note that $\sigma^2(n)\log n \to 0$ as $n \to \infty$ implies (3.1). By part (a), this implies that $X(n) \to 0$ a.s. as $n \to \infty$. Thus (A) implies (C), and (C) obviously implies (B). Finally, we show that (B) implies (A). According to (B) and the proof of Proposition 1 it follows that $\mathbb{P}[\lim_{n\to\infty} \sigma(n)\xi(n+1) = 0] > 0$. However, as $(\xi(n))_{n=1}^{\infty}$ is a sequence of independent random variables, this event is in the tail σ–field of the ξ and as it is of positive probability, by the Kolmogorov zero–one law it is almost sure. Thus by Lemma 4, and (4.1), (4.2) holds for every $\varepsilon \in \mathbb{Q}^+$. Now, fix $\varepsilon \in \mathbb{Q}^+$. Then the sequence $(a_\varepsilon(n))_{n\geq 0}$ defined by $a_\varepsilon(n) = \varepsilon^{-1}|\sigma(n)|\exp\left(-\varepsilon^2/(2\sigma^2(n))\right)$, $n > N_1(\varepsilon)$ is non–increasing. By Lemma 11, $\lim_{n\to\infty} n a_\varepsilon(n) = 0$, or $\lim_{n\to\infty} n|\sigma(n)|\exp\left(-\varepsilon^2/(2\sigma^2(n))\right) = 0$. Thus

$$\lim_{n\to\infty} \sigma^{-2}(n)\left(\sigma^2(n)\log n + 2^{-1}\sigma^2(n)\log\sigma^2(n) - 2^{-1}\varepsilon^2\right) = -\infty.$$

Since $\lim_{x\to\infty} x\log x = 0$, we have $\limsup_{n\to\infty}\left[\sigma^2(n)\log n - \frac{1}{2}\varepsilon^2\right] \leq 0$. However, this inequality holds for every $\varepsilon \in \mathbb{Q}^+$, and so it follows that $\lim_{n\to\infty} \sigma^2(n)\log n = 0$.

4.6. Proof of Corollary 9

To do this suppose first that $\sum_{n=1}^{\infty} |\sigma(n)|^\beta < \infty$. Then $\lim_{n\to\infty} \sigma(n) = 0$. Let ε be any number in $\mathbb{Q}^+$. Since $\varepsilon/\lim_{n\to\infty}\sigma(n) = \infty$,

$$\begin{aligned} &\lim_{n\to\infty} |\sigma(n)|^{-\beta} F(-\varepsilon/|\sigma(n)|) = c_1\varepsilon^{-\beta}, \\ &\lim_{n\to\infty} |\sigma(n)|^{-\beta}[1 - F(\varepsilon/|\sigma(n)|)] = c_2\varepsilon^{-\beta}, \end{aligned} \tag{4.3}$$

and so (2.11) holds. Then by Theorem 6, $\lim_{n\to\infty} X(n) = 0$ a.s. On the other hand, consider the case when $\sum_{n=1}^{\infty} |\sigma(n)|^\beta = \infty$. Suppose that $\limsup_{n\to\infty} |\sigma(n)| > 0$. Then $\limsup_{n\to\infty} |\sigma(n)\xi(n+1)| = \infty$ a.s., and so $\mathbb{P}[\lim_{n\to\infty} X(n) = 0] = 0$. If $\limsup_{n\to\infty}|\sigma(n)| = 0$, we can repeat the calculations above to get (4.3). Since the denominator is not summable, we have $\sum_{n=0}^{\infty}[F(-\varepsilon/|\sigma(n)|) + 1 - F(\varepsilon/|\sigma(n)|)] = \infty$. Thus by Theorem 6, we have $\mathbb{P}[\lim_{n\to\infty} X(n) = 0] = 0$, whence the result.

4.7. Proof of Lemma 10

Suppose to that $\limsup_{n\to\infty}\sigma^2(n) = c_0 > 0$. Then, for every $c \in (0, c_0)$ there is a sequence $m_n^{(c)} \to \infty$ such that $\sigma^2(m_n^{(c)}) > c$ for all $n > N(c)$. Since $\limsup_{n\to\infty}|\xi(m_n^{(c)} + 1)| = \infty$, a.s., we have $\limsup_{n\to\infty}\sigma^2(m_n^{(c)})\xi^2(m_n^{(c)} + 1) = \infty$ a.s., which contradicts the assumption. Hence $\sigma^2(n) \to 0$ as $n \to \infty$.

4.8. Proof of Lemma 11

By the monotonicity of a, the inequalities $2^{j-1}a(2^j) \le \sum_{l=1}^{2^{j-1}} a(2^{j-1} + l)$ hold for each j, so we get

$$\sum_{j=1}^{\infty} a(j) \ge a(1) + \sum_{j=1}^{\infty} 2^{j-1} a(2^j) = a(1) + \frac{1}{2} \sum_{j=1}^{\infty} 2^j a(2^j).$$

Therefore $2^j a(2^j) \to 0$ as $j \to \infty$. This means that for every $\varepsilon > 0$ there is $J = J(\varepsilon)$ such that $j > J$ implies $2^j a(2^j) < \varepsilon$. Consider now $N = 2^{J+1}$. Let $n > N$. Clearly, there is $j > 0$ such that $2^j \le n < 2^{j+1}$. Then $2^j > n/2 > N/2 = 2^J$, so $j > J$. Hence $na(n) < 2^{j+1}a(2^j) < 2\varepsilon$, for all $n > N$.

References

[1] J. A. D. Appleby and M. Riedle, Almost sure asymptotic stability of stochastic Volterra integro-differential equations with fading perturbations, Stoch. Anal. Appl., **24** (2006), 813–826.

[2] J. A. D. Appleby and A. Rodkina, On the asymptotic stability of polynomial stochastic delay differential equations, Funct. Differ. Equ., **12** (2005), 35–66.

[3] J. A. D. Appleby, D. Mackey and A. Rodkina, Almost sure polynomial asymptotic stability of stochastic difference equations. (in Russian), In: Fundamental Directions, Contemp. Math., **17**, 2006, pp. 110–128.

[4] G. Berkolaiko and A. Rodkina, Almost sure convergence of solutions to nonhomogeneous stochastic difference equation, J. Difference Equ. Appl., **12** (2006), 535–553.

[5] T. Chan and D. Williams, An "excursion" approach to an annealing problem, Math. Proc. Cambridge Philos. Soc., **105** (1989), 169–176.

John A. D. Appleby
School of Mathematical Sciences
Dublin City University
Dublin 9
Ireland

Markus Riedle
Humboldt-University of Berlin
10099 Berlin
Germany

Alexandra Rodkina
Dept of Math/CSci
University of the West Indies
Kingston 7
Jamaica

E-mail address: john.appleby@dcu.ie
riedle@mathematik.hu-berlin.de
alexandra.rodkina@uwimona.edu.jm

Advanced Studies in Pure Mathematics 53, 2009
Advances in Discrete Dynamical Systems
pp. 283–290

Local stability of a discrete competition model derived from a nonstandard numerical method

Wendy J. Hernandez-Padilla and Lih-Ing W. Roeger

Abstract.

The continuous two-dimensional Lotka-Volterra competition model is converted to a discrete version using a noncanonical symplectic numerical method. The local stability of the differential equations and difference equations are analyzed and compared. We found that the numerical method preserves the local dynamics of the continuous model. The local stability criteria are the same between the continuous model and the discrete model. The discrete-time model is dynamically consistent with its continuous counterpart.

§1. Introduction

Mathematical models are used to represent phenomena in the biological, ecological, and physical sciences, to name a few. Differential equations are used when the model represents continuous variables. However, when working with discrete variables, difference equations are most appropriate. For example, in ecology, predator-prey models can be formulated as discrete-time mappings. Difference equations are appropriate when organisms have discrete, nonoverlapping generations [1]. Different numerical schemes can be used to convert differential equations into difference equations. If the corresponding difference equations possess the same dynamic behavior as the continuous equations, such as local stability, bifurcations, and/or chaos, then they are said to be *dynamically consistent* [2].

Received November 14, 2006.
Revised October 12, 2007.
2000 *Mathematics Subject Classification.* 34-04.
Key words and phrases. discrete model, competition model, local stability, nonstandard numerical method, dynamically consistent.
This research is supported by the Research Enhancement Fund from College of Arts and Sciences at Texas Tech University

The following is the Lotka-Volterra competition model:

$$\begin{aligned}\frac{dx}{dt} &= x(r_1 - a_{11}x - a_{12}y),\\ \frac{dy}{dt} &= y(r_2 - a_{21}x - a_{22}y),\end{aligned} \tag{1}$$

where

$$r_i > 0 \text{ and } a_{ij} > 0. \tag{2}$$

The parameters r_i are the intrinsic growth rates for species x and y; a_{12} and a_{21} are the interspecific acting coefficients; and $x(t)$ and $y(t)$ represent the number of individuals or population density in species x and y at time t. The dynamics of the model is well-known. We will briefly mention the results of the Lotka-Volterra model.

The system (1) has at most four equilibria. They are the extinct equilibrium $E_0 = (0,0)$; the exclusive equilibria $E_1 = (\frac{r_1}{a_{11}}, 0)$ and $E_2 = (0, \frac{r_2}{a_{22}})$; and the possible coexistence equilibrium $E_3 = (\frac{a_{22}r_1 - a_{12}r_2}{a_{11}a_{22} - a_{12}a_{21}}, \frac{a_{11}r_2 - a_{21}r_1}{a_{11}a_{22} - a_{12}a_{21}})$. Their stability conditions are summarized as follows

(i) E_0 is a repeller, always unstable.

(ii) E_1 is locally asymptotically stable if $\frac{a_{11}}{a_{21}} < \frac{r_1}{r_2}$.

(iii) E_2 is locally asymptotically stable if $\frac{r_1}{r_2} < \frac{a_{12}}{a_{22}}$.

(iv) E_3 is locally asymptotically stable if

$$\frac{a_{11}}{a_{21}} > \frac{r_1}{r_2} > \frac{a_{12}}{a_{22}}$$

and unstable if

$$\frac{a_{11}}{a_{21}} < \frac{r_1}{r_2} < \frac{a_{12}}{a_{22}}.$$

We will introduce a numerical method that converts the continuous model (1) into a discrete-time model. We are inspired by a method that was applied to a continuous predator-prey model which will be explained. The following is a simple normalized Lotka-Volterra predator-prey model:

$$\begin{aligned}\frac{dx}{dt} &= x - xy,\\ \frac{dy}{dt} &= -y + xy.\end{aligned} \tag{3}$$

All solutions of the predator-prey model are mutually stable periodic solutions if the initial conditions are positive, $x(t_0) > 0$ and $y(t_0) > 0$. Standard numerical methods will not produce periodic solutions unless the step size is small enough. Roeger [5] has shown that there are a class of nonstandard numerical methods that will produce periodic solutions for the predator-prey system. However, the methods in [5] do not apply to the Lotka-Volterra models with x^2 or y^2 terms, as it happens in the competition model (1). The simplest numerical method for the predator-prey model (3) that produces periodic solutions and also preserves positivity of the solution is the following method:

$$\begin{aligned} \frac{x(t+h)-x(t)}{h} &= x(t) - x(t+h)y(t), \\ \frac{y(t+h)-y(t)}{h} &= -y(t+h) + x(t+h)y(t). \end{aligned} \tag{4}$$

An example solution is given in Figure 1 when $h = 0.1$.

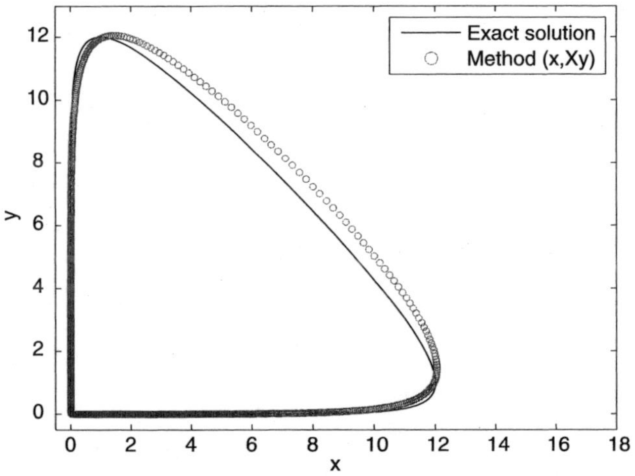

Fig. 1. The method (4) produces periodic solutions for the predator-prey model (3).

Inspired by the method (4), we decide to try the following numerical method on the competition model (1):

$$
\begin{aligned}
\frac{x(t+h)-x(t)}{h} &= r_1x(t) - a_{11}x(t)x(t+h) - a_{12}x(t+h)y(t),\\
\frac{y(t+h)-y(t)}{h} &= r_2y(t+h) - a_{21}x(t+h)y(t) - a_{22}y(t)y(t+h),
\end{aligned} \tag{5}
$$

where the pattern of nonlocal terms on the right hand side equations follows the numerical method (4), except the x^2 and y^2 terms which are replaced by $x(x)x(t+h)$ and $y(t)y(t+h)$ respectively. Since the system has an extra parameter h, further assumptions are needed which will be stated in Section 2. In Section 2, we will discuss the local stability results of the discrete time model (5).

Other different schemes have been used to transform the differential equation (1), into its discrete counterpart [2, 4]. For example, Liu and Elaydi [2] use the following numerical scheme to convert the competition model into its discrete version:

$$
\begin{aligned}
\frac{x(t)(t+h)-x(t)}{\varphi(h)} &= r_1x(t) - a_{11}x(t)x(t)(t+h) - a_{12}x(t)(t+h)y(t),\\
\frac{y(t)(t+h)-y(t)}{\varphi(h)} &= r_2y(t) - a_{21}x(t)y(t)(t+h) - a_{22}y(t)y(t)(t+h),
\end{aligned}
$$

where $\varphi(h) = h + O(h^2)$. Let $x_n = x(t)$, $y_n = y(t)$, $x_{n+1} = x(t+h)$, and $y_{n+1} = y(t+h)$. Its discrete version can be expressed explicitly in terms of x_n and y_n:

$$
\left\{
\begin{array}{rcl}
x_{n+1} &=& \dfrac{[1+r_1\varphi(h)]x_n}{1+\varphi(h)[a_{11}x_n + a_{12}y_n]}\\
y_{n+1} &=& \dfrac{[1+r_2\varphi(h)]y_n}{1+\varphi(h)[a_{21}x_n + a_{22}y_n]}
\end{array}
\right. \tag{6}
$$

Cushing [3] has shown that the dynamics of the discrete system (6) are the same as the continuous model (1).

§2. The main results

We will analyze the local stability of the discrete time system (5). Let $x_n = x(t)$, $y_n = y(t)$, $x_{n+1} = x(t+h)$, and $y_{n+1} = y(t+h)$. The discrete-time competition model (5) can be expressed explicitly in terms of x_n and y_n.

(7)
$$\begin{cases} x_{n+1} = \dfrac{x_n(1+hr_1)}{1+ha_{11}x_n+ha_{12}y_n}, \\[2ex] y_{n+1} = \dfrac{y_n(1-ha_{21}x_{n+1})}{1-hr_2+ha_{22}y_n}. \end{cases}$$

Since this is a discrete-time competition model, we would like to have the solutions to be nonnegative. Therefore, we will assume that

(8)
$$h > 0 \text{ and } 1 - hr_2 > 0.$$

We will show that the local stability conditions for each equilibrium of the discrete-time competitive model (7) are exactly the same as those of the continuous model (1). First, we show that under certain conditions, the solutions for the model (7) are bounded.

Theorem 2.1. *Consider the discrete-time model (7) under the assumptions (2) and (8). If $1 + hr_1 < a_{11}/a_{21}$, then for fixed $h > 0$ the solutions of the system (7) are nonnegative and bounded.*

Proof. It is not difficult to see that x_{n+1} is nonnegative and bounded since

$$0 \le x_{n+1} = \frac{x_n(1+hr_1)}{1+ha_{11}x_n+ha_{12}y_n} \le \frac{x_n(1+hr_1)}{a_{11}hx_n} = \frac{1+hr_1}{a_{11}h}.$$

If we can show that $1 - a_{21}hx_{n+1} > 0$ and since $1 - hr_2 > 0$, then y_{n+1} is nonnegative. We know that $x_{n+1} < \dfrac{1+hr_1}{a_{11}h}$, so $1 - a_{21}hx_{n+1} > 1 - \dfrac{a_{21}}{a_{11}}(1+hr_1) > 0$ if $1 + hr_1 < \dfrac{a_{11}}{a_{21}}$. Then $y_{n+1} \le \dfrac{y_n}{a_{22}hy_n} = \dfrac{1}{a_{22}h}$. y_{n+1} is bounded. Q.E.D.

The system (7) has the same equilibria as the continuous Lotka-Volterra model (1). First, we will show the stability conditions for the equilibria E_0, E_1, and E_2.

Theorem 2.2. *Consider the discrete system (7) under the assumptions (2) and (8). We have the following results.*
(i) E_0 is unstable and a repeller.
(ii) E_1 is locally asymptotically stable if $\dfrac{a_{11}}{a_{21}} < \dfrac{r_1}{r_2}$.
(iii) E_2 is locally asymptotically stable if $\dfrac{r_1}{r_2} < \dfrac{a_{12}}{a_{22}}$.

Proof. We only need to consider the eigenvalues of the Jacobian matrix evaluated at each equilibrium. The Jacobian matrices evaluated at E_0, E_1, and E_2 are

$$J(E_0) = \begin{pmatrix} 1+hr_1 & 0 \\ 0 & \dfrac{1}{1-hr_2} \end{pmatrix}, J(E_1) = \begin{pmatrix} \dfrac{1}{1+hr_1} & \dfrac{r_1 h a_{12}}{a_{11}(1+hr_1)} \\ 0 & \dfrac{a_{11}-hr_1 a_{21}}{a_{11}(1-hr_2)} \end{pmatrix},$$

and

$$J(E_2) = \begin{pmatrix} \dfrac{a_{22}(1+hr_1)}{a_{22}+hr_2 a_{12}} & 0 \\ \dfrac{hr_2 a_{21}(1+hr_1)}{a_{22}+hr_2 a_{12}} & 1-hr_2 \end{pmatrix}$$

respectively. Their eigenvalues appear along the diagonals.

For E_0, since $hr_1 > 0$, $hr_2 > 0$, and $1 - hr_2 > 0$, both eigenvalues of $J(E_0)$ are greater than one, $1 + hr_1 > 1$ and $1/(1 - hr_2) > 1$. E_0 is a repeller and unstable.

For E_1, the first eigenvalue of $J(E_1)$, $1/(1 + hr_1)$, is less than one. Its second eigenvalue is $\lambda_2 = \dfrac{a_{11} - hr_1 a_{21}}{a_{11}(1 - hr_2)}$. We can check that

$$\lambda_2 < 1 \Leftrightarrow \frac{a_{11}}{a_{21}} < \frac{r_1}{r_2}.$$

Therefore, E_1 is stable if $\frac{a_{11}}{a_{21}} < \frac{r_1}{r_2}$. The stability conditions for E_2 can be proved similarly as the case of E_1. Q.E.D.

The positive coexistence equilibrium $E_3 = (x^*, y^*)$ for the system (7) satisfies

$$\begin{aligned} 1 + hr_1 &= 1 + ha_{11}x^* + ha_{12}y^*, \\ 1 - ha_{21}x^* &= 1 - hr_2 + ha_{22}y^*, \end{aligned} \tag{9}$$

which is the same as

$$\begin{aligned} a_{11}x^* + a_{12}y^* &= r_1, \\ a_{21}x^* + a_{22}y^* &= r_2. \end{aligned}$$

Theorem 2.3. *Consider the discrete system (7) under the assumptions (2) and (8). The coexistence equilibrium E_3 is locally asymptotically stable if*

$$\frac{a_{11}}{a_{21}} > \frac{r_1}{r_2} > \frac{a_{12}}{a_{22}}$$

and unstable if

$$\frac{a_{11}}{a_{21}} < \frac{r_1}{r_2} < \frac{a_{12}}{a_{22}}.$$

Proof. The coexistence equilibrium $E_3 = (x^*, y^*) = (\frac{a_{22}r_1 - a_{12}r_2}{a_{11}a_{22} - a_{12}a_{21}}, \frac{a_{11}r_2 - a_{21}r_1}{a_{11}a_{22} - a_{12}a_{21}})$. E_3 exists if $\frac{a_{11}}{a_{21}} > \frac{r_1}{r_2} > \frac{a_{12}}{a_{22}}$ or $\frac{a_{11}}{a_{21}} < \frac{r_1}{r_2} < \frac{a_{12}}{a_{22}}$.

By using the identities in (9), the Jacobian matrix evaluated at E_3 can be simplified to

$$J(E_3) = \begin{pmatrix} \dfrac{1 + ha_{12}y^*}{1 + hr_1} & \dfrac{-ha_{12}x^*}{1 + hr_1} \\ \dfrac{-ha_{21}y^*(1 + ha_{12}y^*)}{(1 + hr_1)(1 - ha_{21}x^*)} & \dfrac{(1 - hr_2)(1 + hr_1) + h^2 a_{12}a_{21}x^*y^*}{(1 + hr_1)(1 - ha_{21}x^*)} \end{pmatrix}.$$

From (9), we know $1 - ha_{21}x^* = 1 - hr_2 + ha_{22}y^* > 0$. Therefore, the entries of the matrix $J(E_3)$ satisfy $J_{11} > 0$, $J_{12} < 0$, $J_{21} < 0$, and $J_{22} > 0$.

To determine the stability of E_3, we conduct the Jury test: E_3 is stable if and only if $|\mathrm{tr}J| < 1 + \det J < 2$. The trace and determinant of $J(E_3)$ are

$$\mathrm{tr}J = J_{11} + J_{22} > 0$$

and

$$\det J = \frac{(1 - hr_2)(1 + ha_{12}y^*)}{(1 + hr_1)(1 - ha_{21}x^*)} > 0.$$

Since $\mathrm{tr}J > 0$, we only need to check two conditions: $1 - \det J > 0$ and $1 + \det J - \mathrm{tr}J > 0$. Using the equations in (9), we can show that

$$\begin{aligned} 1 - \det J &= 1 - \frac{(1 - hr_2)(1 + ha_{12}y^*)}{(1 + hr_1)(1 - ha_{21}x^*)} \\ &= \frac{(1 + hr_1)(1 - hr_2 + ha_{22}y^*) - (1 - hr_2)(1 + hr_1 - ha_{11}x^*)}{(1 + hr_1)(1 - ha_{21}x^*)} \\ &= \frac{ha_{22}y^*(1 + hr_1) + ha_{11}x^*(1 - hr_2)}{(1 + hr_1)(1 - ha_{21}x^*)}, \end{aligned}$$

which is always positive because $1 - hr_2 > 0$ and $1 - ha_{21}x^* > 0$.

For the condition $1 + \det J - \mathrm{tr}J$, using $a_{21}x^* + a_{22}y^* = r_2$, we have

$$\begin{aligned} 1 + \det J - \mathrm{tr}J &= \frac{h^2 y^*(a_{22}r_1 - a_{12}a_{21}x^* - a_{12}a_{22}y^*)}{(1 + hr_1)(1 - ha_{21}x^*)} \\ &= \frac{h^2 y^*(a_{22}r_1 - a_{12}r_2)}{(1 + hr_1)(1 - ha_{21}x^*)}. \end{aligned}$$

Since $1 - ha_{21}x^* > 0$, then $1 + \det J - \operatorname{tr} J$ is positive if $\frac{r_1}{r_2} > \frac{a_{12}}{a_{22}}$ and is negative if $\frac{r_1}{r_2} < \frac{a_{12}}{a_{22}}$. Q.E.D.

§3. Conclusion

We have shown that the local stability conditions between the continuous Lotka-Volterra competition model (1) and the discrete competition model (7) are the same. In other words, the two systems are dynamically consistent.

Numerical methods can be applied to differential equation systems to obtain difference equation systems or discrete-time systems. Comparisons between the two systems help us understand the relationship between them. It will helps us to construct a more appropriate discrete model from continuous system.

The numerical method we used is only one of the methods discussed by Roeger [5]. We will look into other different methods to see if the resulting discrete model is still dynamically consistent with the continuous model.

References

[1] M. Kot, Elements of Mathematical Ecology, Cambridge Univ. Press, Cambridge, 2001.

[2] P. Liu and S. N. Elaydi, Discrete competitive and cooperative models of Lotka-Volterra type, J. Comput. Anal. Appl., **3** (2001), 53–73.

[3] J. M. Cushing, S. Levarge, N. Chitnis and S. M. Henson, Some discrete competition models and the competitive exclusion principle, J. Difference Equ. Appl., **10** (2004), 1139–1151.

[4] L.-I. W. Roeger, Discrete May-Leonard competition models. III, J. Difference Equ. Appl., **10** (2004), 773–790.

[5] L.-I. W. Roeger, Nonstandard finite-difference schemes for the Lotka-Volterra systems: generalization of Mickens's method, J. Difference Equ. Appl., **12** (2006), 937–948.

[6] R. E. Mickens, A nonstandard finite-difference scheme for the Lotka-Volterra system, Appl. Numer. Math., **45** (2003), 309–314.

Department of Mathematics and Statistics
Box 41042, Texas Tech University
Lubbock, TX 79409
USA

E-mail address: `lih-ing.roeger@ttu.edu`

Advanced Studies in Pure Mathematics 53, 2009
Advances in Discrete Dynamical Systems
pp. 291–299

Eventual stability criterion for periodic points of Michio Morishima's example

Seiji Saito

Abstract.

In this paper we discuss the Morishima's example, which implies a kind of eventually asymptotical stability of solutions for a difference equation $x(n+1) = f(x(n))$ for $n = 0, 1, 2, \cdots$. We define new definitions of eventual stability of periodic points in the meaning of the large in the same way as ones of Lakshmikantham et. al. and Yoshizawa. By applying the Lyapunov's second method we give eventual stability criteria in the large of the difference equation. In order to illustrait our main results on eventual stability an example of a set of 2-periodic points for eventual stability is given with an analytical estimation.

§1. Introduction

In 1977 Morishima[3] gave results on the stability, oscillation and chaos of periodic points concerning the following difference equation.

$$x(n+1) = \frac{A(n)}{A(n)+B(n)} \qquad \text{for } n = 0, 1, \cdots \tag{E}$$

and

$$\begin{aligned} A(n) &= \max[\frac{a}{b}x(n) + \{1-(1+a)x(n)\}, 0], \\ B(n) &= \max[(1-x(n))\{\frac{a}{b} - \frac{x(n)(1-(1+a)x(n))}{(1-x(n))^2}\}, 0] \end{aligned}$$

Here a, b are positive parameters. His results[3] with $a = 0.6, b = 1$ were studied concerning the chaos of Eq.(E) independently with Li-Yorke[2] in 1975.

Morishima[4] studied the chaotic behavior and the stability of orbits of

(1) $$x(n+1) = f(x(n)),$$

Received October 12, 2007.

where $f : [0,1] \to [0,1]$ is continuous, $x : \mathbf{Z}_+ = \{0,1,2,\cdots\} \to [0,1]$ is the price of the commodity and also he discussed some type of stability of periodic points, where the stability is not globally uniformly asymptotically stable but every orbit of Eq.(1) has unstable properties in the beginning and the stable behavior from some iterations.

In this paper we show results on the globally asymptotical stability for periodic points of Eq.(1) as well as we discuss the globally eventually asymptotical stability. See Lakshikantham-Leela[1], Yoshizawa[5] concerning the eventual stability for the case of ordinary differential equations.

§2. Notations

Consider difference equation (1) in $I^m \subset \mathbf{R}^m$ with $I = [0,1]$ and positive integer m. Denote $x(n) = (x_1(n), x_2(n), \cdots, x_m(n))^T$, where T means the transpose, is a relative price vector of $m-$commodities, where $0 \le x_j(n) \le 1$ for $j = 1,2,\cdots,m$ and $\sum_{j=1}^{m} x_j(n) = 1$ for $n \in \mathbf{Z}_+$. See [3, 4] in detail. A function $f : I^m \to I^m$ is continuous.

Let k be a positive integer. Denote a set of $k-$peridic points by $P(k) = \{x^* \in I^m\}$. $x^* \in P(k)$ if and only if $f^i(x^*) \neq f^j(x^*)$ for $1 \le i \neq j \le k$ and $f^k(x^*) = x^*$. Denote by $x(n; n_0, x_0)$ a solution of Eq.(1) for $n \ge n_0$ with $x(n_0; n_0, x_0) = x_0$ satisfying the initial condition $(n_0, x_0) \in \mathbf{Z}_+ \times I^m$. Denote by $\| x \|$ a norm of $x \in \mathbf{R}^m$. For $r > 0$ we denote the following neighborhoods: when a point $x_0 \in I^m, B(x_0, r) = \{x \in I^m : \| x - x_0 \| < r\}$; when a subset $P \subset I^m, S(P, r) = \cup_{x \in P} B(x, r)$.

A set of $k-$periodic points $P(k)$ is called eventually uniformly stable [EV-US] if for each $\varepsilon > 0$ there exist $N_0 \in \mathbf{Z}_+$ and $\delta > 0$ such that for every $x_0 \in S(P(k), \delta)$ and every $n_0 \ge N_0$, it holds that each solution $x(n; n_0, x_0) \in S(P(k), \varepsilon)$ for $n \ge n_0, i.e.,$

$$d(x(n; n_0, x_0), P(k)) < \varepsilon.$$

Here a distance between a point $x \in \mathbf{R}^m$ and a subset $S \subset \mathbf{R}^m$ is defined by $d(x, S) = \inf\{\| x - a \| : a \in S\}$. A set of $k-$periodic points $P(k)$ is called eventually uniformly attractive to finite coverings [EV-UA.FC] if each finite covering $\{C_q \subset I^m : \cup_{q=1}^{Q} C_q \supset I^m\}$ and each $\varepsilon > 0$, there exist $N_0 \in \mathbf{Z}_+$ and $T_0 \in \mathbf{Z}_+$ such that for every $1 \le q \le Q$, every $x_0 \in C_q$ and every $n_0 \ge N_0$, it holds that every solution $x(n; n_0, x_0) \in S(P(k), \varepsilon)$ for $n \ge n_0 + T_0, i.e.,$

$$d(x(n; n_0, x_0), P(k)) < \varepsilon.$$

The set of k–periodic points $P(k)$ is callled eventually uniformly asymptotically stable to finite coverings [EV-UAS.FC] if $P(k)$ is [EV-US] and [EV-UA.FC].

§3. Criterion of Evental Stability

Assume that Eq.(1) has a set of k-periodic points

$$P(k) = \{x_1, x_2, \cdots, x_k\}$$

for $k = 1, 2, \cdots$. We show two criterion for eventually uniformly asymptotically stable of $P(k)$ by applying the Lyapunov's second method. In case $k = 1, P(1)$ is a set of fixed point.

Let a set of functions denote

$$CIP = \{a : I \to \mathbf{R}_+ \text{ is continuous, strictly increasing and positively definite}\}$$

and $R_+ = [0, \infty)$. Denote $A - B = \{x \in A : x \notin B\}$ for sets $A, B \subset I^m$.

In the following theorem we prove the eventually uniformly asymptotically stable to finite coverings of $P(k)$.

Theorem. k–periodic points $P(k)$ is eventually uniformly asymptotically stable to finite coverigs under that there exists a function $V : \mathbf{Z}_+ \times I^m \to \mathbf{R}_+$ satisfying the following conditions (a)-(b).

(a) For any $r > 0$ there exist a nonnegative integer $N_0 \geq 0$ and two functions $a_r, b_r \in CIP$ such that

$$a_r(d(x, P(k))) \leq V(n, x) \leq b_r(d(x, P(k))$$

for any $n \geq N_0$ and any $x \in I^m - S(P(k), r)$.

(b) Let $\Delta V(n, x) = V(n+k, f^k(x)) - V(n, x)$ for $(n, x) \in \mathbf{Z}_+ \times I^m$. For any $r > 0$ there exist a nonnegative integer $N_0 \geq 0$ and a function $c_r \in CIP$ such that

$$\Delta V(n, x) \leq -c_r(d(P(k), x))$$

for any $n \geq N_0$ and any $x \in I^m - S(P(k), r)$.

Proof. It is proved that the set $P(k)$ is [EV-US]. At first, we get the following inequalities.

$$\tilde{a_r}(d(x, P(k))) \leq V(n, x) \leq b_r(d(x, P(k))); \tag{2}$$

$$\Delta V(n, x) \leq -\tilde{c_r}(d(x, P(k))). \tag{3}$$

where $\tilde{a_r}(d) = \min[a_r(d), c_r(d)]$ and $\tilde{c_r}(d) = \frac{1}{2}\tilde{a_r}(d)$ for $d > 0$. For a sufficiently large $\alpha_1 > 0$ and small $\alpha_2 > 0$ and any $p_\omega \in P(k)$ it can be seen that $I^m \subset S(P(k), \alpha_1)$ and that

$$\text{(4)} \qquad \text{if } x \in B(p_\omega, \alpha_2), \text{ then } f^k(x) \in B(p_\omega, \alpha_1).$$

For any $\varepsilon > 0$ define

$$\text{(5)} \qquad \phi_\omega(\varepsilon) = \inf\{V(n, x) : \varepsilon \leq \| x - p_\omega \| \leq \alpha_1, n \geq n_0\}.$$

We get

$$\text{(6)} \qquad V(n, x) < \phi_\omega(\varepsilon) \text{ for } x \in B(p_\omega, \delta_\omega),\ n_0 \geq N_0.$$

If not so, it holds that

$$0 < \forall\delta < \varepsilon, \exists n_\delta \geq N_0 : V(n_\delta, x_\delta) \geq \phi_\omega(\varepsilon)$$

for $\exists x_\delta \in B(p_\omega, \delta)$. Inequality (2) means that for $r < \varepsilon$

$$b_r(d(x_\delta, P(k))) \geq \phi_\omega(\varepsilon) > 0.$$

Putting $\delta = 1/j$ for $j = 1, 2, \cdots$, we have a sequence $\{x_j : j = 1, 2, \cdots\} \subset cl(B(p_\omega, \delta))$, the closure of $B(p_\omega, \delta)$, and

$$\lim_{j \to +\infty} \| x_j - p_\omega \| = 0.$$

This imlpies that $0 = \phi_\omega(\varepsilon) > 0$, which leads to a contradiction.

We shall prove that there exist $1 \leq k(1), k(2) \leq k$ and $\delta > 0$ as follows:

$$\text{(7)} \qquad \begin{aligned} &\exists p_{k(1)} \in P(k), 0 < \exists\delta < \delta_\omega : \forall y \in B(p_{k(1)}, \delta), \forall n_0 \geq N_0; \\ &\forall \ell = 1, 2, \cdots, \exists p_{k(2)} \in P(k) : x(n_0 + \ell k; n_0, y)) \in B(p_{k(2)}, \varepsilon). \end{aligned}$$

If not so, then

$$\begin{aligned} &\forall p_\omega \in P(k), 0 < \forall\delta \leq \delta_\omega, \exists y_1 \in B(p_\omega, \delta) : \exists \overline{n_0} \geq N_0, \\ &\exists \ell_1 \geq 1, \forall p_\omega \in P(k), x(\overline{n_0} + \ell_1 k; \overline{n_0}, y_1) \notin B(p_\omega, \varepsilon), \end{aligned}$$

which means that $V(\overline{n_0}, y_1) < \phi_\omega(\varepsilon)$ Dnote $x(\cdot) = x(\cdot; \overline{n_0}, y_1)$. Considering $S(P(k), \alpha_1) \supset I^m$, we get

$$\varepsilon < \| x(\overline{n_0} + \ell_1 k) - p_\omega \| < \alpha_1$$

and $\phi_\omega(\varepsilon) \le V(\overline{n_0} + \ell_1 k, x(\overline{n_0} + \ell_1 k))$. By Condition(b), it holds that

$$\begin{aligned} \phi_\omega(\varepsilon) &\le V(\overline{n_0} + \ell_1 k, x(\overline{n_0} + \ell_1 k)) \\ &= V(\overline{n_0} + \ell_1 k, f^{\ell_1 k}(y_1)) \\ &< V(\overline{n_0} + \ell_1 k - k, f^{(\ell_1 - 1)k}(y_1))) \\ &< \cdots \\ &< V(\overline{n_0}, y_1) < \phi_\omega(\varepsilon). \end{aligned}$$

This leads to a contracition. Hence, $P(k)$ is [EV-US], because for any $0 < \varepsilon < \alpha_2$ there exist a positive $\delta < \min\{\delta_\omega : 1 \le \omega \le k\}$ and an integer $N_0 \ge 0$ such that for any $n_0 \ge N_0$ and any $n \ge n_0$ if $x_0 \in S(P(k), \delta)$, then $x(n; n_0, x_0) \in S(P(k), \varepsilon)$.

It can be seen that Eq.(1) is uniformly bounded as follows:

(8)
$$\forall \alpha > 0,\ \exists \beta(\alpha) > 0 : \forall n_0 \ge 0,$$
$$\| x(n; n_0, x) \| < \beta(\alpha) \text{ for } \| x \| < \alpha, n \ge n_0.$$

If Eq.(1) is not [EV-UA.FC], then there exist a real number $\varepsilon_1 > 0$ and a finite covering $\{C_q \subset I^m$ such that $\cup_{q=1}^{Q} C_q \supset I^m\}$ such that for some $1 \le \tilde{q} \le Q$ and any integers $N, T \in \mathbf{Z}_+$ there exist an initial point $x_0 \in C_{\tilde{q}}$ and integers $n_0 = n_0(N,T) \ge N, n_1 = n_1(N,T) \ge n_0 + T$ such that

(9)
$$d(x(n_1; n_0, x_0), P(k)) \ge \varepsilon_1.$$

Then we have a sequence $\{z_j : \| z_j \| \le \alpha\}$ and $z = \lim_{j \to \infty} z_j \notin P(k)$. For a sufficiently small $\eta > 0$ we get a neighborhood $O(z, \eta)$ of z such that

(10)
$$S(P(k), \varepsilon_1) \cap cl(O(z, \eta)) = \emptyset.$$

From $V(n, x) \ne 0$ on $O(z, \eta)$, we can define $h(n, x) = V(n + k, f^k(x))/V(n, x)$ on $O(z, \eta)$, which is continuous and $h(n, x) < 1$ by Condition (3). By $\tilde{c_r}(d)/\tilde{a_r}(d) = 1/2$, it can be seen that

(11)
$$h(n, x) \le 1 - [\tilde{c_r}(d(x, P(k)))/\tilde{a_r}(d(x, P(k)))] = 1/2.$$

For $\delta_1 > 0$ we have the closure of δ_1-neighborhood at z such that $cl(B(z, \delta_1)) \subset O(z, \eta)$ and also if $x \in cl(B(z, \delta_1))$ then $h(n, x) \le 1/2$. Denote $U = \{u_{i+1} = f^k(u_i) : u_1 = z, i = 1, 2, \cdots\} \in B(z, \delta_1)$.

$$\begin{aligned} V(ik + k, f^k(u_i)) &= h(ik, u_i) V(ik, u_i) \le \frac{1}{2} V(ik, u_i) \\ &\le \frac{1}{2^2} V(ik - k, u_{i-1}) \le \cdots \le \frac{1}{2^i} V(0, z) \end{aligned}$$

Then we have

$$\tilde{a_r}(d(f^k(u_i), P(k))) \leq V(ik, f^k(u_i)) \leq \frac{1}{2^i} b_r(d(z, P(k))).$$

As $r \to 0+, i \to +\infty$ it can be seen that

$$0 \leq \tilde{a_r}(d(f^k(u_i), P(k))) \leq \lim_{i \to \infty} \frac{1}{2^i} b_r(d(z, P(k))) = 0.$$

Hence there exist a subsequence $\{f^k(u_{j_q}) : q = 1, 2, \cdots\} \subset U \subset cl(B(z, \delta_1))$ and a point $v \in I^m$ such that

$$v \in cl(B(z, \delta_1)), \quad f^k(v) \in O(z, \eta), \text{ and } \lim_{q \to \infty} f^k(u_{j_q}) = f^k(v) \in P(k).$$

This implies that $f(v) \in P(k) \cap O(z, \eta)$, which contradicts with the assumtion(9). Therefore $P(k)$ is [EV-UA.FC]. Q.E.D.

In case where $k = 1$ the above theorem leads to an eventual stability theorem of fixed point for Eq.(1).

Corollary. Eq.(1) has a fixed point x^*. The point x^* is eventually uniformly asymptotically stable to finite coverings under that there exists a function $V : \mathbf{Z}_+ \times I^m \to \mathbf{R}_+$ satisfying Condition (a)-(b).

(a) For any $r > 0$ there exist an integer $N_0 \geq 0$ and two functions $a_r, b_r \in CIP$ such that

$$a_r(\| x - x^* \|) \leq V(n, x) \leq b_r(\| x - x^* \|)$$

for any integers $n \geq N_0$ and any initial points $x \in I^m - \{x^*\}$.

(b) Let $\Delta V(n, x) = V(n+1, f(x)) - V(n, x)$ for $(n, x) \in \mathbf{Z}_+ \times I^m$. For any $r > 0$ there exist an intger $N_0 \geq 0$ and a function $c_r \in CIP$ such that

$$\Delta V(n, x) \leq -c_r(\| x - x^* \|)$$

for any $n \geq N_0$ and any $x \in I^m - \{x^*\}$.

§4. Illustration of Theorem

We illustrait Theorem in the case $k = 2$ and $P(2) = \{0.5, 0.7\}$ in the space $\mathbf{R}$ with an analytical result. Consider the Morishima's example as follows.

$$x(n+1) = f(x(n)) = \frac{A(n)}{A(n) + B(n)}.$$

Here $A(n) = \max[x + bE_1(x(n)), 0]$, $B(n) = \max[1 - x + bE_2(x(n)), 0]$ and $a = 0.6$ and $E_1(x) = -x + \frac{1-x}{a}$, $E_2(x) = -\frac{xE_1(x)}{1-x}$. See [3] in detail. Then, in $b = 0.6$, we get

$$f(x) = \frac{1.8x^2 - 4.8x + 3}{9.6x^2 - 13.8x + 6}, \quad f^{'}(x) = \frac{21.24x^2 - 36x + 12.6}{(9.6x^2 - 13.8x + 6)^2}.$$

Let

$$V(x) = d(x, P(2)) = \min[|x - 0.5|, |x - 0.7|]$$

for $x \in I$. Let $a_r(d) = b_r(d) = d$ $(d > 0)$ to any $r > 0$. Then $a_r, b_r \in CIP$ and it holds that Condition(a) of Theorem is satisfied. It can be seen that

$$\begin{aligned}
\Delta V(x) &= \min(|f^2(x) - 0.5|, |f^2(x) - 0.7|) - d(x, P(2)) \\
&= \min(|f^2(x) - f^2(0.5)|, |f^2(x) - f^2(0.7)|) - d(x, P(2)) \\
&= \min_{x^*=0.5,0.7} \left| \int_0^1 \frac{df^2}{dx}(x^* + \theta(x - x^*))(x - x^*)d\theta \right| - d(x, P(2)) \\
&\leq \min_{x^*=0.5,0.7} \max_{x \in I} \left|\frac{df^2}{dx}(x)\right| |x - x^*| - d(x, P(2)) \\
&= \max_{x \in I} \left|\frac{df^2}{dx}(x)\right| d(x, P(2)) - d(x, P(2)) \\
&= (\max_{x \in I} |f^{'}(f(x))f^{'}(x)| - 1)d(x, P(2)).
\end{aligned}$$

It holds that $\Delta V(x) \leq \max_{x \in I}(f^{'}(f(x))f^{'}(x) - 1)V(x)$.

We shall show that $\Delta V(x) \leq -cV(x)$ for $x \in I$ with a real number $c > 0$. Putting $y(x) = f^{'}(f(x))f^{'}(x) - 1$, when $y(x) < 0$, then there exists a positive number c such that

$$\Delta V(x) \leq -cV(x). \tag{12}$$

Putting $C(d) = cd$, we have

$$C \in CIP : \Delta V(x) \leq -C(V(x)).$$

Therefore it holds that Condition (b) of Theorem is satisfied.

Denote

$$p = 21.24x^2 - 36x + 12.6, \quad q = (9.6x^2 - 13.8x + 6)^2,$$

then we have $f^{'} = p/q$ and $\max_{x\in I}|p/q| < 1$. In fact

$$\begin{aligned}
&\frac{p^2-q^2}{q^2}\\
&=\frac{(p-q)(p+q)}{q^2}\\
&=[(21.24x^2-36x+12.6)-(9.6x^2-13.8x+6)^2][21.24x^2-36x\\
&\quad +12.6+(9.6x^2-13.8x+6)^2]/q^2\\
&=[-92.16x^4+264.96x^3-284.4x^2-118.8x-23.4]\\
&\times[21.24(x-(1.18)^{-1})^2+12.6-(9/5.09)+(9.6x^2-13.8x+6)^2]/q^2
\end{aligned}$$

and $12.6-(9/5.09) > 0$, $264.96x^3-284.4x^2 = 264.96x^2(x-284.4/264.96) < 0$ for $0 \le x \le 1$, then we have $|f^{'}(x)| \le \max_{x\in I}\frac{|p|}{|q|} < 1$. Hence it holds that on $x \in [0,1]$

$$y(x) = f^{'}(f(x))f^{'}(x) - 1 \le (\max_{x\in I}\frac{|p|}{|q|})^2 - 1 < 0.$$

Since y is continuous and $[0,1]$ is compact, then there exists a positive number c such that $y(x) \le -c < 0$ on $[0,1]$.

§5. Conclusions

We considered a definition of [EV-UAS.FC] (eventually uniformly asymptotic stability to finite coverings) in the same way as theory of ordinary differential equations.

We proved a theorem for [EV-UAS.FC] of difference equation $x(n+1) = f(x(n))$ by the Lyapunov's second method including an analytical estimation of ΔV.

We illustrated the eventual stability theorem by applying it to the Morishima's example.

References

[1] V. Lakshmikantham and S. Leela, Differential and Integral Inequalities. I, 1969.

[2] T. Y. Li and J. A. Yorke, Period three implies chaos, Amer. Math. Monthly, **82** (1975), 985–992.

[3] M. Morishima, Warlas' Economics, Cambridge Univ. Press, 1977.

[4] M. Morishima, Dynamic Economic Theory, Cambridge Univ. Press, 1997.

[5] T. Yoshizawa, Stability Theory by Liapunov's Second Method, Math. Soc. Japan, 1966.

Faculty of Science and Engineering
Doshisha University
Kyotanabe, Kyoto, 610-0321
Japan

E-mail address: ssaito@mail.doshisha.ac.jp

Advanced Studies in Pure Mathematics 53, 2009
Advances in Discrete Dynamical Systems
pp. 301–309

Detection of periodic driving in nonautonomous difference equations

Timothy Sauer

Abstract.

An algorithm is proposed for determining the periodic behavior of the common driver of a system of nonautonomous difference equations from observations of the equation trajectories only. Methods of attractor reconstruction are used to build a semiconjugacy to a topological version of the driver system. The algorithm is described in detail and implemented on several examples.

§1. Introduction

System identification for nonlinear differential and difference equations has lagged far behind the linear case, due to the greater complexity of the task. Takens' Theorem [10, 1] of 25 years ago was a significant initial foray. The ramifications of this result, in an area that has become known as attractor reconstruction, continue to be worked out.

A particular case we will consider in this article is a system of driven difference equations of general form

$$\begin{aligned} x_t^1 &= f^1(x_{t-1}^1, \ldots, x_{t-n}^1, d_{t-1}) \\ &\vdots \\ x_t^k &= f^k(x_{t-1}^k, \ldots, x_{t-n}^k, d_{t-1}) \\ d_t &= g(d_{t-1}, \ldots, d_{t-p}) \end{aligned} \tag{1}$$

We often refer informally to the g system as the "driver" dynamics. Assume further that the equations $f^1, \ldots, f^k, g$ are unknown to us, and that we can only observe the outputs $x_i^1, \ldots, x_i^k$ as functions of time. We consider the problem of determining the dynamics of g alone.

Takens' Theorem discusses conditions under which such observations allow a topologically accurate reconstruction of the complete system

Received September 16, 2006.

dynamics. Such a reconstruction may not distinguish the g dynamics from the rest of the system. Our present goal goes beyond what is promised by Takens' Theorem, since our goal is to untangle the dynamics of g from the entire system. Moreover, if the theoretical obstructions can be overcome, it is reasonable to expect that the goal can be achieved with a far smaller data requirement than needed to reconstruct the entire system dynamics.

We will use make use of two recent extensions of Takens' Theorem. Stark [9] proved a reconstruction theorem for so-called skew product systems, in which an autonomous subsystem drives the rest of the system. This result was used in [8] to develop a driver reconstruction theorem. An algorithm based on this theorem is discussed below, along with two examples. In the first example, a set of five one-dimensional chaotic logistic maps are driven by a logistic map in a period-six window. The output of the five maps is used (without knowledge of the maps generating the output) to infer the periodicity of the driver. The second example demonstrates a similar reconstruction, using period-four forcing in a system of three two-dimensional difference equations.

It is hoped that this inquiry can further the development of computational techniques for system identification applied to network models of difference equations. The original motivation for this research was the identification of the dynamics of deep brain structures from simultaneous time series collected by surface electrodes. However, the question presents itself rather generally, whenever simultaneous driving of multiple, observable systems occurs.

§2. Background and theoretical results

To begin, we recall a reconstruction theorem for skew products due to Stark [9] (see also Casdagli [2]).

Theorem 1. *(Stark, 1999) Let D and X be compact manifolds, $dim(D) = d$, $dim(X) = k \geq 1$. Let $m \geq 2d + 2k + 1$, and assume the periodic orbits of period $< 2m$ of $g : D \to D$ are isolated and have distinct eigenvalues. Then there exists an open, dense set of C^1 functions $f : D \times X \to X$ and $h : X \to R$ for which the m-dimensional delay map is an embedding.*

The theorem states conditions under which the state space of a system consisting of the combined driver and response can be reconstructed. In the present article, the goal is to use the output of several response units to separate out the dynamics of the driver from the rest of the dynamics, as a way of breaking the system down into its component parts.

We will refer to the systems in (1) as $X_1, \ldots, X_k$ and D, respectively. Apply Stark's theorem to each $D \times X_i$ individually, where D is the ergodic attractor of the driving system and X_i is the state space of the ith nonautonomous system. According to the theorem, for generic dynamics and sufficiently large m, there is a one-to-one correspondence between m-tuples $(x_t^i, \ldots, x_{t-m+1}^i)$ and states of the dynamical attractor A_i in $D \times X_i$. In particular, as a consequence of the one-to-one correspondence guaranteed by Theorem 1, the equality

$$(x_t^i, \ldots, x_{t-m+1}^i) = (x_{t'}^i, \ldots, x_{t'-m+1}^i) \tag{2}$$

implies that $d_t = d_{t'}$. This is the key to identifying individual states of the driver.

Due to continuity, (2) need not hold exactly to provide information. If the difference is small, say in the Euclidean norm, then d_t and $d_{t'}$ must also be close. This motivates choosing m sufficiently large so that Stark's theorem applies to $D \times X_i$ for $i = 1, \ldots, k$. We choose some $\epsilon > 0$ and for a given $(x_t^i, \ldots, x_{t-m+1}^i)$, group the set of times t' for which (2) holds within ϵ. In this way we form a set of equivalence classes.

To simplify the collection of the equivalence classes, one may choose one of the X_i and work in its delay coordinate space. This space contains a one-to-one representation of the attractor A_i in $D \times X_i$, according to Stark's theorem. Grouping the equivalence classes gives a quotient space of A_i, called D^*. Below, we see that D^* is a semiconjugacy with the dynamics of D.

There are three functions that can be defined for the set D^*. First, every equivalence class d^* in D^* has associated with it a unique d in D, so define the function s from D^* to D by $s(d^*) = d$. The function s is *onto*, meaning that the image of s is all of D. (This follows from the fact that the dynamics f is ergodic on D.) Second, there is a natural dynamical rule g^* from the set D^* to itself that is inherited from the dynamics on the delay coordinates. The equivalence class $g^*(d^*)$ is defined to be the one the elements of d^* are mapped to under the system dynamics f_1. In addition to the functions s and g^*, for each $1 \leq i \leq k$, the function p_i from A_i to D^* can be defined by sending each $a_i \in A_i$ to the equivalence class of a_i. The following diagram shows the relation between the functions p_i, s and the new dynamical system g^* on D^*.

$$\begin{array}{ccccc} D & \xleftarrow{\;s\;} & D^* & \xleftarrow{\;p_i\;} & A_i \\ \big\downarrow g & & \big\downarrow g^* & & \big\downarrow f_i^* \\ D & \xleftarrow{\;s\;} & D^* & \xleftarrow{\;p_i\;} & A_i \end{array}$$

Because g^* is onto, the right half of the diagram shows that g^* satisfies the definition of semiconjugacy. The map $g^* : D^* \to D^*$ is said to be *semiconjugate* to the map $f : D \to D$ if there exists an onto map $s : D^* \to D$ satisfying $g \circ s = s \circ g^*$, that is, the left side of the diagram commutes. The analogous statement about the right side of the diagram is also true. The following theorem was introduced in [8]:

Theorem 2. *(Shared Dynamics Reconstruction Theorem.) Assume $g : D \to D$ is ergodic, and in addition drives X_i, $1 \leq i \leq k$ as in (1). Choose m large enough and g, f^i generic such that all skew products $D \times X_i$ are reconstructed in R^m. Define $g^* : D^* \to D^*$ as the map induced as above. Then, generically, (1) the map g^* is semiconjugate to g, and (2) for each i, the induced map is semiconjugate to g^*.*

Roughly speaking, if g_1 is semiconjugate to g_2, then g_1 "contains" the dynamics of g_2. The content of the theorem is that according to the left hand square of the above diagram, D^* captures at least the dynamics of the driver D, and may contain more. However, according to the right side of the diagram for each i, any extra dynamics in D^* must be common to all of the X_i, due to part (2) of the theorem. This is the meaning of "shared dynamics".

Next we show how this theorem leads to an algorithm that extracts shared dynamics of the D using time series data observed from the X_i.

Shared Dynamics Algorithm.

Choose m large enough to unfold the dynamics on each $D \times X_i$, and use delay coordinates to create the reconstructed attractor A_i, which is in one-to-one correspondence with $D \times X_i$. Choose one of the X_i arbitrarily, say X_1. The basis of the algorithm is to group together points in X_1 that lie over the same point in D, the so-called fibers over D. According to the theorem, at each time t when the dynamics returns to the same point in X_j, the state of d in D is the same. With this information, one can search for delay vectors in X_j that are close in the delay reconstruction, and return information to X_1 about points over the same driver state d. A neighborhood size ε must be chosen to decide the meaning of nearly identical, for this purpose. The degree of discretization of the resulting dynamical attractor D^* will depend on this choice.

Using this method of associating points to the fibers over D, one proceeds through all points of the reconstructed attractor X_1 to fit them in an appropriate equivalence class. Choosing representatives for the equivalence classes from a chain of overlapping ε-neighborhoods retains the topological form imposed by the original dynamics. Note that no re-embedding is necessary, since the points of D^* constitute a subset of the reconstructed X_1, which has no self-intersections by assumption.

§3. Examples

In this section we illustrate the use of the Shared Dynamics Algorithm on two examples. In each case, a system of difference equations of form (1) is constructed and the output data x_i are provided to the algorithm for analysis.

Example 1. Consider the system

$$\begin{aligned} x_t^1 &= \lambda_1 x_{t-1}^1(1 - x_{t-1}^1) + 0.5 d_{t-1} \quad \text{mod } 1 \\ &\vdots \\ x_t^5 &= \lambda_5 x_{t-1}^5(1 - x_{t-1}^5) + 0.5 d_{t-1} \quad \text{mod } 1 \\ d_t &= \lambda_0 d_{t-1}(1 - d_{t-1}) \end{aligned} \tag{3}$$

where $\lambda_0 = 3.6266$ represents the period 6 window in the logistic bifurcation sequence, and $\lambda_1 = 3.77, \lambda_2 = 3.775, \lambda_3 = 3.78, \lambda_4 = 3.785, \lambda_5 = 3.79$ are chosen from a chaotic regime. The modulo function was used to keep trajectories from moving outside of the basin of the chaotic attractor.

Figure 1(a) shows the delay coordinate embedding of the reconstructed attractor A_1 of $D \times X_1$, using the embedding dimension $m = 3$. A trajectory of length 3000 time units is shown. The result of the Shared Dynamics Algorithm with $\epsilon = 0.005$, the reconstructed driver D^*, is shown in Figure 1(b). Each point represents an equivalence class of delay coordinate vectors. The topology and dynamics of the period-six attractor from D are efficiently recovered.

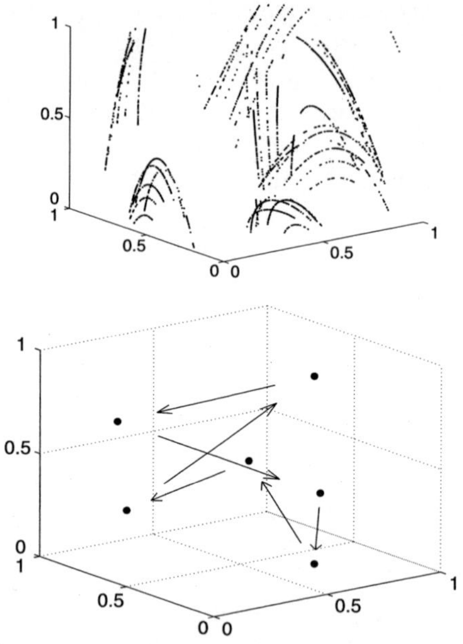

Fig. 1. (a) Delay coordinate reconstruction of one skew product from (3). (b) Reconstructed driver dynamics from the algorithm.

Example 2. The second example is the system

$$
\begin{aligned}
x_t^1 &= a_1 - \tfrac{1}{4}[(x_{t-1}^1)^2 + (y_{t-1}^1)^2] - \tfrac{1}{2}x_{t-1}^1 y_{t-1}^1 \\
&+ \frac{1+b}{2}x_{t-1}^1 + \frac{1-b}{2}y_{t-1}^1 + \alpha d_{t-1} \\
y_t^1 &= a_1 - \tfrac{1}{4}[(x_{t-1}^1)^2 + (y_{t-1}^1)^2] - \tfrac{1}{2}x_{t-1}^1 y_{t-1}^1 \\
&+ \frac{-1+b}{2}x_{t-1}^1 + \frac{-1-b}{2}y_{t-1}^1 \\
x_t^2 &= a_2 - \tfrac{1}{4}[(x_{t-1}^2)^2 + (y_{t-1}^2)^2] - \tfrac{1}{2}x_{t-1}^2 y_{t-1}^2 \\
&+ \frac{1+b}{2}x_{t-1}^2 + \frac{1-b}{2}y_{t-1}^2 + \alpha d_{t-1} \\
y_t^2 &= a_2 - \tfrac{1}{4}[(x_{t-1}^2)^2 + (y_{t-1}^2)^2] - \tfrac{1}{2}x_{t-1}^2 y_{t-1}^2 \\
&+ \frac{-1+b}{2}x_{t-1}^2 + \frac{-1-b}{2}y_{t-1}^2 \\
x_t^3 &= a_3 - \tfrac{1}{4}[(x_{t-1}^3)^2 + (y_{t-1}^3)^2] - \tfrac{1}{2}x_{t-1}^3 y_{t-1}^3 \\
&+ \frac{1+b}{2}x_{t-1}^3 + \frac{1-b}{2}y_{t-1}^3 + \alpha d_{t-1} \\
y_t^3 &= a_3 - \tfrac{1}{4}[(x_{t-1}^3)^2 + (y_{t-1}^3)^2] - \tfrac{1}{2}x_{t-1}^3 y_{t-1}^3 \\
&+ \frac{-1+b}{2}x_{t-1}^3 + \frac{-1-b}{2}y_{t-1}^3 \\
d_t &= a_0 - d_{t-1}^2 - b_0 d_{t-2} \qquad (4)
\end{aligned}
$$

where $a_0 = 0.95, b_0 = 0.4$ generates a period 4 attractor for the Hénon map [3], and $a_1 = 1.26, a_2 = 1.27, a_3 = 1.28, b = 0.3$ represent chaotic trajectories. The drive parameter is $\alpha = 0.05$.

Figure 2(a) shows the delay coordinate embedding of the reconstructed attractor A_1 of $D \times X_1$, using the embedding dimension $m = 3$ and delay coordinate vectors of form $(x_t^1, x_{t-1}^1, x_{t-2}^1)$. These vectors together with vectors of form $(x_t^i, x_{t-1}^i, x_{t-2}^i)$ for $i = 2, 3$ were used in the algorithm. A trajectory of length 2000 time units is shown. The result of the Shared Dynamics Algorithm with $\epsilon = 0.04$, the reconstructed driver D^*, is shown in Figure 2(b). Each point represents an equivalence class of delay coordinate vectors. The topology and dynamics of the period-four attractor from D are efficiently recovered.

§4. Conclusions

This article demonstrates a type of signal processing for difference equations, the idea being to identify system characteristics from output

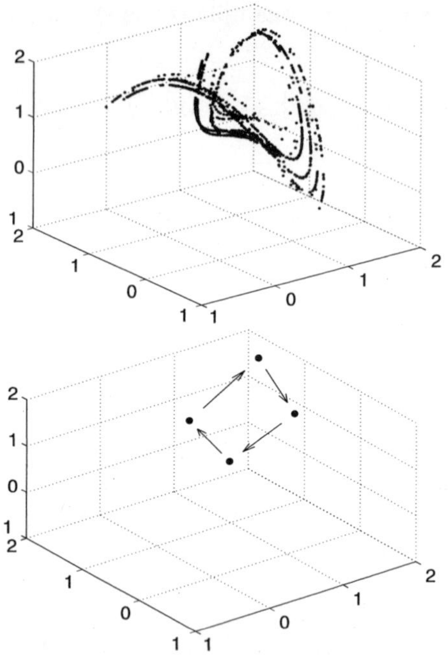

Fig. 2. (a) Reconstruction of one skew product from (4). (b) Reconstructed driver dynamics.

data. The specific goal in this case is to identify the dynamics of the common driver in a system of nonautonomous difference equations.

The examples provided deal with periodic drivers, such as the periodic six orbit driving five logistic maps in Example 1. The algorithm is not limited to periodic drivers, although the job of identifying the result will be more difficult the more complicated the driver dynamics. If the driver is chaotic, for example, sophisticated system identification methods may be needed to analyze and classify the result.

In the case of a chaotic system driving other chaotic systems, for example, although the algorithm introduced here will work in principle, the data requirements may be challenging. For periodic driving, we have shown in the examples that a few thousand data points suffice to determine the driver.

This research was supported by the National Science Foundation.

References

[1] D. Aeyels, Generic observability of differentiable systems, SIAM J. Control Optim., **19** (1981), 595–603.

[2] M. Casdagli, Nonlinear Modeling and Forecasting, (eds. M. Casdagli and S. Eubank), Addison Wesley, Reading, MA, 1992.

[3] M. Hénon, A two-dimensional mapping with a strange attractor, Comm. Math. Phys., **50** (1976), 69–77.

[4] H. Kantz and T. Schreiber, Nonlinear Time Series Analysis, Cambridge Univ. Press, Cambridge, 1997.

[5] E. Ott, T. Sauer and J. Yorke, Coping with Chaos: Analysis of Chaotic Data and the Exploitation of Chaotic Systems, Wiley Interscience, New York, 1994.

[6] N. Packard, J. Crutchfield, J. D. Farmer and R. Shaw, Geometry from a time series, Phys. Rev. Lett., **45** (1980), 712–715.

[7] T. Sauer, J. A. Yorke and M. Casdagli, Embedology, J. Stat. Phys., **65** (1991), 579–616.

[8] T. Sauer, Reconstruction of shared dynamics in a network, Phys. Rev. Lett., **93** (2004), 198701.

[9] J. Stark, Delay embeddings of forced systems I: Deterministic forcing, J. Nonlinear Sci., **9** (1999), 255–332.

[10] F. Takens, Detecting strange attractors in turbulence, In: Dynamical Systems and Turbulence, Lecture Notes in Math., **898**, Springer-Verlag, 1981.

George Mason University
Fairfax, VA 22030
USA

Advanced Studies in Pure Mathematics 53, 2009
Advances in Discrete Dynamical Systems
pp. 311–318

Periodic and chaotic behavior in a class of second order difference equations

Hassan Sedaghat

Abstract.

We discuss the occurrence and nature of periodic and chaotic behavior in a class of nonlinear second order difference equations and present criteria for the attractivity of the periodic solutions.

§1. Introduction

Consider the second order difference equation

$$x_{n+1} = ax_n + bx_{n-1} + f_n(x_n - cx_{n-1}) \tag{1}$$

where $f_n : \mathbb{R} \to \mathbb{R}$ is a given sequence of functions and a, b, c are real constants satisfying the conditions

$$c \neq 0, \;\; b + c(a - c) = 0. \tag{2}$$

Equations bearing some similarity to the autonomous version of (1) have been considered in the literature; see, e.g. [4, Sec.2.5]. These studies have concentrated on the global stability of equilibrium. Equation (1) also generalizes equations introduced in some of the heuristic business cycle models in macroeconomics. For example, special cases of (1) includes Hicks' model (its 2nd order case) and Goodwin's model (its discrete version) as well as Puu's discrete second order equation. A detailed comparative analysis of the mathematics behind these classical models appears in [7].

In this note we take a close look at Equation (1) and obtain criteria for the occurrence of attracting periodic solutions as well as conditions that imply the occurrence of chaotic behavior. We use the fact that under conditions (2), Equation (1) decomposes into a weakly coupled

---

Received November 14, 2006.
Revised October 12, 2007.

system of first order difference equations. Such decompositions are instances of semiconjugacy; see [7] and [8] for some background on this subject.

§2. Main results

Equation (1) may be restated as

$$x_{n+1} - cx_n = (a-c)x_n + bx_{n-1} + f_n(x_n - cx_{n-1}). \tag{3}$$

If $a - c \neq 0$ then we may factor out $a - c$ and using (2) transform (3) into

$$t_{n+1} = (a-c)t_n + f_n(t_n) \doteq g_n(t_n) \tag{4a}$$

$$x_{n+1} = t_{n+1} + cx_n \tag{4b}$$

Note that if $a = c$ then $b = 0$ by (2) so we may still obtain (4a) directly from Equation (1) with $g_n = f_n$.

Equations (4) define a triangular system of first order equations in the sense that the first equation is independent of the second. A general result on the structure of periodic solutions of (4) in terms of the periodic orbits of its two first order equations appears in [1] for the autonomous case, i.e. when $f_n = f_0$ for all n. Here, since the system (4) is specific in its second equation, we derive the needed relationships directly with f_n variable and also establish attractivity when $|c| < 1$.

For a given sequence of real numbers $\{t_n\}$, the general solution of (4b) is

$$x_n = c^n x_0 + \sum_{j=1}^{n} c^{n-j} t_j, \quad n \geq 1. \tag{5}$$

The sum in (5) is of convolution type but here the sequence $\{t_n\}$ is rarely given explicitly.

Lemma 1. *Assume that* $|c| \neq 1$.

(a) *Let* $\{t_n\}$ *be a periodic sequence of real numbers with period* p. *If* $\{\tau_0, \ldots, \tau_{p-1}\}$ *is one cycle of* $\{t_n\}$ *and*

$$\xi_i = \frac{1}{1-c^p} \sum_{j=0}^{p-1} c^{p-j-1} \tau_{(i+j) \bmod p} \quad i = 0, 1, \ldots, p-1 \tag{6}$$

then the solution $\{x_n\}$ *of Eq.(4b) with* $x_0 = \xi_0$ *and* $t_1 = \tau_0$ *has period* p *and* $\{\xi_0, \ldots, \xi_{p-1}\}$ *is a cycle of* $\{x_n\}$.

(b) *If for a given sequence* $\{t_n\}$ *of real numbers Eq.(4b) has a solution* $\{x_n\}$ *of period* p *then* $\{t_n\}$ *is periodic with period* p.

Proof. (a) With $x_0 = \xi_0$ and $t_1 = \tau_0$ we get $x_1 = cx_0 + t_1 = c\xi_0 + \tau_0$. Using (6) for ξ_0 gives

$$x_1 = \frac{c}{1-c^p}\left(\sum_{j=0}^{p-1} c^{p-j-1}\tau_j\right) + \tau_0 = \frac{1}{1-c^p}\left(\sum_{j=0}^{p-2} c^{p-j-1}\tau_{j+1} + \tau_0\right) = \xi_1$$

Proceeding in an inductive fashion, we show in this way that $x_i = \xi_i$ for $i = 0, \ldots, p-1$. Next, we show that $x_p = x_0$. Using (5) we have

$$x_p = c^p\xi_0 + \sum_{j=0}^{p-1} c^{p-j-1}\tau_j = \frac{c^p}{1-c^p}\sum_{j=0}^{p-1} c^{p-j-1}\tau_j + \sum_{j=0}^{p-1} c^{p-j-1}\tau_j = \xi_0 = x_0.$$

Hence $\{x_n\}$ is a solution of (4b) with period p, as claimed.

(b) Suppose that for a given sequence $\{t_n\}$ of real numbers, the corresponding solution of (4b) is periodic with period p. Let $t_1 = x_1 - cx_0$ and from (4b) obtain

$$t_{p+1} = x_{p+1} - cx_p = x_1 - cx_0 = t_1.$$

It follows that $\{t_n\}$ is periodic with period p. Q.E.D.

Theorem 1. (periodic solutions and limit cycles)

(a) *Assume that $|c| \neq 1$ and let $\{t_n\}$ be a periodic solution of the first order equation (4a) with prime period p. If $\{\tau_0, \ldots, \tau_{p-1}\}$ is one cycle of $\{t_n\}$ then (1) has a solution $\{x_n\}$ of prime period p with a cycle $\{\xi_0, \ldots, \xi_{p-1}\}$ given by (6).*

(b) *Assume that the functions f_n are continuous. If $|c| < 1$ and $\{t_n\}$ is an attracting periodic solution of (4a) then $\{x_n\}$ is an attracting periodic solution of (1).*

Proof. (a) In light of Lemma 1(a) we only need to show that p is the prime or minimal period for $\{x_n\}$. Let q be the prime period of $\{x_n\}$ so that $q \leq p$. Then by Lemma 1(b) $\{t_n\}$ has period $q \geq p$ since p is the prime period for $\{t_n\}$. Therefore, $q = p$.

(b) Let $\{\tau_0, \ldots, \tau_{p-1}\}$ be an attracting cycle for (4a) with

$$\lim_{n\to\infty} t_{pn+i} = \tau_{i-1}, \quad i = 1, 2, \ldots, p$$

Let $s_n = \sum_{j=1}^{n} c^{n-j} t_j$. Then by rearranging terms in the summation we find that

$$\begin{aligned} s_{pn} &= c^{pn-1}t_1 + c^{pn-2}t_2 \cdots + c^{pn-p}t_p \\ &+ c^{pn-p-1}t_{p+1} + c^{pn-p-2}t_{p+2} \cdots + c^{pn-2p}t_{2p} \\ &+ \cdots \\ &+ c^{p-1}t_{p(n-1)+1} + c^{p-2}t_{p(n-1)+2} \cdots + c^{pn-pn}t_{p(n-1)+p} \end{aligned}$$

$$\begin{aligned} &= c^{p-1}(c^{pn-p}t_1 + c^{pn-2p}t_{p+1} + \cdots + t_{pn-p+1}) \\ &+ c^{p-2}(c^{pn-p}t_2 + c^{pn-2p}t_{p+2} + \cdots + t_{pn-p+2}) \\ &+ \cdots \\ &+ c^{pn-p}t_p + c^{pn-2p}t_{2p} + \cdots + t_{pn} \\ &= \sum_{i=1}^{p} c^{p-i} \sum_{k=0}^{n-1} (c^p)^{n-k-1} t_{pk+i}. \end{aligned}$$

Now for $i = 1, 2, \ldots, p$ define

$$\sigma_n^i = \sum_{k=0}^{n-1} (c^p)^{n-k-1} t_{pk+i} \quad \text{and} \quad \gamma_n^i = \sum_{k=0}^{n-1} (c^p)^{n-k-1} \tau_{i-1} = \tau_{i-1} \sum_{k=0}^{n-1} c^{pk}.$$

Notice that

$$\begin{aligned} \left|\sigma_n^i - \frac{\tau_{i-1}}{1-c^p}\right| &\le |\sigma_n^i - \gamma_n^i| + \left|\gamma_n^i - \frac{\tau_{i-1}}{1-c^p}\right| \\ &\le \sum_{k=0}^{n-1} |c^p|^{n-k-1} |t_{pk+i} - \tau_{i-1}| + \left|\gamma_n^i - \frac{\tau_{i-1}}{1-c^p}\right| \end{aligned}$$

Clearly the second term on the right hand side approachs 0 as $n \to \infty$. As for the first term, let $m \ge 1$ and define

$$\delta = \max_{1 \le i \le p} \left\{ \sup_{k \ge 1} |t_{pk+i} - \tau_{i-1}| \right\} < \infty, \quad \delta_m^i = \sup_{k \ge m} |t_{pk+i} - \tau_{i-1}|$$

and observe that for $m < n$

$$\sum_{k=0}^{n-1} |c^p|^{n-k-1}|t_{pk+i} - \tau_{i-1}|$$
$$= \sum_{k=0}^{m-1} |c^p|^{n-k-1}|t_{pk+i} - \tau_{i-1}| + \sum_{k=m}^{n} |c^p|^{n-k-1}|t_{pk+i} - \tau_{i-1}|$$
$$\leq |c^p|^{n-m}\,\delta \sum_{k=0}^{m-1} |c^p|^k + \delta_m^i \sum_{k=m}^{n} |c^p|^{n-k-1}$$

By taking n and m sufficiently large, each of the last two terms above can be made arbitrarily small. Therefore,

$$\lim_{n\to\infty} \sigma_n^i = \frac{\tau_{i-1}}{1-c^p} \qquad i = 1, \ldots, p.$$

It follows that

$$\lim_{n\to\infty} x_{pn} = \lim_{n\to\infty} s_{pn} = \sum_{i=1}^{p} \frac{c^{p-i}\tau_{i-1}}{1-c^p} = \frac{c^{p-1}\tau_0 + \cdots + c\tau_{p-2} + \tau_{p-1}}{1-c^p}.$$

Therefore, $x_{pn} \to \xi_0$ as $n \to \infty$ with ξ_0 as in Lemma 1. From this and (4b) we obtain

$$\lim_{n\to\infty} x_{pn+1} = \lim_{n\to\infty} (t_{pn+1} + cx_{pn}) = \tau_0 + c\xi_0 = \xi_1.$$

Inductively, we find that $x_{pn+i} \to \xi_i$ for $i = 0, 1, \ldots, p-1$. This implies that $\{x_n\}$ is an attracting periodic solution of (1). Q.E.D.

Examples. 1. Consider the difference equation

(7a) $$x_{n+1} = cx_n + \alpha_n(x_n - cx_{n-1})^q$$

(7b) $$0 < |c|, |q| < 1,\ \alpha_{2m} = \alpha_0 > 0,\ \alpha_{2m+1} = \alpha_1 > 0,\ m = 0, 1, \ldots$$

In this case, $g_n(t) = \alpha_n t^q$ for $t > 0$ and straightforward calculations show that all positive solutions of the first order equation $t_{n+1} = \alpha_n t_n^q$ converge to the 2-cycle

$$t_{2n} \to \alpha_0^{q/(1-q^2)}\alpha_1^{1/(1-q^2)} = \tau_0, \quad t_{2n+1} \to \alpha_0^{1/(1-q^2)}\alpha_1^{q/(1-q^2)} = \tau_1$$

Therefore, every solution of (7a) in the invariant region $\{(x, y) : x > cy\}$ converges to the attracting cycle

$$\xi_0 = \frac{c\tau_0 + \tau_1}{1-c^2}, \quad \xi_1 = \frac{c\tau_1 + \tau_0}{1-c^2}.$$

Note that in this example α_n can be taken as a sequence with any period p with slightly more calculating effort.

2. For the difference equation

$$x_{n+1} = c^2 x_{n-1} + \alpha(x_n - cx_{n-1})^q \tag{8}$$
$$\alpha > 0,\ 0 < |c| < 1,\ q = 2j/(2k+1) < 1,\ k \geq j \geq 1$$

we have $g_n(t) = g(t) = -ct + \alpha t^q$ for all real t and straightforward calculations show that all nonzero solutions of the first order equation $t_{n+1} = -ct_n + \alpha t_n^q$ converge to its unique fixed point

$$\bar{t} = \left(\frac{\alpha}{1+c}\right)^{1/(1-q)} = \left(\frac{\alpha}{1+c}\right)^{\frac{2k+1}{2(k-j)+1}}.$$

Hence the fixed point $\bar{x} = \bar{t}/(1-c)$ of (8) is globally attracting.

Now we consider conditions that imply chaotic behavior. For difference equations a chaotic solution is typically a non-periodic, oscillatory solution that is sensitive to initial values. See [2], [5], [6] and [7] for some background on this concept. For first order difference equations a more refined definition of chaotic solutions was given in [3]. We first give conditions for solutions of (1) to be uniformly bounded.

Lemma 2. (boundedness) *Let $|c| < 1$. If $\{t_n\}$ is a bounded sequence with $|t_n| \leq B$ for some $B > 0$, then the corresponding solution $\{x_n\}$ for Eq.(4b) is also bounded and there is a positive integer N such that*

$$|x_n| < |c| + \frac{B}{1-|c|} \quad \textit{for all } n \geq N. \tag{9}$$

Proof. From (5) we obtain

$$|x_n| \leq |c|^n |x_0| + \sum_{j=1}^{n} |c|^{n-j} B < |c|^n |x_0| + B \sum_{k=0}^{\infty} |c|^k = |c|^n |x_0| + \frac{B}{1-|c|}.$$

Q.E.D.

Now if n is large enough, then $|c|^n |x_0| \leq |c|$ from which (9) follows.

In light of Lemma 2, the following result is easy to prove using Theorem 3.3.3 in [7].

Theorem 2. (chaotic behavior) *Assume that $|c| < 1$ and let $f_n = f$ for all n where f is continuous on an invariant closed interval $[\mu, \nu]$ on the line. If the first order equation (4a) is chaotic within $[\mu, \nu]$ then the* *second order equation (1) is chaotic in the following invariant compact, convex set in the plane:*

$$\{(x,y) : cx+\mu \le y \le cx+\nu\} \cap \left[-|c| - \frac{\max\{|\mu|,|\nu|\}}{1-|c|}, |c| + \frac{\max\{|\mu|,|\nu|\}}{1-|c|}\right]^2.$$

A straightforward example for illustration is the autonomous equation

$$x_{n+1} = cx_n + \alpha(x_n - cx_{n-1})(1 - x_n + cx_{n-1}) \tag{10}$$

where we have picked $a = c \in (-1,1)$ and $f(t) = \alpha t(1-t)$ in (1). As α varies in the interval [0,4] the familiar behavior of $f(t)$ on the interval [0,1] is translated via Theorems 1 and 2 into the analogous behavior for the solutions of (10) in the compact invariant region

$$\{(x,y) : cx \le y \le cx+1\} \cap \left[-|c| - \frac{1}{1-|c|}, |c| + \frac{1}{1-|c|}\right]^2$$

in the plane. Similar observations apply to

$$x_{n+1} = c^2 x_{n-1} + (x_n - cx_{n-1})(\alpha - x_n + cx_{n-1}),\ 0 < |c| < 1 \le \alpha \le 4.$$

A less routine example is the one parameter family of autonomous rational equations

$$x_{n+1} = \frac{6x_n^2 - 5x_n x_{n-1} + x_{n-1}^2 + 4}{4x_n - 2x_{n-1}} - \alpha, \quad 0 < \alpha < 2 \tag{11}$$

obtained from (1) by setting $a = 3/2$, $c = 1/2$, $f(t) = 1/t - \alpha$ and $g(t) = t + f(t)$. For $\alpha > \sqrt{2}$ it can be shown that all iterates of g will eventually enter and remain in the invariant interval $[2-\alpha, g(2-\alpha)]$. With increasing value of α a sequence of bifurcations of periodic orbits ensues that progresses through the Sharkovski ordering. This behavior can then be translated into the analogous behavior for (11) using Theorems 1 and 2.

§3. Extensions and future directions

The results of the previous section can be readily extended to the equation

$$x_{n+1} = a_n x_n + b_n x_{n-1} + f_n(x_n - cx_{n-1}) \tag{12}$$

where the sequences $\{a_n\}$ and $\{b_n\}$ satisfy

$$c \ne 0,\ b_n + c(a_n - c) = 0,\ \text{for all } n. \tag{13}$$

This would lead to essentially the same type of triangular system as (4) but with a slightly greater range of possibilities. Theorems and Lemmas 1 and 2 apply to Equation (12) essentially as they are presented above. A more distant generalization is to higher order equations of the following type:

$$x_{n+1} = c_n x_{n-k+1} + f_n(x_n - c_{n-1}x_{n-k}) \tag{14}$$

where k is a fixed positive integer, $\{c_n\}$ is a given sequence of real numbers and $\{f_n\}$ is a sequence of real valued functions all defined on a given interval I. Equation (14) is equivalent to the triangular system

$$t_{n+1} = f_n(t_n) \tag{15a}$$

$$x_{n+1} = t_{n+1} + c_n x_{n-k+1} \tag{15b}$$

Equation (15b) produces different results from those seen in Theorems 1 and 2 above, especially if c_n does not converge to a limit.

References

[1] L. Alseda and J. Llibre, Periods for triangular maps, Bull. Austral. Math. Soc., **47** (1993), 41–53.

[2] P. Collet and J.-P. Eckmann, Iterated Maps on the Interval as Dynamical Systems, Birkhäuser, Boston, 1980.

[3] R. L. Devaney, An Introduction to Chaotic Dynamical Systems, 2nd ed., Addison-Wesley, Redwood City, 1989.

[4] V. L. Kocic and G. Ladas, Global Behavior of Nonlinear Difference Equations of Higher Order with Applications, Kluwer, Dordrecht, 1993.

[5] T.-Y. Li and J. A. Yorke, Period three implies chaos, Amer. Math. Monthly, **82** (1975), 985–992.

[6] F. R. Marotto, Snap-back repellers imply chaos in $\mathbb{R}^n$, J. Math. Anal. Appl., **63** (1978), 199–223.

[7] H. Sedaghat, Nonlinear Difference Equations: Theory with Applications to Social Science Models, Kluwer, Dordrecht, 2003.

[8] H. Sedaghat, Order-reducing form symmetries and semiconjugate factorizations of difference equations, arXiv:0804.3579.

Department of Mathematics
Virginia Commonwealth University
Richmond, Virginia 23284-2014
USA

E-mail address: `hsedagha@vcu.edu`

Advanced Studies in Pure Mathematics 53, 2009
Advances in Discrete Dynamical Systems
pp. 319–331

Uniqueness of polymorphism for a discrete, selection-migration model with genetic dominance

James F. Selgrade[1] and James H. Roberds

Abstract.

The migration into a natural population of a controlled population, e.g., a transgenic population, is studied using a one island selection-migration model. A 2-dimensional system of nonlinear difference equations describes changes in allele frequency and population size between generations. Biologically reasonable conditions are obtained which guarantee the existence and uniqueness of a polymorphic equilibrium in the cases of complete dominance and no dominance in fitness. This model may provide some useful information about the migration of transgenes into a natural population.

§1. Introduction

Migration and natural selection are two evolutionary processes that are important determinants of genetic composition and demographic properties of populations [3]. Effects produced by the interaction of these factors can significantly influence genetic variability in populations as well as affect mean population fitness. Here we study the impact of these two agencies within the framework of a one-island migration model. In this context, genes and individuals are viewed as migrating from a large donor population into an island or receiving population in which density-dependent selection takes place. Based upon this model, we describe the changes in allele frequency and population size that occur over generations in an island population in terms of a system of two-dimensional difference equations. With certain assumptions this system is applicable to the investigation of the fate of a transgene that has migrated into a nontargeted natural population. Considered

---

Received October 12, 2007.
2000 *Mathematics Subject Classification.* 39A11, 92D25.
[1]Research partially supported by the USDA-Forest Service, Southern Research Station, Southern Institute of Forest Genetics, Saucier, MS.

in this light, a population composed of individuals genetically modified through transgenics is understood to be the donor population, and a natural population is viewed as the island or recipient population. In this paper we demonstrate that a unique polymorphic equilibrium exists for this system when density-dependent fitness is regarded to be under the control of a single genetic locus with alleles expressed according to either a complete dominance or a no dominance mode [1].

The one-island model studied here could potentially be used to explore incorporation of transgenes into natural tree populations. In forest biology, there is much concern about ecological consequences that might result from such gene transfer. As a case in point, genetic engineering is being seriously considered as a method for improving wood properties of loblolly pine (*Pinus taeda*), a timber tree grown extensively in the southeastern United States and several South American countries (see Sedjo [7]). This biotechnology has the potential to produce trees with greater quantities of useful fiber and fiber that can more efficiently be processed into wood pulp. But in the southern United States, loblolly pine commercial stands composed of trees carrying certain transgenes could pose an ecological hazard to nearby natural pine forests because of the possibility of transgene transfer (see Williams [13]). Uncertainties abound about effects that will accrue from the spread of transgenes in natural forest environments and this heightens unease about the risks that could accompany use of this technology. As a result, there is interest in studying the possible fate of transgenes that have entered nontargeted populations.

§2. Model background

Density-dependent selection may be studied by considering a diploid population with two alleles, A and a, at a single autosomal locus. Here, a population of density x consists of individuals with one of three genotypes, AA, Aa, or aa. Let p denote the frequency of the A allele, where $0 \leq p \leq 1$, and hence $1-p$ is the frequency of the a allele. The effects of natural selection determine an average per capita replacement rate or *fitness* f_{ij} for the ij-genotype, where $i,j = A, a$, which measures fertility and viability of that genotype. Allele fitnesses f_A and f_a are linear combinations of genotype fitnesses weighted by allele frequency and are defined by $f_A \equiv pf_{AA} + (1-p)f_{Aa}$ and $f_a \equiv pf_{Aa} + (1-p)f_{aa}$. The population mean fitness f is given by $f \equiv pf_A + (1-p)f_a$. For this study we assume density-dependent selection, i.e., genotypic fitnesses $f_{ij}(x)$ depend on population density x and are independent of the allele frequency p.

When population density is small, each genotype does well but genotype fitness decreases as population density increases because of the detrimental effects of crowding. Such functions f_{ij} are referred to as *pioneer* fitness functions [8]. Exponential (see Ricker [4]), rational (see Hassell and Comins [2]), and linear (see Selgrade and Roberds [9, 10]) functions have been used as pioneer fitnesses in modeling populations. More specifically, we assume that each $f_{ij}(0) > 1$ and that $f_{ij}(x) \searrow 0$ as $x \to \infty$. This guarantees that for each fixed $p \in [0,1]$, there is a population density $x > 0$ such that the population equilibrates, i.e., $f(p,x) = 1$. Henceforth, for $i, j = A, a$ we assume:

(A1) $f'_{ij}(x) < 0$ **for all** $x > 0$**,** $f_{ij}(0) > 1$ **and** $f_{ij}(x) \to 0$ **as** $x \to \infty$.

To study the migration of a population of constant allele frequency into another population, we appeal to a one-island model discussed by Roberds and Selgrade, see [5] and [11]. Such a model might represent the migration into a natural population of a controlled population, e.g., a transgenic population. Let x and p denote the island population density and allele frequency, respectively. Following selection in each generation, assume gametes are contributed to the island population by immigration from a population with allele frequency the constant q with $0 \le q \le 1$. Random mating occurs following migration so that the number of additional zygotes in the next generation produced as a consequence of immigration is denoted by y. In [5] and [11], we derived the following system of difference equations that describes changes in allele frequency and population size between generations:

$$\begin{aligned} p_{n+1} &= \frac{p_n\, x_n\, f_A + q\, y}{x_n\, f + y} \\ x_{n+1} &= x_n\, f + y\,. \end{aligned} \tag{1}$$

Here p_{n+1} and x_{n+1} represent allele frequency and population size in the next generation.

It is convenient both mathematically and biologically to introduce a per capita migration rate for $x > 0$ given by

$$h(x) \equiv \frac{y}{x}\,,$$

which measures the per capita migration per generation relative to the island population size x. It is biologically reasonable to assume that $h(x)$ is a nonincreasing function of population size, i.e., $h'(x) \le 0$. There are two standard examples of migration in this form. The simplest case occurs when the same amount of migration takes place each generation so that y is constant and $h'(x) = -y/x^2 < 0$. The second type occurs when the amount of migration increases linearly with x in each generation as discussed in Selgrade and Roberds [11], i.e., $y = m\,x$ where m is a constant. For this study, we consider migration of the former type, i.e., we assume that y is constant.

After replacing y by $xh(x)$ in (1), the transition equations become

$$
\begin{aligned}
p_{n+1} &= \frac{p_n\, f_A + q\, h}{f + h} \\
x_{n+1} &= x_n\, (f + h)\,.
\end{aligned}
\tag{2}
$$

In (2), $f + h$ denotes the per capita transition function for the island population. For $x > 0$, systems (1) and (2) are equivalent so, henceforth, we study (2).

§3. Properties of equilibria

The phase space for system (2) is the slot in the (p, x)-plane designated by

$$\mathcal{S} \equiv \{(p, x) : 0 \le p \le 1, 0 < x\}\,.$$

When $y = 0$ (i.e., $h = 0$), the boundary lines of $\mathcal{S}$, $\{p = 0\}$ and $\{p = 1\}$, represent allele fixation and, therefore, are invariant. If $y > 0$ (i.e., $h > 0$) and $0 < q < 1$, points on the vertical boundaries of $\mathcal{S}$ are mapped into the interior of $\mathcal{S}$.

An equilibrium E is an allele frequency $\overline{p}$, $0 \le \overline{p} \le 1$, and a population density $\overline{x} > 0$ which remain constant across generations, i.e., $p_n = \overline{p}$ and $x_n = \overline{x}$ for all n. Such an E is said to be a *polymorphism* if $0 < \overline{p} < 1$. From (2), an equilibrium $E = (\overline{p}, \overline{x})$ satisfies the system:

$$
\begin{aligned}
p &= p\, f_A + q\, h \\
1 &= f + h\,.
\end{aligned}
\tag{3}
$$

Since the frequency of the allele a is $1 - p$ and is constant at equilibrium, the equation $1 - p_{n+1} = 1 - p_n$ implies that the following equation must also be satisfied at equilibrium

$$1 - p = (1 - p)\, f_a + (1 - q)\, h\,. \tag{4}$$

Hence, a polymorphism is a point of intersection of the three isocline curves:

$$\begin{aligned}
&\mathcal{C} \equiv \{(p,x) : f(p,x) + h(x) = 1\} \\
&\mathcal{C}_A \equiv \{(p,x) : p\,[f_A(p,x) - 1] + q\,h(x) = 0\} \quad \text{and} \\
&\mathcal{C}_a \equiv \{(p,x) : (1-p)\,[f_a(p,x) - 1] + (1-q)\,h(x) = 0\}\,.
\end{aligned} \tag{5}$$

Any pair of equations in (5) will determine E.

Since genotype fitnesses are decreasing then $\partial f_A/\partial x < 0$, $\partial f_a/\partial x < 0$ and $\partial f/\partial x < 0$. With $h'(x) < 0$, it follows from the implicit function theorem that the curves defined in (5) may be considered as the graphs of x as functions of p, which will be denoted by $\tilde{x}(p)$, $\tilde{x}_A(p)$ and $\tilde{x}_a(p)$, respectively. Using the implicit function theorem and the fact that $\frac{\partial f}{\partial p} = 2\,(f_A - f_a)$, we compute that

$$\frac{d\tilde{x}}{dp} = \frac{-2\,(f_A - f_a)}{\frac{\partial f}{\partial x} + h'}\,, \tag{6a}$$

$$\frac{d\tilde{x}_A}{dp} = \frac{1 - f_A - p\,\frac{\partial f_A}{\partial p}}{p\,\frac{\partial f_A}{\partial x} + q\,h'}\,, \tag{6b}$$

$$\frac{d\tilde{x}_a}{dp} = \frac{f_a - 1 - (1-p)\,\frac{\partial f_a}{\partial p}}{(1-p)\,\frac{\partial f_a}{\partial x} + (1-q)\,h'}\,. \tag{6c}$$

From (A1) and the fact that $h(x) \searrow 0$ as $x \to \infty$, it follows that for each value of $p \in [0,1]$, there is an x so that $f(p,x) + h(x) = 1$ and so the function $\tilde{x}(p)$ exists for all p and the curve $\mathcal{C}$ separates $\mathcal{S}$ into two subsets. Similar arguments show that the functions $\tilde{x}_A(p)$ and $\tilde{x}_a(p)$ exist for $0 < p < 1$. In addition, it follows that as $p \to 0$ the function $\tilde{x}_A(p)$ approaches infinity and that as $p \to 1$ the function $\tilde{x}_a(p)$ approaches infinity, see [5]. By rewriting the equation that defines $\mathcal{C}$ as

$$p\,[f_A(p,x) - 1] + q\,h(x) + (1-p)\,[f_a(p,x) - 1] + (1-q)\,h(x) = 0\,,$$

it follows that the graph of the function $\tilde{x}(p)$ is exactly between the graphs of $\tilde{x}_A(p)$ and $\tilde{x}_a(p)$. In fact, Roberds and Selgrade [5] showed that these isoclines must cross at least once and so there is at least one polymorphism.

Theorem 3.1. Existence of a polymorphism [5]. *Fix* $0 < q < 1$. *Assume that each genotype fitness* f_{ij} *satisfies (A1) and that* $h(x) = y/x$ *where* y *is a constant. Then (2) has at least one polymorphic equilibrium* $E = (\overline{p}, \overline{x})$, *i.e.,* $0 < \overline{p} < 1$ *and* $\overline{x} > 0$.

§4. Dominance assumptions

Here we introduce two classical notions of dominance from genetics that describe genotypic effects on fitness, e.g., see [1] or [6]. First we assume that the presence of the allele A confers maximal fitness on a genotype and consider the case where genotypic fitnesses exhibit *complete dominance* (CD). Thus, for all $x > 0$, we assume that

$$f_{AA}(x) = f_{Aa}(x) > f_{aa}(x)\,. \tag{CD}$$

With complete dominance, the allele and mean fitnesses become

$$\begin{aligned} f_A &= f_{AA} \\ f_a &= p\, f_{AA} + (1-p)\, f_{aa} \\ f &= p\,(2-p)\, f_{AA} + (1-p)^2\, f_{aa}\,. \end{aligned} \tag{7}$$

For the second case, we assume that the heterozygote fitness is the average of the homozygote fitnesses, i.e., for all $x > 0$

$$f_{Aa}(x) = (f_{AA}(x) + f_{aa}(x))/2\,. \tag{ND}$$

In this case the fitnesses are said to be *additive* or to exhibit *no dominance* (ND). Allele and mean fitnesses then become

$$\begin{aligned} f_A &= p\,(f_{AA} - f_{aa})/2 + (f_{AA} + f_{aa})/2 \\ f_a &= (1-p)\,(f_{aa} - f_{AA})/2 + (f_{AA} + f_{aa})/2 \\ f &= p\, f_{AA} + (1-p)\, f_{aa}\,. \end{aligned} \tag{8}$$

It follows that $f_A - f_a = (1-p)\,(f_{AA} - f_{aa})$ if (CD) holds and $f_A - f_a = 0.5\,(f_{AA} - f_{aa})$ if (ND) holds.

§5. Uniqueness of polymorphism

From Theorem 3.1, (2) always has at least one polymorphism. In the case of complete dominance (CD), we show here that this polymorphism is unique. Also, we obtain conditions for uniqueness if (ND) holds. To do this we study the geometry of the isoclines (5).

The curve $\mathcal{C}$ in $\mathcal{S}$ is the graph the function $\tilde{x}(p)$. So for each fixed $p \in [0,1]$, the x-coordinate of a point on $\mathcal{C}$ is determined by the intersection of the two functions of x, i.e., $f(p,x)$ and $1 - y/x$. Since $f_{ii}(0) > 1$ and $f_{ii}(x) \searrow 0$ as $x \to \infty$ for $i = A, a$, $f_{AA}(x)$ crosses $1 - y/x$ at some $x_A > y$ and $f_{aa}(x)$ crosses $1 - y/x$ at some $x_a > y$, see Figure 1. If $x_A = x_a$ then for all $p \in [0,1]$ it follows from (7) or (8) that

$$f(p, x_A) = f_{AA}(x_A) = 1 - \frac{y}{x_A}$$

and $\mathcal{C}$ is the horizontal line $\{(p,x) : x = x_A = x_a\}$. If $x_A \neq x_a$ then define $m = \min\{x_a, x_A\}$ and $M = \max\{x_a, x_A\}$. Clearly for all $x \in [m, M]$, one of the genotype fitnesses is larger than the other fitness (Figure 1) and, for each p, $f(p,x)$ is a decreasing function of x which lies in between the curves $f_{AA}(x)$ and $f_{aa}(x)$. Hence $f(p,x)$ crosses $1 - y/x$ uniquely at x-value given by $\tilde{x}(p) \in [m, M]$. If $m = x_a$ (which is the case for (CD)) then $f_{AA}(x) - f_{aa}(x) > 0$ for all $x \in [m, M]$ and (6a) implies that $\tilde{x}(p)$ is an increasing function of p. If $m = x_A$ then $f_{AA}(x) - f_{aa}(x) < 0$ for all $x \in [m, M]$ and (6a) implies that $\tilde{x}(p)$ is an decreasing function of p. We have established the following result:

Proposition 5.1. *Assume (A1) and (CD) or (ND) for all $x > 0$. Then one of the following three cases applies:*
(i) $\mathcal{C}$ is the graph of the increasing function $\tilde{x}(p)$ from $\tilde{x}(0) = x_a$ to $\tilde{x}(1) = x_A$ and $f_{AA}(x) - f_{aa}(x) > 0$ for all $x \in [m, M] = [x_a, x_A]$.
(ii) $\mathcal{C}$ is the graph of the decreasing function $\tilde{x}(p)$ from $\tilde{x}(0) = x_a$ to $\tilde{x}(1) = x_A$ and $f_{AA}(x) - f_{aa}(x) < 0$ for all $x \in [m, M] = [x_A, x_a]$.
(iii) $\mathcal{C}$ is the horizontal line $\{(p,x) : x = x_A = x_a\}$ and $f_{AA}(x) - f_{aa}(x) = 0$ for all $x \in [m, M]$.

If $E = (\overline{p}, \overline{x})$ is a polymorphism then multiplying the second equation in (3) by $\overline{p}$ and subtracting from the first equation in (3) gives

$$h(\overline{x})\,(\overline{p} - q) = \begin{cases} p\,(1-p)^2\,(f_{AA}(\overline{x}) - f_{aa}(\overline{x})) & \text{if (CD) holds} \quad (9a) \\[2ex] 0.5\,p\,(1-p)\,(f_{AA}(\overline{x}) - f_{aa}(\overline{x})) & \text{if (ND) holds} \quad (9b) \end{cases}$$

Hence, the terms $(f_{AA}(\overline{x}) - f_{aa}(\overline{x}))$ and $(\overline{p} - q)$ have the same sign. If case (iii) of Prop. 5.1 holds then $\overline{x} = x_A = x_a$. From (9), It follows that $\overline{p} = q$ and so the polymorphism $E = (q, x_A)$ is unique. If (CD) holds then case (i) of Prop. 5.1 applies and $\tilde{x}(p)$ is an increasing function of

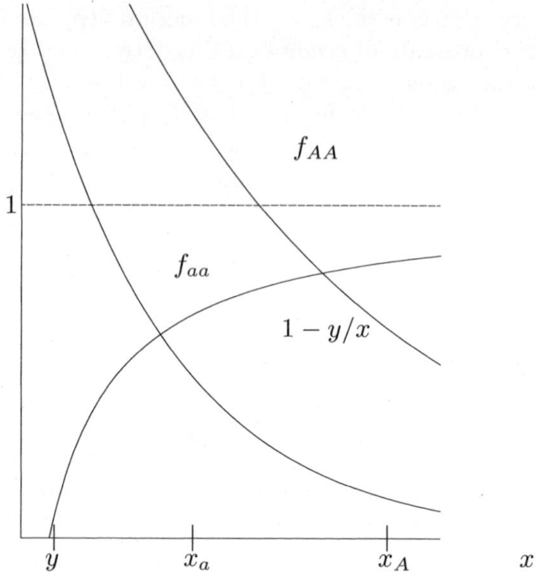

Fig. 1. Graphs of $f_{AA}(x)$, $f_{aa}(x)$ and $1-y/x$ when $x_a < x_A$.

p. Using (7), the numerator of (6b) is

$$1 - f_A - p\,\frac{\partial f_A}{\partial p} = 1 - f_{AA}(x)\,.$$

which is positive along $\mathcal{C}_A$. Hence, $\tilde{x}_A(p)$ is a decreasing function of p and may intersect $\tilde{x}(p)$ at most once. With Theorem 3.1 and (9a), we have the result:

Theorem 5.2. Uniqueness of polymorphism for (CD). *Assume that each genotype fitness f_{ij} satisfies (A1) and (CD) for all $x > 0$. If $0 < q < 1$ then there exist a unique polymorphic equilibrium $E = (\overline{p}, \overline{x})$ where $\overline{x} > 0$ and $q < \overline{p} < 1$.*

The situation is more difficult when (ND) holds because $\tilde{x}_A(p)$ and $\tilde{x}_a(p)$ may not be monotone functions. This problem is avoided by assuming that both homozygote fitnesses $f_{ii}(x)$ decrease faster than $h(x) = y/x$ for $i = A, a$. Many biologically reasonable fitnesses behave this way. Linear fitnesses may not but are not typically used in discrete models because population sizes may become negative. The precise assumption we make for $i = A, a$ is:

(A2) $\quad f'_{ii}(x) < h'(x) = -y/x^2$ **for all** $x \in [m, M]$.

We proceed by considering cases (i) and (ii) of Prop. 5.1 where (ND) holds.

For $E = (\overline{p}, \overline{x})$ in case (i), we have $x_a \leq \overline{x} \leq x_A$ and $f_{AA}(x) - f_{aa}(x) > 0$ for all $x \in [x_a, x_A]$. From (9b) it follows that $\overline{p} > q$. We use the equations for $\mathcal{C}$ and $\mathcal{C}_A$ in (5) to find two functions for p in terms of x whose intersections determine $E = (\overline{p}, \overline{x})$. Then we show that these curves intersect only once if $q \geq 0.5$.

Multiply the equation for $\mathcal{C}$ by p and combine the result with the equation for $\mathcal{C}_A$ to eliminate the quadratic term in p. The resulting equation may be solved for p in terms of x giving

$$p = \frac{2\,q\,h(x)}{1 + h(x) - f_{AA}(x)} \equiv F(x)\,. \tag{10}$$

Let the right side of (10) define the function $F(x)$ for $x \in [x_a, x_A]$ where the denominator is not zero. Because of (A2), the denominator is increasing on $[x_a, x_A]$ and positive at any equilibrium $\overline{x}$ value. Hence, $F(x)$ is defined on a subinterval of $[x_a, x_A]$ containing all the x-coordinates of equilibria. Also $\mathcal{C}$ determines p in terms of x by solving $f(p, x) = 1 - h(x)$ to define the function $p = G(x)$ as

$$p = \frac{1 - h(x) - f_{aa}(x)}{f_{AA}(x) - f_{aa}(x)} \equiv G(x)\,. \tag{11}$$

The points (p, x) of intersection of the F and G curves are the equilibria for $0 < p < 1$ and $x \in [x_a, x_A]$. Setting $F(x) = G(x)$ yields the condition

$$0 < \frac{2\,q\,h(x)}{1 + h(x) - f_{AA}(x)} = \frac{1 - h(x) - f_{aa}(x)}{f_{AA}(x) - f_{aa}(x)} < 1\,. \tag{12}$$

For $E = (\overline{p}, \overline{x})$ the first ratio in (12) implies that

$$0 < 2\,q\,h(\overline{x}) < 1 + h(\overline{x}) - f_{AA}(\overline{x})$$

or

$$(2\,q - 1)\,h(\overline{x}) < 1 - f_{AA}(\overline{x})\,. \tag{13}$$

Hence, if $q \geq 0.5$ then $0 < 1 - f_{AA}(\overline{x})$. Clearly

$$F'(\overline{x}) = \frac{2\,q\,h'(\overline{x})\,[1 - f_{AA}(\overline{x})] + 2\,q\,h(\overline{x})\,f'_{AA}(\overline{x})}{[1 + h(\overline{x}) - f_{AA}(\overline{x})]^2}\,, \tag{14}$$

so $F'(\overline{x}) < 0$ if $q \geq 0.5$. Since $f_{AA}(\overline{x}) - f_{aa}(\overline{x}) > 0$ then the second ratio in (12) gives

$$0 < 1 - h(\overline{x}) - f_{aa}(\overline{x}) < f_{AA}(\overline{x}) - f_{aa}(\overline{x})\,, \tag{15}$$

which implies that

$$1 - h(\overline{x}) - f_{AA}(\overline{x}) < 0\,. \tag{16}$$

Computing

$$G'(\overline{x}) = \frac{-h'(\overline{x})[f_{AA}(\overline{x}) - f_{aa}(\overline{x})] - f'_{AA}(\overline{x})[1 - h(\overline{x}) - f_{aa}(\overline{x})]}{[f_{AA}(\overline{x}) - f_{aa}(\overline{x})]^2} + \frac{f'_{aa}(\overline{x})[1 - h(\overline{x}) - f_{AA}(\overline{x})]}{[f_{AA}(\overline{x}) - f_{aa}(\overline{x})]^2} \tag{17}$$

and using (15) and (16), we see that $G'(\overline{x}) > 0$. Hence, if $q \geq 0.5$, for each $\overline{x}$ where the F and G curves intersect we must have

$$F'(\overline{x}) < 0 \quad \text{and} \quad G'(\overline{x}) > 0\,. \tag{18}$$

Since F and G are continuous on the subinterval of $[x_a, x_A]$ containing all $\overline{x}$, an application of the intermediate value theorem implies that (18) may hold at only one $\overline{x}$. Hence, the polymorphism $E = (\overline{p}, \overline{x})$ is unique.

Finally, for E in case (ii) of Prop. 5.1, we have $x_A \leq \overline{x} \leq x_a$ and $f_{AA}(x) - f_{aa}(x) < 0$ for all $x \in [x_A, x_a]$. From (9b) it follows that $\overline{p} < q$. We use the equations for $\mathcal{C}$ and $\mathcal{C}_a$ in (5) to find two functions for p in terms of x whose intersections determine $E = (\overline{p}, \overline{x})$. Then we show that these curves intersect only once if $q \leq 0.5$.

Multiply the equation for $\mathcal{C}$ by $1-p$ and combine the result with the equation for $\mathcal{C}_a$ to eliminate the term with factor $p(1-p)$. The resulting equation may be solved for $1-p$ in terms of x giving

$$1 - p = \frac{2\,(1-q)\,h(x)}{1 + h(x) - f_{aa}(x)}\,. \tag{19}$$

Since $0 < 1 - p < 1$, from (19) we see that for $E = (\overline{p}, \overline{x})$

$$0 < 2\,(1-q)\,h(\overline{x}) < 1 + h(\overline{x}) - f_{aa}(\overline{x})$$

which implies that

$$(1 - 2\,q)\,h(\overline{x}) < 1 - f_{aa}(\overline{x})\,. \tag{20}$$

Hence, if $q \leq 0.5$ then $0 < 1 - f_{aa}(\overline{x})$. Solve (19) for p in terms of x to obtain the function

$$H(x) \equiv 1 - \frac{2\,(1-q)\,h(x)}{1 + h(x) - f_{aa}(x)}\,. \tag{21}$$

As with $F(x)$ using (A2), we see that the denominator in (21) is increasing on $[x_A, x_a]$ and positive at any equilibrium $\overline{x}$ value. Hence, $H(x)$ is defined on a subinterval of $[x_A, x_a]$ containing all the x-coordinates of equilibria. Compute

$$H'(\overline{x}) = \frac{-2\,(1-q)\;[h'(\overline{x})\,(1 - f_{aa}(\overline{x})) + h(\overline{x})\,f'_{aa}(\overline{x})]}{[1 + h(\overline{x}) - f_{aa}(\overline{x})]^2} \tag{22}$$

and notice that $H'(\overline{x}) > 0$ if $q \leq 0.5$. Recall $\mathcal{C}$ determines $G(x)$ defined in (11) and the points (p, x) of intersection of the H and G curves are the equilibria for $0 < p < 1$ and $x \in [x_A, x_a]$. Since $f_{AA}(x) - f_{aa}(x) < 0$ and $0 < G(\overline{x}) < 1$, from (11) we conclude that

$$f_{AA}(x) - f_{aa}(x) < 1 - h(\overline{x}) - f_{aa}(\overline{x}) < 0$$

and so

$$0 < 1 - h(\overline{x}) - f_{AA}(\overline{x})\,.$$

With these inequalities, it is clear that each term in the numerator in (17) is negative and hence $G'(\overline{x}) < 0$. Hence, if $q \leq 0.5$, for each $\overline{x}$ where the G and H curves intersect we must have

$$G'(\overline{x}) < 0 \quad \text{and} \quad H'(\overline{x}) > 0\,, \tag{23}$$

which may occur at most once on a closed interval.

Thus we have established the following result where the three cases correspond to those in Proposition 5.1.

Theorem 5.3. Uniqueness of polymorphism for (ND). *Assume (A1), (A2) and no dominance (ND) for all $x > 0$. Suppose that $E = (\overline{p}, \overline{x})$ is a polymorphism, i.e., $0 < \overline{p} < 1$. Then one of the following three statements holds:*
(i) $x_a < \overline{x} < x_A$ and $q < \overline{p}$. In addition, if $q \geq 0.5$ then E is unique.
(ii) $x_A < \overline{x} < x_a$ and $\overline{p} < q$. In addition, if $q \leq 0.5$ then E is unique.
(iii) $x_a = \overline{x} = x_A$ and $E = (q, x_A)$.

§6. Conclusion

Selgrade and Roberds [12] present an example of transgenic immigration with exponential fitnesses satisfying (CD). If $q = 0.9$, the unique polymorphism $E = (0.906, 1.80)$ appears globally attracting. In fact, if the A allele is absent in the island population, numerical results indicate that within five generations of immigration the island population evolves so that p is at least 0.9.

In general, the stability of the unique polymorphism is not known. In [5] where immigration depends on the population size x, examples of saddle point equilibria are presented for (CD). We conjecture similar behavior for the case of constant immigration studied here and this will be a topic of future investigation. Also, whether or not the bound on q in Theorem 5.3 is necessary for equilibrium uniqueness needs to be determined.

References

[1] D. S. Falconer and T. F. C. Mackay, Introduction to Quantitative Genetics, 4th ed., Longman, Essex, 1996.

[2] M. P. Hassell and H. N. Comins, Discrete time models for two-species competition, Theor. Pop. Biol., **9** (1976), 202–221.

[3] P. W. Hedrick, Genetics of Populations, 3rd ed., Jones and Barlett Publ., Sudbury, MA, 2005.

[4] W. E. Ricker, Stock and recruitment, J. Fish. Res. Bd. Can., **11** (1954), 559–623.

[5] J. H. Roberds and J. F. Selgrade, Dynamical analysis of density-dependent selection in a discrete one-island migration model, Math. Biosciencs, **164** (2000), 1–15.

[6] J. Roughgarden, Theory of Population Genetics and Evolutionary Ecology: An Introduction, MacMillan, New York, 1979.

[7] R. A. Sedjo, Global agreements and US forestry: Genetically modified trees, J. Forestry, **103** (2005), 109–113.

[8] J. F. Selgrade and G. Namkoong, Stable periodic behavior in a pioneer-climax model, Nat. Res. Modeling, **4** (1990), 215–227.

[9] J. F. Selgrade and J. H. Roberds, Lumped-density population models of pioneer-climax type and stability analysis of Hopf bifurcations, Math. Biosciences, **135** (1996), 1–21.

[10] J. F. Selgrade and J. H. Roberds, Period-doubling bifurcations for systems of difference equations and applications to models in population biology, Nonlinear Anal., **29** (1997), 185–199.

[11] J. F. Selgrade and J. H. Roberds, Equilibrium and nonequilibrium attractors for a discrete, selection-migration model, Can. Appl. Math. Q., **11** (2003), 195–211.

[12] J. F. Selgrade and J. H. Roberds, Global attractors for a discrete selection model with periodic immigration, J. Difference Equ. Appl., **13** (2007), 275–287.

[13] C. G. Williams, Genetically modified pines at the interface of private and public lands: A case study approach, Paper presented at USDA Forest Service, Washington, DC, 2004.

James F. Selgrade
Department of Mathematics and Biomathematics Program
North Carolina State University
Raleigh, NC 27695-8205
USA

James H. Roberds
USDA Forest Service, Southern Research Station
Southern Institute of Forest Genetics, Saucier, MS 39574
USA

E-mail address: `selgrade@math.ncsu.edu`
`jroberds@fs.fed.us`

Advanced Studies in Pure Mathematics 53, 2009
Advances in Discrete Dynamical Systems
pp. 333–337

Periodicity in the May's host parasitoid equation

Walter Sizer

Abstract.

In this paper we consider the May's host parasitoid equation,

$$x_{n+1} = \frac{\alpha x_n^2}{(1+x_n)x_{n-1}}, \alpha > 1. \tag{1}$$

We show that with initial conditions $x_{-1} = x_0 = 1$ there are values of α giving periodic solutions of prime period n for all integers $n \geq 7$. There are no non-equilibrium periodic solutions of periods 2, 3, 4, 5 or 6.

§1. Introduction

May's host parasitoid equation (1) shows some rhythmic behavior, and some results about its long term behavior are given in [2]. In this paper we indicate that in the case $x_{-1} = x_0 = 1$, for different values of α there are periodic solutions of prime period n for all natural numbers $n \geq 7$. A key observation is that for appropriate values of α there are in this situation arbitrarily long decreasing sequences

$x_0 > x_1 > \ldots > x_n$. The techniques used are based on those of [3] and [4].

§2. Getting periodic solutions

With equation (1) and initial conditions $x_{-1} = x_0 = 1$, the value of x_n will be a rational function of α. We will often write $x_n(\alpha)$ to emphasize this dependence. Direct computation (with simplification of fractions) gives

$x_1(\alpha) = \frac{\alpha}{2}$,
$x_2(\alpha) = \frac{\alpha^3}{4+2\alpha}$,

Received October 12, 2007.
2000 *Mathematics Subject Classification.* 39A11.

$x_3(\alpha) = \frac{\alpha^6}{(2+\alpha)(4+2\alpha+\alpha^3)}$,
$x_4(\alpha) = \frac{2\alpha^{10}}{(4+2\alpha+\alpha^3)(8+8\alpha+2\alpha^2+2\alpha^3+\alpha^4+\alpha^6)}$,
etc..

We will later need the observation that the functions $x_i(\alpha)$ are rational functions in α with non-negative coefficients in both numerator and denominator, hence are continuous for $\alpha > 1$. This can be formally established by induction on i.
The existence of periodic solutions of given period will eventually follow from the following theorem.

Theorem 1. *With May's host parasitoid equation with $x_{-1} = x_0 = 1$, and for $k \geq 0$,*

(i) *α gives a periodic solution of period $2k+2$ if and only if $x_k(\alpha) = x_{k+1}(\alpha)$;*

(ii) *α gives a periodic solution of period $2k+3$ if and only if $x_k(\alpha) = x_{k+2}(\alpha)$.*

Proof. We prove (i); the proof of (ii) is similar and is omitted.
Equation (1) can be rewritten as

$$x_{n-1} = \frac{\alpha x_n^2}{(1+x_n)x_{n+1}}. \tag{2}$$

If: Assume that $x_k(\alpha) = x_{k+1}(\alpha)$, $k \geq 0$. We can generate the sequence in both directions starting with $x_k = x_{k+1}$. The similarity of equations (1) and (2) gives us equations $x_{k-1} = x_{k+2}$, $x_{k-2} = x_{k+3}$, and in general $x_{k-i} = x_{k+i+1}$. This means we get $x_0 = x_{2k+1} = 1$ and $x_{-1} = x_{2k+2} = 1$. So $x_{-1} = x_{2k+1}$, $x_0 = x_{2k+2}$. Since (1) is a second order difference equation it follows that the seuence $\{x_n\}$ is periodic of period $2k+2$.

Only if: Assume the sequence is periodic of period $2k+2$. Then $x_{-1} = x_{2k+1} = 1$, and $x_0 = x_{2k+2} = 1$. Using (1) and (2) and working both ways, we get $x_1 = x_{2k}$, $x_2 = x_{2k-1}$, . . ., $x_k = x_{k+1}$. Q.E.D.

We can use theorem 1 to show the existence of some periodic solutions algebraically. For example, setting the expressions for $x_3(\alpha)$ and $x_4(\alpha)$ that we have above equal and simplifying we get the equation $f(\alpha) = 8 + 8\alpha + 2\alpha^2 + 2\alpha^3 - 3\alpha^4 - 2\alpha^5 + \alpha^6 = 0$.

Descartes' Rule of Signs says there are zero or two positive solutions. We know (and can check) that $\alpha = 2$ is a solution (giving the equilibrium $x_n = 1$ for all n). Since $f'(2) = -24$ (in particular, $f'(2) \neq 0$), we know that $\alpha = 2$ is not a repeated root. In fact, since $f(\alpha)$ has a positive leading coefficient we know the other root is greater than 2. (In fact, the value of α giving a period 8 solution is approximately 2.43). Similarly we could show that a periodic solution of period 7 comes when α is approximately 4.71, and a period 9 solution comes when α is approximately 1.89.

We can also show there is no non-equilibrium solution of period 6 in this manner. Setting $x_2(\alpha) = x_3(\alpha)$ and simplifying gives the equation $\alpha^3 - 2\alpha - 4 = 0$. Descartes' Rule of Signs says there is one positive root. It is $\alpha = 2$, giving the equilibrium solution.

Computationally this algebraic approach to proving the existence of periodic solutions becomes unmanageable as k increases, and another, non-algebraic approach (still using theorem 1) is needed.

§3. Long decreasing sequences

Another observation which plays a key role in establishing periodicity is contained in the following theorem.

Theorem 2. *With the May's host parasitoid equation with initial conditions $x_{-1} = x_0 = 1$, for any positive integer N there is a value of α greater than 1, $\alpha = c_N$, with*

$$x_0(c_N) > x_1(c_N) > \ldots > x_N(c_N).$$

Proof. We first show that if $1 \geq x_k \geq \frac{1}{2^{2^k-1}}$ for $k = 0, 1, \ldots n-1$ and $\alpha < 2$, then $x_n \geq \frac{1}{2^{2^n-1}}$. This is done, of course, using induction on n: The fact is clearly true for $n = 1$, where $1 \geq x_1 = \frac{\alpha}{2} \geq \frac{1}{2} = \frac{1}{2^{2^1-1}}$. So assume it is true for $n-1$. Then

$$x_n = \frac{\alpha x_{n-1}^2}{(1+x_{n-1})x_{n-2}} > \frac{x_{n-1}^2}{(1+x_{n-1})x_{n-2}} \geq \frac{x_{n-1}^2}{(1+x_{n-1})} \geq \frac{x_{n-1}^2}{2} \geq$$
$$\left(\frac{1}{2^{2^{n-1}-1}}\right)^2 \cdot \frac{1}{2} = \frac{1}{2^{(2^{n-1}-1)2}} \cdot \frac{1}{2} = \frac{1}{2^{2^n-2+1}} = \frac{1}{2^{2^n-1}},$$

as desired.

Next we show that if $\alpha < 2$ and if $1 + x_k \geq \alpha$ for $0 \leq k \leq n$, then

$x_0 > x_1 > \ldots > x_n$. Again the proof uses induction. For $n = 1$, the conclusion says $x_0 = 1 > x_1 = \frac{\alpha}{2}$, which is true as $\alpha < 2$. Assume $n > 1$ and the claim is true for all $k < n$. Then $x_n = \frac{\alpha}{1+x_{n-1}} \cdot \frac{x_{n-1}}{x_{n-2}} \cdot x_{n-1}$. Since $1 + x_{n-1} \geq a$ and, by induction, $x_{n-2} > x_{n-1}$, we get $x_{n-1} > x_n$, as desired.

Now, take $\alpha = c_N = \frac{1}{2^{2^N-1}}$. By the first observation we get $1+x_k \geq \alpha$ for all $k \leq N$ (assuming $x_k \leq 1$ for $k \leq N$), and so by the second observation

$x_0 > x_1 > \ldots > x_N$ (and indeed $x_k \leq 1$ for $k \leq N$). Q.E.D.

With values of α decreasing towards 1 we get arbitrarily long decreasing sequences $x_{-1} = x_0 = 1 > x_1 > x_2 > \ldots > x_N$. It is shown in [2] that none of these sequences have limit 0, in contrast to the situation where $\alpha \leq 1$ (treated in [2], and in the case $\alpha = 1$ also covered in [1]).

§4. The existence of periodic solutions

We are now in a position to prove the existence of periodic solutions to the May's host parasitoid equation.

Theorem 3. *For every positive integer $n \geq 7$, there exists a value of $\alpha > 1$ so that the May's host parasitoid equation (1) has a periodic solution of prime period n and with initial conditions $x_{-1} = x_0 = 1$.*

Proof. The cases of $n = 7$, 8, and 9 were dealt with previously.

Continuing the notation introduced in theorem 2, let $c_N \in (1, 2)$ satisfy the inequalities $x_0(c_N) > x_1(c_N) > \ldots > x_N(c_N)$. Following Theorem 1 we want to find an α so that (for odd periodicity) $x_0(\alpha) > x_1(\alpha) > \ldots > x_{N+1}(\alpha)$ and $x_N(\alpha) = x_{N+2}(\alpha)$, $(N \geq 4)$. Define a set A_N by

$A_N = \{\alpha | \alpha \geq c_{N+2}$ and $x_0(c) > x_1(c) > \ldots > x_{N+1}(c)$ and $x_N(c) \geq x_{N+2}(c)$ for all c in $[c_{N+2}, \alpha]\}$.

The set A_N is non-empty, as $c_{N+2} \in A_N$. Also, A_N is bounded above by the value b of α giving a period 9 solution, since $x_0(b) > x_1(b) > x_2(b) > x_3(b) > x_4(b)$ but $x_3(b) = x_5(b)$, so $b \notin A_N$ $(N \geq 4)$. Denote by a_N the least upper bound of A_N.

We claim that

$x_0(a_N) > x_1(a_N) > \ldots > x_{N+1}(a_N)$ and that $x_N(a_N) = x_{N+2}(a_N)$.

Because of the continuity of the functions $x_k(\alpha)$, we get

(*) $\quad x_0(a_N) \geq x_1(a_N) \geq \ldots \geq x_{N+1}(a_N)$ and $x_N(a_N) \geq x_{N+2}(a_N)$.

If $x_0(a_N) = x_1(a_N)$, then $x_1(a_N) = 1$ and we have the equilibrium solution, This means $a_N = 2$, but this can't happen since A_N has an upper bound $b < 2$ and a_N is supposed to be the least upper bound of A_N. If $x_k(a_N) = x_{k+1}(a_N)$ for any $k = 1, 2, \ldots, N-1$, take k to be minimal with this property. The as before, using equations (1) and (2), we get $x_{k+2} = x_{k-1}$. This contradicts $x_{k-1}(a_N) > x_k(a_N) \geq x_{k+1}(a_N) \geq x_{k+2}(a_N)$, so we can't have $x_k(a_N) = x_{k+1}(a_N)$. Thus inequalities are strict through $x_{N-1} > x_N$. If $x_N = x_{N+1}$, then (again using equations (1) and (2)) we get $x_{N+2} = x_{N-1} > x_N$, contradicting (*). Finally, if $x_N > x_{N+2}$, again using continuity we can find a $\delta > 0$ so that $a_N + \delta \in A_N$, contradicting the fact that a_N is an upper bound for A_N. Then by theorem 1, a_N gives a periodic solution of period $2N + 3$.

Similar arguments using the sets

$B_N = \{\alpha | \alpha \geq c_{N+2}$ and $x_0(c) > x_1(c) > \ldots > x_{N-1}(c) \geq x_N(c)$ for all c in $[c_{N+2}, a]\}$

give even period periodic solutions of period greater than or equal to 10.

The constructions involving the sets A_N and B_N guarantee that the indicated period of the solution is its prime period, and in fact the period is made up of just one cycle of the sequence. Q.E.D.

Note in the proof of theorem 3 that it was essential to have an upper bound b of the sets A_N and B_N which was less than 2 to rule out the possibility that $a_N = 2$ and we got the equilibrium solution. The rest of the argument would work for A_1, except that in this case a_1 is 2 and our period 5 solution is the equilibrium solution.

Problems:

1. Is there only one value of α for which equation (1) has a one cycle period n solution ($n \geq 7$)?
2. Show multicycle period n solutions exist (for some values of n).
3. Give periodicity results for arbitrary x_{-1} and x_0.
4. Show that in general solutions are bounded.
5. Give an invariant for equation (1).

References

[1] G. Bastien and M. Rogalski, On some algebraic difference equations $u_{n+2}u_n = \psi(u_{n+1})$ in $\mathbb{R}_*^+$, related to families of conics or cubics: generalization of the Lyness' sequences, J. Math. Anal. Appl., **300** (2004), 303–333.

[2] G. Ladas, A. Tovbis and G. Tzanetopoulos, On May's host parasitoid model, J. Differ. Equations Appl., **2** (1996), 195–204.

[3] W. S. Sizer, Some periodic solutions of the Lyness equation, In: New Trends in Difference Equations, Proceedings of the Fifth International Conference on Difference Equations, (eds. S. Elaydi et al.), Taylor & Francis, New York, 2002, pp. 267–272.

[4] W. S. Sizer, Periodicity in the Lyness equation, Math. Sci. Res. J., **7** (2003), 366–372.

Minnesota State University Moorhead

E-mail address: `sizer@mnstate.edu`

Advanced Studies in Pure Mathematics 53, 2009
Advances in Discrete Dynamical Systems
pp. 339–346

On variational methods for second order discrete periodic problems

Petr Stehlík

Abstract.

We use some basic results from the theory of variational methods to prove the exisetnce and uniqueness of periodic solutions to second-order nonlinear discrete problems. Our method has no continuous counterpart since it is based on the finite dimension of the corresponding function space. Our main tools include matrix theory and the Fundamental theorem of calculus for Lebesgue Integral.

§1. Introduction

In recent years, the number of publications concerning difference equations and corresponding boundary value problems has sharply risen. Most of the existence and multiplicity results are based on the corresponding statements from differential equations or are even directly proven in the time scales settings, i.e. in a general scheme which encompasses both continuous and discrete problems (see e.g. Bohner, Peterson [6] for more details).

In this short note we show that some purely discrete approaches can yield results which can't be obtained in a traditional, continuous-like, way. Namely, we reformulate the second-order nonlinear periodic boundary value problem as a nonlinear equation in $\mathbb{R}^N$. Then we take advantage of the finite dimension of the considered problem and use some basic properties of positive (semi)definite matrices and Lebesgue integration. In contrast to traditional results, we are able to prove existence and uniqueness theorems for problems with non-continuous right-hand sides.

2000 *Mathematics Subject Classification.* Primary 39A12; Secondary 34B15.

Key words and phrases. Boundary value problem, Difference equations, Variational methods.

Throughout this paper, we consider the second order nonlinear periodic problem:

(1)
$$\begin{cases} \Delta\left(p(k-1)\Delta x(k-1)\right)) + q(k)x(k) = g(k, x(k)), & k = 1, 2, \ldots, N, \\ x(0) = x(N), \quad p(0)\Delta x(0) = p(N)\Delta x(N), \end{cases}$$

where $p : \{0, 1, \ldots, N-1, N\} \to \mathbb{R}$ is such that $p(k) > 0$ and $q : \{1, \ldots, N-1, N\} \to \mathbb{R}$ satisfies either

(Q^0) $$q(k) \equiv 0,$$

or

(Q$^+$) $q(k) \geq 0$ for all k, $q(k_1) > 0$ for some $k_1 \in \{1, 2, \ldots, N\}$.

By the solution to (1) we understand a vector

$$[x(0), x(1), \ldots, x(N), x(N+1)]^T,$$

such that the equation in (1) holds for every $k = 1, 2, \ldots, N$, and the boundary conditions there are satisfied as well.

Similar problems have been studied recently in different settings. Firstly, Atici, Cabada [3] considers the problem with $p \equiv 1$ and uses the method of lower and upper solutions to prove, for instance, that if g is continuous and nondecreasing in x and q satisfies (Q$^+$), then the problem (1) has a solution. Furthermore, if $q(k) > 0$ for all k and g is stricly increasing in x then the solution is unique. Bereanu and Mawhin [5] lists discrete results which correspond to Ambrosetti-Prodi and Landesman-Lazer problems for differential equations. Those results are proven for continuous right-hand sides as well.

The main contribution of this new technique consists in the facts that

(i) we prove the existence and uniqueness results also for non-continuous right-hand sides of (1), and
(ii) we prove the existence and uniqueness also in the case (Q^0).

§2. Matrix formulation

Taking into account the finite dimension of the space in which we seek a solution of (1), we rewrite now the periodic nonlinear problem (1) into a nonlinear equation in $\mathbb{R}^N$. First, let us consider the case (Q^0), i.e. the case without the term $q(k)x(k)$. Exploiting the difference relation

$\Delta x(k) = x(k+1) - x(k)$ and the boundary value problems in (1) we can reformulate (1) as

$$\overline{A}_{pq}x = G(x), \tag{2}$$

where $x = [x(1), x(2), \ldots, x(N)]^T$, $\overline{A}_{pq}$ is an $N \times N$ matrix

$$\overline{A}_{pq} = \begin{bmatrix} p(0)+p(1) & -p(1) & & & -p(0) \\ -p(1) & p(1)+p(2) & -p(2) & & \\ & \ddots & \ddots & \ddots & \\ & & -p(N-2) & p(N-2)+p(N-1) & -p(N-1) \\ -p(0) & & & -p(N-1) & p(N-1)+p(0) \end{bmatrix},$$

and $G : \mathbb{R}^N \to \mathbb{R}^N$ is a nonlinear function defined by ($x(i)$ is abbreviated as x_i)

$$G(x) = \begin{bmatrix} g_1(x_1) \\ g_2(x_2) \\ g_3(x_3) \\ \vdots \\ g_{N-1}(x_{N-1}) \\ g_N(x_N) \end{bmatrix} := \begin{bmatrix} g(1, x_1) \\ g(2, x_2) \\ g(3, x_3) \\ \vdots \\ g(N-1, x_{N-1}) \\ g(N, x_N) \end{bmatrix}. \tag{3}$$

Furthermore, if we consider the general case with $q \geq 0$, then we can rewrite the problem (1) into a nonlinear equation in $\mathbb{R}^N$

$$A_{pq}x = G(x), \tag{4}$$

with $A_{pq} = \overline{A}_{pq} + Q$ where Q is a diagonal matrix with entries $q(1), q(2), \ldots, q(N)$.

Obviously, if we find a solution $x_0 \in \mathbb{R}^N$ of the problem (4) then the vector

$$[x(N), x(1), x(2), \ldots, x(N-1)x(N), x(1)]^T$$

is the desired solution of (1).

§3. Definiteness of matrices A_{pq}

In this section we show some basic properties of matrices A_{pq}. We follow the standard notation of Lütkepohl [14]. The interested reader can find more about matrices which are connected to discrete operators in Agarwal [1].

Lemma 1. *Let us suppose that* (Q^0) *holds, then the matrix* $A_{pq} = \overline{A}_{pq}$ *is positive semidefinite and* $\lambda = 0$ *is eigenvalue. If the condition* (Q^+) *is satisfied then the matrix* A_{pq} *is positive definite*

.

Proof. Firstly, let us notice that in all cases the matrices A_{pq} are symmetric. Therefore the matrix A_{pq} is positive (semi)definite if and only if its eigenvalues are positive (non-negative in semidefinite case) (e.g. [9, Theorem 9.12]).

Secondly, let us realize that symmetric matrix has only real eigenvalues (e.g. [9, Corollary 9.11]). Therefore, the Gerschgorin Circle Theorem (e.g. [9, Theorem 9.13]) yields that all eigenvalues of A_{pq} lies in the interval

$$\left[0, \max\left\{\max_{i=1,2,\ldots,N-1} 2\left(p(i-1)+p(i)\right); 2\left(p(0)+p(N-1)\right)\right\}\right].$$

Let us consider first the case (Q^0), i.e. $A_{pq} = \overline{A}_{pq}$. All negative off-diagonal entries are counterbalanced by the occurence of their positive counterparts on the diagonal. Therefore, $\overline{A}_{pq}v = 0$ for $v = t\,[1, 1, \ldots, 1]^T$ with $t \in \mathbb{R}$. This implies that $\lambda = 0$ is eigenvalue of $\overline{A}_{pq}$.

Furthermore, if (Q^+) holds then

$$x^T A_{pq} x = x^T(\overline{A}_{pq} + Q)x = x^T\overline{A}_{pq}x + x^T Q x \geq 0, \tag{5}$$

i.e. A_{pq} is at least positive semidefinite. But the equality in (5) could have occurred only if both terms $x^T\overline{A}_{pq}x$ and $x^T Q x$ had vanished for the same vector $x \in \mathbb{R}^N$. However, the former vanishes if and only if $v = t\,[1, 1, \ldots, 1]^T$ and $t \in \mathbb{R}$. But for such a v (now considering only $t \neq 0$) we get

$$v^T Q v = t^2 \sum_{k=1}^{N} q(k) > 0,$$

since at least one $q(k)$ is positive. This implies that A_{pq} is positive definite. Q.E.D.

§4. Application of variational methods

At this stage, we are ready to show that even the simplest conlusions of the theory of variational methods can provide very strong statements if we consider the above matrix formulation of discrete nonlinear problems.

Namely, we first use the following result:

Theorem 2. [8, Theorem 6.2.8.] *Let H be a Hilbert space. Let $F : H \to \mathbb{R}$ be a weakly sequentially lower semi-continuous and weakly coercive functional. Then F is bounded from below on H and there exists $x_0 \in H$ such that $F(x_0) = \min F(x)$. Moreover, if the Fréchet derivative $F'(x_0)$ exists then*

$$F'(x_0) = o.$$

If we work with locally integrable functions $g_k(u)$ then there exists a functional $\widetilde{G} : \mathbb{R}^N \to \mathbb{R}$ such that $\nabla\widetilde{G} = G$. Thanks to the symmetry of A_{pq}, there exists also a functional $F : \mathbb{R}^N \to \mathbb{R}$ defined by ($\langle\cdot,\cdot\rangle$ denotes a scalar product in $\mathbb{R}^N$.)

$$F(x) := \frac{1}{2}\langle A_{pq}x, x\rangle - \widetilde{G}(x), \tag{6}$$

whose local extremum $x_0 \in \mathbb{R}^N$ is a solution of (4). Therefore, we are looking for extrema of the functional (6).

Theorem 3. *Let (Q^0) or (Q^+) be satisfied. Let us suppose that $g : \{1, 2, \ldots, N\} \times \mathbb{R} \to \mathbb{R}$ is a function such that for each $k = 1, 2, \ldots, N$:*

(I) $g_k \in L^1_{loc}(\mathbb{R})$,

(L) *there exists $M > 0$ such that*

$$\lim_{u\to\infty} g_k(u) \le -M, \qquad \textit{and}$$
$$\lim_{u\to-\infty} g_k(u) \ge M.$$

Then BVP (1) has a solution.

Proof. Taking into account the assumption (I) and the matrix procedure from the above sections, we can transform the problem (1) into the problem of existence of an extremum of the funcional F defined in (6). Moreover, the integrability of $g_k(u)$ ensures that the functional F is continuous (this is a consequence of Fundamental theorem of calculus for Lebesgue Integral, see e.g. [13, Theorem 23.4]). This implies, thanks to the finite dimension of $\mathbb{R}^N$, that F is also weakly sequentially semi-continuous. In order to apply Theorem 2 it suffices to show that the functional is weakly coercive.

Condition (I) implies that for each $k = 1, 2, \ldots, N$, there exists an antiderivative $\widetilde{g}_k(u) = \int\limits_0^u g_k(\tau)\mathrm{d}\tau$. If we take into account the assumption (L) we obtain

$$\lim_{u\to\infty} \widetilde{g}_k(u) = -\infty,$$
$$\lim_{u\to-\infty} \widetilde{g}_k(u) = -\infty.$$

Since $\widetilde{G}(x) = \sum_{k=1}^{N} \widetilde{g}_k(x_k)$ and A_{pq} is at least a positive semidefinite matrix ($\langle A_{pq}x, x\rangle \geq 0$ for each nonzero vector $x \in \mathbb{R}^N$), the following inequality holds

$$\lim_{\|x\|\to\infty} F(x) \geq \lim_{\|x\|\to\infty} -\widetilde{G}(x) = +\infty.$$

Thus F is a weakly coercive functional.

Consequently, Theorem 2 yields that the functional (6) has a minimum $x_0 \in \mathbb{R}^N$, in other words $A_{pq}x_0 = G(x_0)$ which completes the proof. Q.E.D.

In order to obtain a uniqueness of the solution we suppose additionally that the functional in Theorem 2 is strictly coercive. In this case the minimum of the functional is unique which is summarized in the following theorem.

Theorem 4. [8, Theorem 6.2.12]: *If, in addition to assumptions of Theorem 2, the functional F is continuous and strictly convex, then x_0 is uniquely determined.*

As usual, it is the monotonicity of right-hand sides that ensures the convexity of the functional F in our case.

Theorem 5. *Let us suppose that $g : \{1, 2, \ldots, N\} \times \mathbb{R} \to \mathbb{R}$ is a function such that for each $k = 1, 2, \ldots, N$ the assumptions* (I), (L) *and*

(M) *g_k is nonincreasing,*

are satisfied. If (Q^+) *holds, then the solution of* (1) *is unique. Moreover, if*

(M') *g_k is strictly decreasing,*

then the solution is unique also in the case (Q^0).

Proof. Similarly as in the proof of Theorem 3, one can show that F exists and is weakly coercive and continuous. In order to apply Theorem 4 it suffices to prove the strict convexity.

First, let us realize that the assumption (M) is sufficient for the concavity of $\widetilde{g}_k(u) = \int_0^u g_k(\tau)\mathrm{d}\tau$ for each $k = 1, 2, \ldots, N$ (e.g. [15, Theorem 12A]). Thus for each $x, y \in \mathbb{R}^N$:

$$
\begin{aligned}
\widetilde{G}(tx+(1-t)y) &= \sum_{k=1}^{N} \widetilde{g}_k(tx_k+(1-t)y_k) \\
&\geq t\sum_{k=1}^{N} \widetilde{g}_k(x_k) + (1-t)\sum_{k=1}^{N} \widetilde{g}_k(y_k) \\
&= t\widetilde{G}(x) + (1-t)\widetilde{G}(y),
\end{aligned}
$$

and therefore $\widetilde{G}$ is concave, i.e. $-\widetilde{G}$ is a convex functional. Since A_{pq} is at least positive semidefinite, we obtain that $\frac{1}{2}\langle A_{pq}x, x\rangle$ is convex. Hence, F is a convex functional.

Thus the functional F has a minimum. Moreover, it is unique if the assumption (Q^+) is satisfied. This is caused by the fact that in this case A_{pq} is positive definite and thus F is not only convex, but also strictly convex.

Finally, if g_k are strictly decreasing for each $k = 1, 2, \ldots, N$, then $-\widetilde{G}$ is strictly convex and, consequently, F is strictly convex also in the case (Q^0). Therefore, Theorem 4 yields a unique solution even in this case. Q.E.D.

Acknowledgements. The author gratefully acknowledges the support by the Ministry of Education, Youth and Sports of the Czech Republic, Research Plan No. MSM 4977751301.
Moreover, he is also obliged to Pavel Drábek and Petr Přikryl for their valuable hints and suggestions.

References

[1] Ravi P. Agarwal, *Difference equations and inequalities*, second ed., Monographs and Textbooks in Pure and Applied Mathematics, vol. 228, Marcel Dekker Inc., New York, 2000, Theory, methods, and applications.

[2] Ravi P. Agarwal, Kanishka Perera, and Donal O'Regan, *Multiple positive solutions of singular and nonsingular discrete problems via variational methods*, Nonlinear Anal. **58** (2004), no. 1-2, 69–73.

[3] F. M. Atici and A. Cabada, *Existence and uniqueness results for discrete second-order periodic boundary value problems*, Comput. Math. Appl. **45** (2003), no. 6-9, 1417–1427, Advances in difference equations, IV.

[4] F. Merdivenci Atici and G. Sh. Guseinov, *Positive periodic solutions for nonlinear difference equations with periodic coefficients*, J. Math. Anal. Appl. **232** (1999), no. 1, 166–182.

[5] C. Bereanu and J. Mawhin, *Existence and multiplicity results for periodic solutions of nonlinear difference equations.*, J. Difference Equ. Appl. **12** (2006), no. 7, 677–695.

[6] Martin Bohner and Allan Peterson, *Dynamic equations on time scales*, Birkhäuser Boston Inc., Boston, MA, 2001, An introduction with applications.

[7] Klaus Deimling, *Nonlinear functional analysis*, Springer-Verlag, Berlin, 1985.

[8] Pavel Drábek and Jaroslav Milota, *Lectures on nonlinear analysis.*, Plzen: Vydavatelský Servis. , 2004.

[9] J.Douglas Faires and Richard L. Burden, *Numerical methods.*, Andover, Hampshire: PWS Publishing Co., 1993.

[10] Robert Gaines, *Difference equations associated with boundary value problems for second order nonlinear ordinary differential equations*, SIAM J. Numer. Anal. **11** (1974), 411–434.

[11] Walter G. Kelley and Allan C. Peterson, *Difference equations*, second ed., Harcourt/Academic Press, San Diego, CA, 2001, An introduction with applications.

[12] Gabriel Klambauer, *Problems and propositions in analysis*, Lecture Notes in Pure and Applied Mathematics, vol. 49, Marcel Dekker Inc., New York, 1979.

[13] Jaroslav Lukeš and Jan Malý, *Measure and integral.*, Prague: Matfyzpress. , 1995.

[14] H. Lütkepohl, *Handbook of matrices*, John Wiley & Sons Ltd., Chichester, 1996.

[15] A. Wayne Roberts and Dale E. Varberg, *Convex functions*, Academic Press [A subsidiary of Harcourt Brace Jovanovich, Publishers], New York-London, 1973, Pure and Applied Mathematics, Vol. 57. MR MR0442824 (56 #1201)

Department of Mathematics
University of West Bohemia
Univerzitní 23, Plzeň
Czech Republic 30614

E-mail address: pstehlik@kma.zcu.cz

Advanced Studies in Pure Mathematics 53, 2009
Advances in Discrete Dynamical Systems
pp. 347–361

Time evolution with and without remote past

Yoichiro Takahashi

Abstract.

We usually discuss the time evolution from the present to the future or from the past, precisely, from some fixed initial time in the past, to the present or to the future. But we sometimes consider the time evolution from the remote past to the remote future as in the theory of stationary stochastic processes or dynamical systems. In the present paper we consider time evolutions governed by noise driven automorphism on locally compact abelian groups and give a necessary and sufficient condition for the time evolution to admit remote past. It turns out that *to admit the remote past is fairly restrictive.*

§1. Introduction

Let G be a locally compact group and φ be an automorphism of G. Consider the random time evolution governed by a stochastic equation

(1.1) $$\eta_n = \xi_n \varphi(\eta_{n-1})(n \in \underline{Z})$$

on G where (ξ_n) is a noise in the sense that
(a) the random variables ξ_n are mutually independent and subject to a common probability distribution, say μ, and
(b) for each n the random variable ξ_n is independent of the random variables η_k with $k < n$.

The point is that ξ_n and η_n are indexed for all integer n, including negative n. We say that the time evolution admits remote past if there exists a solution (ξ_n, η_n) of (1.1).

Received October 12, 2007.
2000 *Mathematics Subject Classification.* Primary 60G50; Secondary 37H10, 60J05.
Key words and phrases. random time evolution, stochastic difference equations, random walks on groups, remote past, automorphisms of groups, non-strong solutions.

It is immediate (cf. [1]) to see that the above equation (1.1) is reduced to the convolution equation

$$\lambda_n = \mu_n * \lambda_{n-1} \qquad (n \in \mathbb{Z}) \tag{1.2}$$

where λ_n's are unknown probability measures on G which stand for the probability distribution of $\varphi^{-n}\eta_n$ and μ_n's are known probability measures which come from $\varphi^{-n}\xi_n$. The equation (1.2) for arbitrarily given μ_n's will be discussed in Section 3. We remark that the equation (1.2) always admits the *trivial solution* (λ_n) where each λ_n is the normalized Haar measure ω_G on G and that the set of the solutions (λ_n) of (1.2) is a convex set in the infinite product space of copies of the space of probability Borel measures on G.

If the automorphism is the identity, i.e., if $\varphi = \text{id.}$, then the solution of (1.1) is nothing but a random walk on the group G with remote past. There is a long history in the study of random walks on groups (cf., e.g., [3]) and the asymptotic behavior as time goes to infinity is studied in detail. Due to [3] the theory of random walks on groups goes back to the paper [4] by Y. Kawada and K. Itô and our Theorem 1.1 stated below may be regarded as its improvement from the viewpoint of "with remote past".

On the other hand, our motivation of the study of the time evolution with remote past has the background in the theory of stochastic differential equation which was invented by Kiyosi Itô in 1942. He gave the definition of Itô's stochastic integral and found the Itô's formula. He solved the stochastic differential equation by generalizing the successive iteration method. Those solutions he and his pupils gave on the first stage are now called "strong" solutions because the solution process is adapted to the given Brownian filtration: in other words, the solution up to time t is a functional of the noise up to time t. In 1960's "pathological" solutions are found: there exist nonstrong solutions. A typical example is Tanaka's stochastic differential equation:

$$dX(t) = \text{sgn}(X(t))dB(t), \ \text{sgn}(x) = 1(x \geq 0); = -1(x < 0).$$

It has various solutions:
1) $X(t) = |B(t)|$ is a solution(strong solution).
2) At any zero of $B(t)$ one can switch $\text{sgn}(X(t))$ to obtain a new solution which are not strong because extra randomness is put in there. Recently, Le Jan et al. characterized the solution set of Tanaka's equation (cf., [5]).

In 1975 Tsirelson ([7]) has constructed examples of nonstrong solutions by reducing stochastic differential equations into stochastic difference equations on the one dimensional torus $G = \mathbb{T}^1$. Later he studies the isomorphism problem of noises: white noise, black noise, etc.

In 1990's M. Yor ([8]) formulated Tsirelson's argument in the form (1.1) and studied the equation on $G = \mathbb{T}^1$ for general noise. In 2006 Akahori et al. ([1]) have studied it on general compact groups G and showed the structure of the extremal solution set, i.e., the extremal set of the solution set which is a compact convex set:

$$ex(\{solutions\}) \cong G/H \quad \text{for some subgroup} \quad H \subset G.$$

Now we state the results in the random walk case on a locally compact abelian group G. We denote the characteristic group of G by Γ.

Theorem 1.1. *Assume that the noise is stationary and the automorphism is the identity:*

$$\mu_n = \mu \text{ for any } n \in \mathbb{Z} \text{ and } \varphi = \text{id}. \tag{1.3}$$

Set

$$\Gamma_\mu = \{\chi \in \Gamma : |\mu(\chi)| = 1\} \tag{1.4}$$

and

$$G_\mu = \{g \in G : \chi(g) = 1 \text{ for all } \chi \in \Gamma_\mu\}. \tag{1.5}$$

Then there exists a unique element $\alpha(\mu)$ in G/G_μ such that solutions (λ_n) of (1.2) *are characterized by the following two properties:*
(a) *Each λ_n is G_μ-invariant.*
(b) *The projections $\widehat{\lambda}_n$ of λ_n to G/G_μ evolves by the Weyl transformation (or the translation) $\tau_{\alpha(\mu)}$ by $\alpha(\mu)$:*

$$\widehat{\lambda}_n = \tau_{\alpha(\mu)} \widehat{\lambda}_{n-1} \qquad (n \in \mathbb{Z}) \tag{1.6}$$

where $\tau_\alpha \beta = \beta + \alpha$ for $\alpha, \beta \in G/G_\mu$.

To illustrate the idea, we give examples in the case where $G = \mathbb{T}^1$. Here we identify $\mathbb{T}$ with the interval $[0, 1)$.

Example 1.2 (cf. [8, Section 7, Lemma 5]). For $p = 1, 2, \ldots$, we denote $\mathbb{Z}_p = \{k/p : k = 0, 1, \ldots, p-1\}$.
(i) Assume that $\mu(x + \mathbb{Z}_p) < 1$ for any $p = 1, 2, \ldots$ and any $x \in \mathbb{T}$. Then $G_\mu = \mathbb{T}$. In this case the solution of (1.2) is only the trivial solution.
(ii) Assume that $\mu(\{x\}) = 1$ for some $x \in \mathbb{T}$. Then $G_\mu = \{0\}$ and $\alpha(\mu) = \{x\}$. In this case every solution (λ_n) of (1.2) evolves by the translation by x.
(iii) Otherwise, we can choose $x \in \mathbb{T}$ and $p = 2, 3, \ldots$ such that p is

minimum among the pairs (x,p) with $\mu(x+\mathbb{Z}_p)=1$. Then $G_\mu=\mathbb{Z}_p$ and $\alpha(\mu)=x+\mathbb{Z}_p$.

In particular, we consider the case where the support of μ consists of two points: $\mu=p_1\delta_{x_1}+p_2\delta_{x_2}$ for some $p_1,p_2>0$, $p_1+p_2=1$ and $x_1,x_2\in\mathbb{T}$, $x_1\neq x_2$.
(i)′ Assume that x_2-x_1 is rational under identification $G\simeq[0,1)$. We can express $x_2-x_1=r/p$ for some $p=2,3,\dots$ and $r\in\mathbb{N}$ where p and r are coprime. Then $G_\mu=\mathbb{Z}_p$ and $\alpha(\mu)=x_1+\mathbb{Z}_p$ $(=x_2+\mathbb{Z}_p)$.
(ii)′ Otherwise, $G_\mu=\mathbb{T}$.

The following result also shows that the existence of nontrivial solution is fairly restrictive.

Theorem 1.3. *Assume that G has a countable basis. Let μ be a probability measure on G and φ an automorphism of G. Set*

$$\Gamma_\mu=\left\{\chi\in\Gamma:\prod_{k=m}^{\infty}|\mu(\chi\circ\varphi^k)|>0 \text{ for some } m\in\mathbb{Z}\right\}\tag{1.7}$$

and

$$G_\mu=\{g\in G:\chi(g)=1 \text{ for all } \chi\in\Gamma_\mu\}.\tag{1.8}$$

Then there exists an element $\alpha(\mu)\in G/G_\mu$ such that $\mu(\cap_{\chi\in\Gamma_\mu}W^s(a,\chi,\varphi))=1$ for any $a\in\alpha(\mu)$ where $W^s(a,\chi,\varphi)$ is the "stable set of a in direction χ":

$$W^s(a,\chi,\varphi)=\left\{x\in G:\lim_{k\to\infty}\chi(\varphi^k x)/\chi(\varphi^k a)=1\right\}.\tag{1.9}$$

Theorem 1.4. *Under the same assumption and notations as in Theorem 1.3, the following statements hold:*
(a) *Every solution (λ_n) of (1.2) consists of G_μ-invariant measures λ_n.*
(b) *There exists a sequence (ν_n) of G_μ-invariant probability measures on G such that every extremal solution (λ_n) of the convolution equation (1.2) corresponds to a unique element $\gamma\in G/G_\mu$ by the relation*

$$\lambda_n=\nu_n*\delta_{-\sum_{j=0}^{-n-1}\varphi^j a*\delta_c}\qquad(n\in\mathbb{Z})\tag{1.10}$$

where a and c are arbitrary elements of the cosets $\alpha(\mu)$ and γ, respectively, and the sum $\sum_{j=0}^{-n-1}$ is interpreted as 0 for $n=0$ and $-\sum_{j=-n}^{-1}$ for positive n.

In the case of nonabelian groups we use the unitary representation theory. We keep Tannaka's duality theorem in mind, though it will not be explicitly stated below. We denote the totality of unitary representations of a compact group G by $\widehat{G}$. Thus, each $\rho \in \Gamma_\mu$ is a unitary operator acting on some finite dimensional Hilbert space $U(\rho)$.

Theorem 1.5. *For a given noise probability distribution μ set*

$$\Gamma_\mu = \{\rho \in \widehat{G}; \|\mu(\rho)\|_{op} = 1\} \tag{1.11}$$

where $\mu(\rho) = \int_G \mu(dx)\rho(x)$ as before and $\|\cdot\|_{op}$ denotes the operator norm. Then each $\rho \in \Gamma_\mu$ has the component ρ_{11} of operator norm exactly 1 which is necessarily the identity operator on some subspace $U_1(\rho)$ of $U(\rho)$. Set

$$G_\mu = \{g \in G; \rho_{11}(g) = \mathrm{id} \quad \textit{for all} \quad \rho \in \Gamma_\mu\}. \tag{1.12}$$

Then G_μ is a subgroup of G and, if (λ_n) is a solution of (1.1), each λ_n is G_μ-invariant and is obtained by n times translation of λ_0 by some element $\alpha(\mu)$ in G/G_μ.

The proofs of Theorem 1.1-4 is given below following the joint work with K. Yano [6] where G is assumed to be compact but the proof of Theorem 1.5 will be published elsewhere.

In Section 5 we give a further property in the special case where G is a finite-dimensional torus and $\Gamma_\mu = \Gamma$.

We must also refer to a work of Brossard and Leuridan ([2]). They have studied rather general Markov chains and investigate the uniqueness problem and the behavior of sample paths at the remote past. But they imposed a restrictive assumption that the one-step transition probability is absolute continuous with respect to a measure. Consequently, the case which involves the Weyl transform is excluded.

§2. Random walks on abelian groups

In this section we assume that G is abelian, the noise is stationary and the automorphism is the identity:

$$\mu_n = \mu \text{ for any } n \in \mathbb{Z} \text{ and } \varphi = \mathrm{id}. \tag{2.1}$$

Then the convolution equation (1.2) takes the form

$$\lambda_n = \mu * \lambda_{n-1} \qquad (n \in \mathbb{Z}) \tag{2.2}$$

and we obtain

$$\lambda_n(\chi) = \mu(\chi)^m \lambda_{n-m}(\chi) \qquad n \in \mathbb{Z},\ m \in \mathbb{N},\ \chi \in \Gamma. \tag{2.3}$$

Definition 2.1. Set

$$\Gamma_\mu = \{\chi \in \Gamma : |\mu(\chi)| = 1\}. \tag{2.4}$$

Lemma 2.2. *If $|\mu(\chi)| < 1$, then $\lambda_n(\chi) = 0$ for any $n \in \mathbb{Z}$.*

Proof. By (2.3), we have $|\lambda_n(\chi)| = |\mu(\chi)|^m |\lambda_{n-m}(\chi)|$. Since $|\lambda_{n-m}(\chi)| \le \lambda_{n-m}(|\chi|) \le 1$, we obtain $|\lambda_n(\chi)| \le |\mu(\chi)|^m$ for $n \in \mathbb{Z}$ and $m \in \mathbb{N}$. Letting m tend to infinity, we obtain $\lambda_n(\chi) = 0$. Q.E.D.

Lemma 2.3. *Let λ be a probability measure on G and Γ_0 be a subset of Γ. Assume that $\lambda(\chi) = 0$ whenever $\chi \notin \Gamma_0$. Then, the measure λ is G_0-invariant where G_0 is the annihilator of Γ_0:*

$$G_0 = \{x \in G : \chi(x) = 1 \text{ for all } \chi \in \Gamma_0\}. \tag{2.5}$$

Proof. Let T_g be the translation by $g \in G_0$. Then, $(T_g\lambda)(\chi) = \chi(g)\lambda(\chi) = \lambda(\chi)$ if $\chi \in \Gamma_0$ by the definition of G_0. Otherwise, $(T_g\lambda)(\chi) = \chi(g)\lambda(\chi) = 0 = \lambda(\chi)$ by the assumption on λ. Hence, $T_g\lambda = \lambda$. Q.E.D.

Let us denote the annihilator of Γ_μ by G_μ.

Lemma 2.4. *If $|\mu(\chi)| = 1$, then $\chi(x)$ is constant μ-a.e. In particular, $\chi(x) = \mu(\chi)$ for μ-a.e. x.*

Proof. Since $|\chi(x)| = 1$ μ-a.e., one obtains

$$\begin{aligned} 0 &\le \int_X |\chi(x) - \mu(\chi)|^2 \mu(dx) \\ &\le \int_X \int_X |\chi(x) - \chi(y)|^2 \mu(dx)\mu(dy) = 2(1 - |\mu(\chi)|^2). \end{aligned} \tag{2.6}$$

Hence, if $|\mu(\chi)| = 1$, then $\chi(x) = \mu(\chi)$ μ-a.e. Q.E.D.

The above lemma can have various versions, for instance, for vectors in Hilbert spaces, for unitary operators etc.

Proposition 2.5. *The following statements hold:*
(i) *Γ_μ is a subgroup of the character group Γ.*
(ii) *For any $\chi_1, \chi_2 \in \Gamma_\mu$,*

$$\mu(\chi_1\overline{\chi_2}) = \mu(\chi_1)\overline{\mu(\chi_2)}. \tag{2.7}$$

In other words, the restriction $\mu|_{\Gamma_\mu}$ is a character of Γ_μ.
(iii) *There exists a unique element α_μ in G/G_μ such that $\mu(\chi) = \chi(a)$ for any $a \in \alpha_\mu$ and any character $\chi \in \Gamma_\mu$.*

Proof. Let $\chi_1, \chi_2 \in \Gamma_\mu$. By Lemma 2.4, we see that $\chi_1(x) = \mu(\chi_1)$ and $\chi_2(x) = \mu(\chi_2)$ for μ-a.e. $x \in G$. Then we have $(\chi_1\overline{\chi_2})(x) = \mu(\chi_1)\overline{\mu(\chi_2)}$ for μ-a.e. $x \in G$, and, hence, we obtain $\mu(\chi_1\overline{\chi_2}) = \mu(\chi_1)\overline{\mu(\chi_2)}$. This implies (ii) and also (i).

Note that Γ_μ is identified with the character group of G/G_μ. By Pontryagin's duality theorem, the character $\mu|_{\Gamma_\mu}$ of Γ_μ obtained in Proposition 2.5 can be identified with an element of G/G_μ. We identify it with a coset $\alpha(\mu)$. Then, for any $a \in \alpha(\mu)$ and any character $\chi \in \Gamma_\mu$, we obtain $\mu(\chi) = \chi(a)$. Q.E.D.

Proof of Theorem 1.1. We already proved (a) in Lemmas 2.2 and 2.3.

It then follows from Proposition 2.5 (iii) and from a similar argument in the proof of Lemma 2.3 that $\lambda_n(\chi) = \chi(a)\lambda_{n-1}(\chi)$ for all $n \in \mathbb{Z}$, $a \in \alpha(\mu)$ and $\chi \in \Gamma$. Consequently, $\lambda_n = T_a\lambda_{n-1}$. Since each λ_n is G_μ-invariant, we obtain (b). Q.E.D.

Remark 2.6. Assume, in addition, that G has a countable basis. Since each λ_n is G_μ-invariant, there exists a probability measure $\widehat{\lambda}_n$ on the quotient group G/G_μ such that

$$\int_G \lambda_n(dx)f(x) = \int_{G/G_\mu} \widehat{\lambda}_n(dh) \int_{G_\mu} \nu(dy)f(h.y) \tag{2.8}$$

for any continuous function f on G where ν is the normalized Haar measure on G_μ and $h.y$ stands for an element of $h \in G/G_\mu$ identified with a coset.

§3. Nonstationary noise on abelian groups

In this section we continue to assume that $\varphi = \text{id}$ but we consider the case where μ_n may depend on n. Now the convolution equation (1.2) takes the original form

$$\lambda_n = \mu_n * \lambda_{n-1} \qquad n \in \mathbb{Z} \tag{3.1}$$

and, hence, we obtain

(3.2)
$$\lambda_n(\chi) = \mu_n(\chi)\mu_{n-1}(\chi)\cdots\mu_{n-m+1}(\chi)\lambda_{n-m}(\chi) \quad n \in \mathbb{Z},\ m \in \mathbb{N},\ \chi \in \Gamma.$$

Lemma 3.1. *If* $\prod_{k=1}^{\infty} \mu_{-k}(\chi) = 0$, *then* $\lambda_n(\chi) = 0$ *for any* $n \in \mathbb{Z}$.

Proof. By (3.2), we have $|\lambda_n(\chi)| = \prod_{k=1}^{m-1} |\mu_{n-k}(\chi)||\lambda_{n-m}(\chi)|$. Since $|\lambda_{n-m}(\chi)| \leq \lambda_{n-m}(|\chi|) \leq 1$, we obtain $|\lambda_n(\chi)| \leq \prod_{k=1}^{m-1} |\mu_{n-k}(\chi)|$ for $n \in \mathbb{Z}$ and $m \in \mathbb{N}$. Letting m tend to infinity, we obtain $\lambda_n(\chi) = 0$. Q.E.D.

Definition 3.2. Set

$$\Gamma_\mu = \left\{ \chi \in \Gamma : \prod_{k=m}^{\infty} |\mu_{-k}(\chi)| > 0 \text{ for some } m \right\}. \tag{3.3}$$

Remark 3.3. In the case considered in Section 2 the two definitions of Γ_μ given by (2.4) and (3.3) coincide.

Lemma 3.4. *The inequality* $\prod_{k=m}^{\infty} |\mu_{-k}(\chi)| > 0$ *holds if and only if* $\mu_{-k}(\chi) \neq 0$ *for any* $k \geq m$ *and*

$$\sum_{k=m}^{\infty} \int_G \int_G \mu_{-k}(dx)\mu_{-k}(dy)|\chi(x) - \chi(y)|^2 < \infty. \tag{3.4}$$

Proof. Notice that

$$\int_G \int_G \mu_{-k}(dx)\mu_{-k}(dy)|\chi(x) - \chi(y)|^2 = 2(1 - |\mu_{-k}(\chi)|^2). \tag{3.5}$$

Hence the assertion follows from the fact that the infinite product $\prod_{k=1}^{\infty} c_k$ of $0 \leq c_k \leq 1$ converges to a positive limit if and only if c_k's are positive and $\sum_{k=1}^{\infty}(1 - c_k) < \infty$. Q.E.D.

Proposition 3.5. Γ_μ *is a subgroup of the character group* Γ.

Proof. Let $\chi_1, \chi_2 \in \Gamma_\mu$. Then it follows from Lemma 3.4 that, for sufficiently large m,

(3.6)

$$\begin{aligned} &\sum_{k=m}^{\infty} \int_G \int_G \mu_{-k}(dx)\mu_{-k}(dy)|(\chi_1\overline{\chi_2})(x) - (\chi_1\overline{\chi_2})(y)|^2 \\ &\leq \sum_{k=m}^{\infty} \int_G \int_G \mu_{-k}(dx)\mu_{-k}(dy) 2 \left\{ |\chi_1(x) - \chi_1(y)|^2 + |\chi_2(x) - \chi_2(y)|^2 \right\} \\ &< \infty. \end{aligned}$$

Moreover,

(3.7)

$$\sum_{k=m}^{\infty} \int_G \mu_{-k}(dx)|(\chi_1\overline{\chi_2})(x) - \mu_{-k}(\chi_1\overline{\chi_2})|^2 < \infty.$$

Hence, $\mu_{-k}(\chi_1\overline{\chi_2}) \neq 0$ except for finitely many k. Consequently, again by Lemma 3.4 we conclude that $\chi_1\overline{\chi_2} \in \Gamma_\mu$. Since we assume that G is compact, the character group Γ is discrete. Hence, an algebraic subgroup of Γ is a (topological) subgroup. Q.E.D.

Remark 3.6. Lemma 2.3 in the previous section works here, too. Thus, each λ_n of a solution (λ_n) is G_μ-invariant.

Here we stop the preliminary discussion on nonstationary noises and we proceed in the next section to the case where the noise is stationary and the automorphism is arbitrary.

§4. Noise-driven automorphisms of locally compact abelian group

Let φ be an automorphism of a compact abelian group G. We assume that the noise (ξ_n) is stationary so that the random variables ξ_n's are independent and subject to a common probability distribution μ. So we consider the stochastic equation (1.1) stated in Section 1 where μ_n is the probability distribution of $\varphi^{-n}\xi_n$ and λ_n is the probability distribution of $\varphi^{-n}\eta_n$. We denote by $\widetilde{\lambda}_n$ the probability distribution of η_n itself.

The automorphism φ of G induces an automorphism φ^* of the character group Γ: $\varphi^*\chi(x) = \chi(\varphi x)$, $x \in G$, $\chi \in \Gamma$. For $\chi \in \Gamma$ define

(4.1)

$$W_2^s(\chi, \varphi) = \left\{ (x, y) \in G \times G : \sum_{k=0}^{\infty} |(\varphi^{*n}\chi)(x) - (\varphi^{*n}\chi)(y)|^2 < \infty \right\}$$

and, for $x \in G$,

$$\begin{aligned} (4.2) \quad W_2^s(x; \chi, \varphi) &= \{y \in G : (x, y) \in W_2^s(\chi, \varphi)\} \\ &= \left\{ y \in G : \sum_{k=0}^{\infty} |(\varphi^{*n}\chi)(y) - (\varphi^{*n}\chi)(x)|^2 < \infty \right\}. \end{aligned}$$

We may call $W_2^s(x; \chi, \varphi)$ the *ℓ^2-stable set of x in the direction χ with respect to φ.*

Remark 4.1. We have the obvious relation $W_2^s(x; \chi, \varphi) \subset W^s(x; \chi, \varphi)$ where $W^s(x; \chi, \varphi)$ is defined in (1.9).

Lemma 4.2. *The set $W_2^s(0; \chi, \varphi)$ is a φ-invariant subgroup of G.*

Proof. Obvious. Q.E.D.

Now Lemma 3.4, Proposition 3.5 and Remark 3.6 can be restated as follows.

Proposition 4.3. *Assume that* (λ_n) *solves the equation* (1.2). *Then the following statements hold.*
(i) *If* $\chi \in \Gamma_\mu$, *then,* $(\mu \otimes \mu)(W_2^s(\chi, \varphi)) = 1$.
(ii) Γ_μ *is* φ^**-invariant.*
(iii) G_μ *is a* φ*-invariant subgroup.*
(iv) λ_n *is* G_μ*-invariant and so is* $\widetilde{\lambda}_n$.

Proof. (ii)-(iv) are obvious restatements. To see (i) it suffices to note that

$$\int_G \int_G \mu(dx)\mu(dy) \sum_{k=m}^{\infty} |(\varphi^{*k}\chi)(x) - (\varphi^{*k}\chi)(y)|^2 \tag{4.3}$$
$$= \sum_{k=m}^{\infty} \int_G \int_G \mu_{-k}(dx)\mu_{-k}(dy)|\chi(x) - \chi(y)|^2.$$

Q.E.D.

Remark 4.4. If $\varphi = \mathrm{id}$, then, $W_2^s(x; \chi, \varphi) = \{y : \chi(y) = \chi(x)\}$. Hence, if we assume, in addition, that G is metrizable or that Γ_μ is countable, then we can apply Fubini's theorem and the assertion (i) of Proposition 4.2 shows

$$(\mu \otimes \mu)\{(x, y) \in G : \chi(x) = \chi(y) \text{ for all } \chi \in \Gamma_\mu\} = 1. \tag{4.4}$$

This implies that the support of μ consists of a single coset in G/G_μ, which is nothing but the element $\alpha(\mu)$ introduced in Section 2.

Proof of Theorem 1.2. Obvious from (i) of Proposition 4.2. Q.E.D.

Proof of Theorem 1.3. Let (λ_n) be a solution of (1.2) and $a \in \alpha(\mu)$. Set

$$\mu_n^\circ = T_{-\varphi^{-n}a}\mu_n \qquad (n \in \mathbb{Z}) \tag{4.5}$$

and

$$\lambda_n^\circ = T_{\sum_{j=0}^{-n-1} \varphi^j a}\lambda_n \qquad (n \in \mathbb{Z}) \tag{4.6}$$

for each n. Here we interpret $\sum_{j=0}^{-1} = 0$ and $\sum_{j=0}^{-n-1} = -\sum_{j=-n}^{-1}$ for positive n. Then they satisfy

$$\lambda_n^\circ = \mu_n^\circ * \lambda_{n-1}^\circ \qquad (n \in \mathbb{Z}), \tag{4.7}$$

and, hence,

$$\lambda_n^\circ = \mu_n^\circ * \mu_{n-1}^\circ * \cdots * \mu_{n-k}^\circ * \lambda_{n-k-1}^\circ \qquad (n \in \mathbb{Z},\ k \in \mathbb{N}). \tag{4.8}$$

Recall that the totality of probability measures on a compact metrizable space is compact in the weak topology. Here we say that $\mu_n \to \mu$ weakly if $\mu_n(f) \to \mu(f)$ for any continuous function f. This topology is called the *weak* topology* in the context of the functional analysis.

Now we can choose an increasing sequence of integers $m_j \to \infty$ such that the weak limit

$$\nu_n^\circ = \lim_{j\to\infty} \mu_n^\circ * \mu_{n-1}^\circ * \cdots * \mu_{n-m_j}^\circ \qquad (n \in \mathbb{Z}) \tag{4.9}$$

exists. Since

$$\nu_n^\circ(\chi) = \lim_{j\to\infty} \prod_{i=0}^{m_j} \mu_{n-i}^\circ(\chi) \qquad (n \in \mathbb{Z}) \tag{4.10}$$

for any $\chi \in \Gamma$, we see that $\nu_n^\circ(\chi)$ for each $n \in \mathbb{Z}$ is not equal to 0 for any $\chi \in \Gamma_\mu$ and is equal to 0 for any $\chi \notin \Gamma_\mu$. Thus we conclude that ν_n° is G_μ-invariant by Lemma 2.3. Note that (ν_n) is not uniquely determined from (μ_n), but, for each choice of a sequence m_j, the limit $(\nu_n(\chi))$ is uniquely determined up to a multiplicative constant of modulus 1.

Take a limit point of the sequence $(\lambda_{-m_j}^\circ)$ and denote it by $\lambda_{-\infty}^\circ$. Recall that $\mu_n \to \mu$ and $\nu_n \to \nu$ weakly imply $\mu_n * \nu_n \to \mu * \nu$ weakly. In fact, it is obvious that the product measure $\mu_n \otimes \nu_n \to \mu \otimes \nu$ weakly and that the pullback of any continuous function under the map $(x, y) \mapsto x + y$ is again a continuous function on the product space. Letting k tend to infinity in (4.8), we obtain

$$\lambda_n^\circ = \nu_n^\circ * \lambda_{-\infty}^\circ. \tag{4.11}$$

Then (λ_n) is expressed as

$$\begin{aligned} \lambda_n &= T_{-\sum_{j=0}^{-n-1} \varphi^j a}(\nu_n^\circ * \lambda_{-\infty}^\circ) \\ &= \nu_n * T_{-\sum_{j=0}^{-n-1} \varphi^j a}(\lambda_{-\infty}^\circ). \end{aligned} \tag{4.12}$$

Here we denote $\nu_n = T_{-\sum_{j=0}^{-n-1} \varphi^j a}(\nu_n^\circ)$, which is also G_μ-invariant.

Consequently, an extremal solution (λ_n) is expressed as

$$\lambda_n = \nu_n * \delta_{-\sum_{j=0}^{-n-1} \varphi^j a + c} \tag{4.13}$$

for some $c \in G$ and the element c is unique modulo G_μ. Q.E.D.

§5. The case of toral automorphisms

By the definition the probability measures ν_n° satisfy the equation

$$\nu_n^\circ = \mu_n^\circ * \nu_{n-1}^\circ \qquad n \in \mathbb{Z}. \tag{5.1}$$

Consequently, if we take a sequence of independent random variables ξ_n° subject to μ_n°, then, we may formally understand that each ν_n° is the probability distribution of the infinite sum $\sum_{k=-\infty}^{n} \xi_k^\circ$.

In some special cases the convergence of $\sum_{k=-\infty}^{n} \xi_k^\circ$ is justified and an explicit formula for the solution (η_n) of the equation (1.1) is obtained.

Theorem 5.1. *Let G be a finite dimensional torus, say $G = \mathbb{T}^d$ and assume that $\Gamma_\mu = \mathbb{Z}^d$. Let d be a distance in $\mathbb{T}^d$. Then*

$$\sum_{k=-n}^{\infty} \varphi^k(\xi_{-k}^\circ) \text{ converges almost surely} \tag{5.2}$$

and

$$E\left[\sum_{k=-n}^{\infty} d(\varphi^k(\xi_{-k}^\circ), 0)^2\right] < \infty \tag{5.3}$$

for each $n \in \mathbb{Z}$. Moreover, the extremal solution (η_n) is given by the formula

$$\varphi^{-n}(\eta_n) = c + \sum_{k=-n}^{\infty} \varphi^k(\xi_{-k}^\circ) + \sum_{k=0}^{-n-1} \varphi^k(a) \quad \text{for} \quad n \in \mathbb{Z} \tag{5.4}$$

with $c \in \mathbb{T}^d$ and $a \in \alpha(\mu)$ where $\alpha(\mu)$ is defined in Theorem 1.4.

Lemma 5.2. *There exists a constant r with $0 < r < 1$ such that*

$$\begin{aligned} \cap_{\chi\in\Gamma} W^s(a;\chi,\varphi) &= \cap_{\chi\in\Gamma} W_2^s(a;\chi,\varphi) \\ &= \{x \in G : d(\varphi^k(x), \varphi^k(a)) \le Cr^n \\ &\qquad \text{for any } k \in \mathbb{N} \text{ and for some constant } C\}. \end{aligned} \tag{5.5}$$

Proof. Let us identify $\mathbb{T}^d$ with the unit cube $[-1/2, 1/2)^d$ in $\mathbb{R}^d$ and measures on $\mathbb{T}^d$ with those on $[-1/2, 1/2)^d$. Then the automorphism φ is regarded as an automorphism on $[-1/2, 1/2)^d$ and is defined by a matrix A as

$$\varphi(x) \;=\; Ax \bmod \mathbb{Z}^d. \tag{5.6}$$

Under the identification stated above, $\varphi^k(x) \to 0$ as $k \to \infty$ in $\mathbb{T}^d$ if and only if $A^k x \to 0$ as $k \to \infty$ in $\mathbb{R}^d$. Since A is a finite dimensional matrix, it means that the vector x in $\mathbb{R}^d$ belongs to the linear span of eigenvectors of A corresponding to eigenvalues of modulus less than 1. Take a constant r which is less than 1 and is greater than the maximum modulus of such eigenvalues. Then for any norm $\|\cdot\|$ there holds the inequality $\|A^k x\| \le Cr^k$ for some constant C. Consequently, for any distance d on $\mathbb{T}^d$ there holds the inequality $d(0, \varphi^k(x)) \le Cr^k$ for some constant C depending on x (which may be different from the previous C). Hence follows the desired assertion. Q.E.D.

Remark 5.3. If $x \neq 0$ and $A^k x \to 0$ as $k \to \infty$, then $\|A^{-k}x\| \to \infty$ as $k \to \infty$ but the converse is not true.

To prove Theorem 5.1 we want to apply a well-known convergence theorem: if X_k, $k = 0, 1, \ldots$, are independent $\mathbb{R}^d$-valued random variables and if they are square integrable with $\sum_{k=0}^{\infty} E[\|X_k - E[X_k]\|^2] < \infty$, then $\sum_{k=0}^{\infty}(X_k - E[X_k])$ converges almost surely and in L^2 sense. Here appear two obstacles:

(a) Absence of the notion of mean for group elements.

(b) The sequence $\chi(\varphi^k(x)) - \int_G \mu(dy)\chi(\varphi^k(y))$ is square summable for μ-a.e. x but is generally not summable.

Lemma 5.2 above shows that the assumptions of Theorem 5.1 eliminates (b). Indeed, the sequence $\chi(\varphi^k(x)) - \int_G \mu(dy)\chi(\varphi^k(y))$ decreases exponentially.

Lemma 5.4. *Under the identification of $\mathbb{T}^d$ with $[-1/2, 1/2)^d$, we obtain*

$$E\left[\sum_{k=0}^{\infty} \|\varphi^k(\xi^{\circ}_{-k})\|^2\right] < \infty. \tag{5.7}$$

Proof. We start with the following restatement of (3.4):

$$\int_G \mu(dx) E\left[\sum_{k=0}^{\infty} |\chi(\varphi^k(\xi_{-k})) - \chi(\varphi^k(x))|^2\right] < \infty. \tag{5.8}$$

Thus, for μ-a.e. x,

$$E\left[\sum_{k=0}^{\infty} |\chi(\varphi^k(\xi_{-k})) - \chi(\varphi^k(x))|^2\right] < \infty. \tag{5.9}$$

Therefore, $\xi^\circ_{-k} = \xi_{-k} - \varphi^k(a)$ satisfies

(5.10) $$E\left[\sum_{k=0}^{\infty} |\chi(\varphi^k(\xi^\circ_{-k})) - 1|^2\right] < \infty.$$

Now let χ_j, $j = 1, 2, \ldots, d$, be the standard generators of Γ:

(5.11) $$\chi_j(x) = \exp(2\pi\sqrt{-1}x_j) \text{ for } x = (x_1, \ldots, x_d) \in [-1/2, 1/2)^d.$$

Note that $|\exp(2\pi\sqrt{-1}x) - 1| \geq c|x|$ for $x \in [-1/2, 1/2)$. Hence it follows from (5.10) with $\chi = \chi_j$, $j = 1, 2, \ldots, d$, in $\mathbb{T}^d$ that

(5.12) $$E\left[\sum_{k=0}^{\infty} \|\varphi^k(\xi^\circ_{-k})\|^2\right] < \infty$$

in $\mathbb{R}^d$. Q.E.D.

Proof of Theorem 5.1. It follows from Lemma 5.4 that

(5.13) $$E\left[\sum_{k=0}^{\infty} \left\|\varphi^k(\xi^\circ_{-k}) - E[\varphi^k(\xi^\circ_{-k})]\right\|^2\right] < \infty.$$

Hence, the sum

(5.14) $$\sum_{k=0}^{\infty} \left\{\varphi^k(\xi^\circ_{-k}) - E[\varphi^k(\xi^\circ_{-k})]\right\}$$

converges almost surely. On the other hand, we already know that

(5.15) $$\sum_{k=0}^{\infty} (E[\varphi^k(\xi^\circ_{-k})] - 1)$$

converges absolutely. Consequently, the sum $\sum_{k=0}^{\infty} \varphi^k(\xi^\circ_{-k})$ converges almost surely. Q.E.D.

Added in proofs

The author would like to dedicate this article to the memory of Kiyosi Itô who passed away on Nov. 10, 2008.

References

[1] J. Akahori, C. Uenishi and K. Yano, Stochastic equation on compact groups in discrete negative time, Preprint RIMS-1535, 2006, available at RIMS web page, to appear in Probab. Theory Related Fields.

[2] J. Brossard and C. Leuridan, Chaînes de Markov indexées par$-\mathbf{N}$: existence et comportement, Ann. Probab., **29** (2001), 1033–1046.

[3] U. Grenander, Probabilities on algebraic structures, Wiley, New York, 1963.

[4] Y. Kawada and K. Itô, On the probability distribution on a compact group. I, Proc. Phys. Math. Soc. Japan, **22** (1940), 977–998.

[5] Y. Le Jan and O. Raimond, Flows associated to Tanaka's SDE, ALEA Lat. Am. J. Probab. Math. Stat., **1** (2006), 21–34.

[6] Y. Takahashi and K. Yano, Time Evolution with and without Remote Past (I): Noise Driven Automorphisms of Compact Abelian Groups, Preprint RIMS-1557, 2006, available at RIMS web page.

[7] B. S. Tsirel'son, An example of a stochastic differential equation having no strong solution, Translated from Russian, Theory Probab. Appl., **20** (1975), 416–418.

[8] M. Yor, Tsirel'son's equation in discrete time, Probab. Theory Related Fields, **91** (1992), 135–152.

Research Institute for Mathematical Sciences
Kyoto University
Japan

E-mail address: `takahasi@kurims.kyoto-u.ac.jp`

Advanced Studies in Pure Mathematics 53, 2009
Advances in Discrete Dynamical Systems
pp. 363–370

The Friedman rule under habit formation

Yoshimasa Aoki and Yasunobu Tomoda

Abstract.

This paper studies the Friedman rule on the optimal quantity of money in a money in the utility model with habit formation. If habits are formed only in consumption, or if habit formations in both consumption and real money holdings are symmetric, the Friedman rule is optimal. However, if habit forming is assumed in real money holdings, then the Freidman rule may not be optimal.

§1. Introduction

In the seminal paper "The Optimum Quantity of Money" (1969), Milton Friedman claimed that government should supply as large an amount of money as possible to the market. This follows since the amount of money supply is optimal if the marginal utility of real money holdings is equal to the marginal cost of money supply and the production cost of money is negligible. This optimal rule of money supply implies that the nominal interest rate is zero in market and that government should not tax the real money holdings of households.

Recently, a considerable number of studies have been made on the Friedman rule, including models with distortional taxes. In particular, Chari et al (1996) and Correia and Teles (1999) demonstrate that, in money in the utility models, the Friedman rule is optimal if the utility function is homothetic in consumption and real money holdings and separable in leisure[1].

In this study, we extend their models by introducing habit formation which arises from consumption and real balance holdings[2]. And we show that if habits are only formed in consumption, the Freidman rule is

Received November 14, 2006.
Revised October 12, 2007.
We appreciate Professor Kazuo Nishimura for his valuable comments and suggestions. All remaining errors are, of cause, our responsibility.

[1]Chari and Kehoe (1999) surveys the literature of this field.

[2]Habit formation is formalized by Ryder and Heal (1973).

optimal. If habits are symmetrically formed in both consumption and real balance holdings, the Friedman rule is still optimal. However, if habits are formed only in real money holdings, the Freidman rule is not necessarily optimal. We conclude that the optimality of the Friedman rule depends only on habits formed from real balance holdings with the exception of the symmetric habit parameter case.

This paper is organized as follows. In section 2, we set up the basic model with habit formation. In section 3, we show steady state solution and consider the optimal quantity of money. A summary is given section 4.

§2. The model

We consider an infinite horizon model in which a representative agent maximizes utility. Utility is a function of consumption, real balance holdings, and leisure. Consumption and real balance holdings include habit from past periods. The representative agent maximizes the discounted sum of utility over an infinite horizon:

$$\sum_{t=0}^{\infty} \beta^t U(X_t, M_t, l_t), \text{ with } U(\cdot) = \nu(X_t, M_t) + u(l_t) \tag{1}$$

where β represents the discount factor, l_t is leisure, and X_t and M_t are consumption and real balance holdings in relative amounts. $\nu(\cdot)$ and $u(\cdot)$ possess the following properties.

Assumption 1. $\nu(\cdot)$ is strictly concave, homogeneous of degree k (≥ 1), differentiable, and $\lim\limits_{M_t \to \infty} \left(\frac{\partial U}{\partial M_t}\right) = 0$.

Assumption 2. $u(\cdot)$ is strictly concave and differentiable.

We also assume that X_t and M_t have the following forms.

$$X_t = \frac{x_t}{(h_t)^{\eta}}, \tag{2}$$

$$M_t = \frac{m_t}{(b_t)^{\sigma}}, \tag{3}$$

where $\eta \in [0, 1)$ and $\sigma \in [0.1)$. x_t and m_t represent the consumption good and real balance holdings in the current period, respectively. Let h_t and b_t represent the weighted average of past consumption as in the following: $h_t = \rho \sum_{t=0}^{\infty} (1-\rho)^j x_{t-1-j}$ and $b_t = \omega \sum_{t=0}^{\infty} (1-\omega)^j m_{t-1-j}$, where $\rho \in (0, 1)$ and $\omega \in (0.1)$. These equations are equivalent as

$$(4) \qquad h_{t+1} = h_t + \rho \cdot (x_t - h_t),$$

$$(5) \qquad b_{t+1} = b_t + \omega \cdot (m_t - b_t).$$

In period t, households trade money, bonds, and the consumption good in each market. The budget constraint of households is

$$(6) \quad (1-\tau_t)P_t(1-l_t)+n_t+(1+i_t)B_t \geq P_t x_t + n_{t+1} + B_{t+1}, t = 0, 1, \cdots$$

where τ_t, $(1 - l_t)$, n_t, i_t, B_t, and P_t represent, respectively, the income tax, labor supply, nominal money holdings, the nominal interest rate, bond holdings from period t to $t+1$, and the price of the consumption good in units of money.

Next, we introduce the resource constraint. Households are endowed with one unit of time in each period. For simplicity, one unit of labor produces one unit of the consumption good. Thus, the resources constraint which the economy faces in period t is

$$(7) \qquad 1 - l_t \geq x_t + g,$$

where g is a given level of government expenditures that is constant over time.

§3. The Friedman rule

Under a given sequence $\{P_t, i_t, \tau_t\}_{t=0}^{\infty}$, the representative household maximizes the utility function (1) subject to budget constraints (6), the initial condition $n_{-1} = B_{-1} = 0$, and a no Ponzi games condition $\lim_{t\to\infty} I_t(n_t + B_t) \geq 0$, with $I_t = \frac{1}{\Pi_{s=0}^{t}(1+i_s)}$. The set of budget constraints can be written as a unique intertemporal budget constraint;

$$(8) \qquad \sum_{t=0}^{\infty} I_t P_t (1-\tau_t)(1-l_t) \geq \sum_{t=0}^{\infty} I_t P_t x_t + \sum_{t=0}^{\infty} I_t i_t P_t m_t.$$

Therefore, the household maximizes (1) subject to (8). The first order conditions of this problem, for $t \geq 0$, are

$$\beta^t \left(U_{X_t} \cdot (h_{t-1})^{-\eta} + \sum_{T=t+1}^{\infty} \beta^{T-t} U_{X_t} \frac{\partial X_T}{\partial x_t} \right) = \lambda I_t P_t, \tag{9}$$

$$\beta^t \left(U_{M_t} \cdot (b_{t-1})^{-\sigma} + \sum_{T=t+1}^{\infty} \beta^{T-t} U_{M_t} \frac{\partial M_T}{\partial m_t} \right) = \lambda I_t i_t P_t, \tag{10}$$

$$\beta^t U_{l_t} = \lambda I_t P_t (1 - \tau_t), \tag{11}$$

where λ is the Lagrange multiplier of the intertemporal budget constraint. Terms $U_{X_t} \cdot (h_{t-1})^{-\eta}$ and $U_{M_t} \cdot (b_{t-1})^{-\sigma}$ in (9) and (10) represent the current marginal utilities, which are positive terms. On the other hand, terms of $\sum_{T=t+1}^{\infty} \beta^{T-t} U_{X_t} \frac{\partial X_T}{\partial x_t}$ and $\sum_{T=t+1}^{\infty} \beta^{T-t} U_{M_t} \frac{\partial M_T}{\partial m_t}$ represent the marginal effects of habit formation. Because consumption and real balance holdings in period t reduce the utilities of future periods, these terms are negative. We assume that the sums of the left hand terms of (9) and (10) are non-negative. To simplify notation, we define $U_{x_t} = U_{X_t} \cdot (h_{t-1})^{-\eta}$, $U_{m_t} = U_{M_t} \cdot (b_{t-1})^{-\sigma}$, $V_{x_t} = \sum_{T=t+1}^{\infty} \beta^{T-t} U_{X_t} \frac{\partial X_T}{\partial x_t}$, and $V_{m_t} = \sum_{T=t+1}^{\infty} \beta^{T-t} U_{M_t} \frac{\partial M_T}{\partial m_t}$.

From (9), (10), and (11), we can derive the following conditions:

$$\left(U_{x_t} + V_{x_t} \right) \cdot i_t = \left(U_{m_t} + V_{m_t} \right), \tag{12}$$

$$\left(U_{m_t} + V_{m_t} \right) (1 - \tau_t) = U_{l_t}, \tag{13}$$

From (12), the Friedman rule $i_t = 0$ is equivalent to $(U_{m_t} + V_{m_t}) = 0$. Moreover, from (9), we have following difference equation:

$$U_{x_t} + V_{x_t} = \frac{\beta(1 + i_{t+1}) P_t}{P_{t+1}} \left(U_{x_{t+1}} + V_{x_{t+1}} \right). \tag{14}$$

Since consumption causes a habitual effect represented by V_{x_t} and $V_{x_{t+1}}$, (14) includes terms for not only period t and *t+1* but also all periods after period t.

A benevolent government chooses a policy assuming that households' maximize their utilities. The first order condition (12), (13), and (14) and the budget constraint (8) represent the sequence of households' behaviors on markets. From these equations, the implementability constraint of this problem is

(15) $$\sum_{t=0}^{\infty} \beta^t \left[(U_{x_t} + V_{x_t})\, x_t + (U_{m_t} + V_{m_t}) m_t - U_{l_t}(1 - l_t) \right] \leq 0.$$

Government chooses a $\{c_t, m_t, l_t\}_{t=0}^{\infty}$ that maximizing (1) subject to the implementability constraints (15) and the resource constraints (7). First order conditions are

(16)
$$(U_{x_t}+V_{x_t})-\mu[(U_{x_t}+V_{x_t})+(U_{x_t x_t}+V_{x_t x_t})x_t+(U_{x_t m_t}+V_{x_t m_t})m_t]-\Psi_t = 0,$$

(17)
$$(U_{m_t}+V_{m_t})-\mu[(U_{m_t}+V_{m_t})+(U_{x_t m_t}+V_{x_t m_t})x_t+(U_{m_t m_t}+V_{m_t m_t})m_t] = 0,$$

(18) $$U_{l_t} - \mu[U_{l_t} - U_{l_t l_t}(1 - l_t)] - \Psi_t = 0,$$

where $\Psi_t = \frac{\psi_t}{\beta^t}$, and μ and ψ_t are multipliers of the implementability constraint and resource constraints. From the first order conditions (16), (17), and (18), and constraints (7) and (15), five variables, i.e. x_t, m_t, l_t, μ, and Ψ_t, are determined. Since these conditions are independent of t, the solution is stationary in all variables. For the remainder of this section, we omit time subscript t. Now, we check whether $i = 0$ is the optimal policy in the steady state. First, we have the following proposition.

Proposition 1. *The Friedman rule, $i = 0$, is the optimal policy iff the following equation is satisfied.*

(19) $$(U_{xm} + V_{xm})\, x + (U_{mm} + V_{mm})\, m = 0.$$

Moreover, when $\sigma\omega\beta \neq 1 - \beta(1 - \omega)$, $i = 0$ iff $U_M m^{-\sigma} = 0$.

Proof. From (12), $i = 0$ is equivalent to $(U_m + V_m) = 0$ since $(U_x + V_x) \neq \infty$. Substituting $(U_m + V_m)$ into (17), we have (19) since we consider an interior solution, i.e. $\mu \neq 0$. Moreover, since, in the steady state, $x = h$ and $m = b$ from (4) and (5), we have $U_m + V_m = U_M m^{-\sigma}\left(1 - \frac{\sigma\omega\beta}{1-\beta(1-\omega)}\right)$. Therefore, if $\sigma\omega\beta \neq 1 - \beta(1 - \omega)$, $U_m + V_m = 0 \Longleftrightarrow U_M m^{-\sigma} = 0$. Q.E.D.

From the definitions of U_j and V_j $(j = x_t, m_t)$, we calculate the terms in equation (19). Since $x = h$ and $m = b$ in the steady state, from

(4) and (5) we have $X = x^{1-\eta}$, and $M = m^{1-\sigma}$.Thus, (19) is rewritten as

$$(20) \qquad A \cdot U_{XM} X m^{-\sigma} + B \cdot U_{MM} M m^{-\sigma} + C \cdot U_M m^{-\sigma} = 0,$$

where $A = 1 + \frac{\beta\eta\sigma\rho\omega}{1-\beta(1-\rho)(1-\omega)}$, $B = 1 + \frac{\beta\sigma^2\omega^2}{1-\beta(1-\omega)^2}$, and $C = \frac{\beta\sigma(\sigma+1)\omega^2}{1-\beta(1-\omega)^2}$. When $i = 0$, equation (20) is not always established. Therefore, $i = 0$ is not always the solution of this problem in the steady state. We consider several parameter cases to obtain a better description of the optimal money supply.

Case 1. *(Chari et al. 1996)*
If $\eta = \sigma = 0$ (households do not depend on habit formation), the Friedman rule is optimal.

Proof. Substituting $\eta = \sigma = 0$ into (20), because of $A = B = 1$ and $C = 0$, we have $U_{XM}X + U_{MM}M = 0$. Since the utility function is homogeneous of degree k, this equation is rewritten as $(k-1)U_M = 0$ using Euler's formula. This result stems from assumption 1. Q.E.D.

This case is identical to Chari et al. (1996). Thus, if preferences do not have habit formation, the Freidman rule is the optimal policy when preferences are separable and homogeneous of degree k.

Case 2. *If $\eta = \sigma$ and $\rho = \omega$ (households have symmetric habit parameters), the Friedman rule is optimal.*

Chari et al. (1996) proved the optimality of Freidman rule under homothetic preferences. If there is symmetric habit formation in both consumption and real balance holdings, preference is homothetic for x and m as well as X and M. Thus, Case 2 above is a special case of Chari et al. (1996).

Case 3. *If $\eta \neq 0$ and $\sigma = 0$ (i.e. only consumption is habit forming), the Friedman rule is optimal.*

Faria (2001) obtained a similar result. However, there is a big difference between Case 3 and Faria's result. In Faria's case, there is no constraint on the government's supply of money. On the other hand, we solved the problem in which government has to impose a distortional tax and/or an inflation tax since government needs financial sources for expenditure g in (8). Therefore, we demonstrate the optimality of the Friedman rule under a different setting.

Case 4. *Suppose $\eta = 0$ and $\sigma \neq 0$ (i.e. only real money holdings are habit forming). If $\sigma > \frac{1}{2}$, the Friedman rule is the optimal policy.*

Proof. Since, by substituting $\eta = 0$ into (20), we obtain $A = 1$, (20) is rewritten as $U_{XM}Xm^{-\sigma} + U_{MM}Mm^{-\sigma}B + U_M m^{-\sigma}C = 0$. The first and third terms in the left hand side of this equation are $\lim_{m\to\infty}(U_{XM}Xm^{-\sigma}) = \lim_{m\to\infty}(U_M m^{-\sigma}C) = 0$. However, since $M = m^{1-\sigma}$, the second term in the left hand side is rewritten as $U_{MM}m^{1-2\sigma}B$. Therefore, if $1 - 2\sigma < 0$, then $\lim_{m\to\infty}(U_{MM}m^{1-2\sigma}B) = 0$. Q.E.D.

This result contrasts with that of Case 3. In Case 3, if habits are only formed from consumption, the Friedman rule is independent of consumption based habit formation. However, Case 4 claims that, if habits are only formed from real balance holdings, the optimality of the Friedman rule depends on habits formed from real balance holdings.

From the proof of Case 4, we can extend Case 4 for cases where $\eta \neq 0$. Then, we obtain a more general proposition than the above cases.

Proposition 2. *If $\sigma = 0$, $\sigma > \frac{1}{2}$, or $\eta = \sigma$ and $\rho = \omega$, the Friedman rule is the optimal policy.*

We omit the proof because it is trivial given the above cases. In our study, we analyzed a model with habit formation arising from consumption and real money holdings. We introduced four habitual parameters, i.e. η, σ, ρ, and ω. However, we conclude that the optimality of the Friedman rule in habit forming models depends only on the habitual parameter of money holdings σ, with the exception of the symmetric parameter case, i.e. $\eta = \sigma$ and $\rho = \omega$.

§4. Summary

We studied the optimality of the Friedman rule in a model with habit formation in real balance holdings as well as consumption. If the functions of habit formation are symmetric, the Friedman rule is optimal. If habit formation is assumed only in consumption, the Friedman rule is again optimal. Moreover, if the habit parameter of real balance holdings σ is greater than one-half, the Friedman rule is optimal regardless of the habit parameter of consumption. We introduced four habitual parameters. However, we conclude that the optimality of the Friedman rule in our model only depend on the habit parameter of money holdings σ with the expection of the symmetric parameter case.

References

[1] V. V. Chari, L. J. Christiano and P. J. Kehoe, Optimality of the Friedman rule in economics with distorting taxes, J. Monetary Econom., **37** (1996), 203–223.

[2] V. V. Chari and P. J. Kehoe, Optimal fiscal and monetary policy, In: Handbook of Macroeconomics Vol. 1C, (eds. J. Taylor and M. Woodford), Elsevier North-Holland, Amsterdam, 1999, pp. 1671–1745.

[3] I. Correia and P. Teles, The optimal inflation tax, Rev. Econ. Dyn., **2** (1999), 325–346.

[4] J. R. Faria, Habit formation in a monetary growth economy, Econom. Lett., **73** (2001), 51–55.

[5] M. Friedman, The optimum quantity of money, In: The Optimum Quantity of Money and other Essays, (ed. M. Friedman), Aldine, Chicago, 1969, pp. 1–50.

[6] H. E. Ryder and G. M. Heal, Optimal growth with intertemporally dpendent preference, Rev. Econom. Stud., **40** (1973), 1–31.

Yoshimasa Aoki
Department of Economics
Sapporo Gakuin University

Yasunobu Tomoda
Institute of Economic Research
Kyoto University, JSPS Research Fellow

E-mail address: tomoda@kier.kyoto-u.ac.jp

Advanced Studies in Pure Mathematics 53, 2009
Advances in Discrete Dynamical Systems
pp. 371–385

Subharmonic bifurcation from relative equilibria in reversible systems with rotation symmetry

André Vanderbauwhede

Abstract.

In this paper we study the bifurcation of subharmonic tori from branches of symmetric relative equilibria in reversible systems with an additional S^1-symmetry. The analysis is based on a detailed study of the Poincaré map, which appears to be generated by a reduced vectorfield on the section. The results are applied to a simple example system, but the bifurcation behaviour described by our theoretical results can also be observed in, for example, the spherical pendulum and the Furuta pendulum.

§1. Introduction

Consider a system

$$\dot{x} = f(x), \tag{1}$$

with $x \in \mathbb{R}^{2n}$ and $f : \mathbb{R}^{2n} \to \mathbb{R}^{2n}$ a C^∞-smooth vectorfield. Such system is *reversible* if there exists a linear involution $R \in \mathcal{L}(\mathbb{R}^{2n})$ such that $f(Rx) = -Rf(x)$ for all $x \in \mathbb{R}^{2n}$; we will assume that $\dim(\mathrm{Fix}(R)) = \dim(\mathrm{Fix}(-R)) = n$. Denoting by $\tilde{x}(t,x)$ the flow of (1) we have then that $\tilde{x}(t,Rx) = R\tilde{x}(-t,x)$, and if $\tilde{x}(t_0,x_0) \in \mathrm{Fix}(R)$ for some $(t_0,x_0) \in \mathbb{R} \times \mathbb{R}^{2n}$ then the orbit $\{\tilde{x}(t,x_0) \mid t \in \mathbb{R}\}$ is invariant under R; we call such orbits *symmetric*. If for some $t_0 < t_1$ we have $\tilde{x}(t_0,x_0) \in \mathrm{Fix}(R)$, $\tilde{x}(t_1,x_0) \in \mathrm{Fix}(R)$ and $\tilde{x}(t,x_0) \notin \mathrm{Fix}(R)$ for $t_0 < t < t_1$, then $\tilde{x}(t,x_0)$ is both symmetric and periodic, with minimal period $T = 2(t_1 - t_0)$. Such symmetric periodic orbits appear typically in one-parameter families, since they are determined by the intersection of the n-dimensional subspace $\mathrm{Fix}(R)$ with the $(n+1)$-dimensional manifold $\{\tilde{x}(t,x) \mid t \in \mathbb{R}, x \in \mathrm{Fix}(R)\}$. When along such one-parameter family one finds an orbit which has a (non-trivial) pair of multipliers which are q-th roots of unity ($q \geq 3$) then one expects at this orbit the bifurcation of one or

Received October 12, 2007.

more branches of q-subharmonics. The precise conditions for such subharmonic bifurcation and the detailed form of the bifurcating branches have been studied in [8]; see also [1] and [2].

Typically, two branches of q-subharmonic orbits will bifurcate from the primary family; the two branches are close to each other, like the boundaries of an Arnold tongue (i.e. at a distance ρ from the bifurcation point the separation between the two branches is of the order $(q-2)/2$). Along one of the two branches there is a pair of near-critical multipliers (i.e. close to 1) on the unit circle, on the other branch the near-critical multipliers are on the real axis. To see this typical subharmonic branching behaviour three conditions must be satisfied: (1) the critical periodic orbit along the primary family must have 1 as a multiplier with multiplicity 2, together with exactly one pair of simple multipliers which are q-th roots of unity; (2) this pair of multipliers must move with non-zero speed through the root of unity as one moves along the primary family (using an appropriate intrinsic parametrization); (3) some higher order coefficient (of order $q-1$) in the normal form of the Poincaré-map must be different from zero. In [3] it is shown what happens when the critical multipliers are not simple (condition (1) is not satisfied), and in a forthcoming paper we will study the case where the transversality conditions (2) fails. In this note we describe a particular class of systems where due to additional symmetries the condition (3) is not satisfied.

We will assume that the system (1) has an S^1-symmetry which is compatible with the reversibility: there exists some $J_0 \in \mathcal{L}(\mathbb{R}^{2n})$ such that $J_0^2 = -I$, $J_0 R = -RJ_0$ and

$$f(e^{J_0\theta}x) = e^{J_0\theta}f(x), \qquad \forall\theta \in \mathbb{R},\ \forall x \in \mathbb{R}^{2n}. \tag{2}$$

Observe that $(\theta, x) \in \mathbb{R} \times \mathbb{R}^{2n} \mapsto e^{J_0\theta}x \in \mathbb{R}^{2n}$ defines an S^1-action on $\mathbb{R}^{2n}$, since $e^{2\pi J_0} = I$. Without loss of generality we can assume that both R and J_0 are orthogonal (this implies that R is symmetric and J_0 anti-symmetric). A symmetric *relative equilibrium* is a symmetric orbit of (1) (generated by some $x_0 \in \mathrm{Fix}(R)$) which is at the same time an orbit under the S^1-action: $\{\tilde{x}(t, x_0) \mid t \in \mathbb{R}\} = \{e^{J_0\theta}x_0 \mid \theta \in \mathbb{R}\}$. A necessary and sufficient condition for $x_0 \in \mathrm{Fix}(R)$ to generate such symmetric relative equilibrium is that there exists some $\Omega_0 \in \mathbb{R}$ such that

$$f(x_0) = \Omega_0 J_0 x_0. \tag{3}$$

By imposing the condition $\Omega_0 \neq 0$ we exclude actual equilibria; replacing J_0 by $-J_0$ if necessary we can w.l.o.g. assume that $\Omega_0 > 0$. The corresponding orbit is then periodic, with minimal period $T_0 := 2\pi/\Omega_0$, and $\tilde{x}(t, x_0) = e^{J_0 \Omega_0 t} x_0$. In the subsequent sections we will study subharmonic bifurcation from such symmetric relative equilibrium.

Observe that since both f and J_0 map $\text{Fix}(R)$ into $\text{Fix}(-R)$, we can rewrite the condition (3) for symmetric relative equilibria as $F(x, \Omega) = 0$, with $F : \text{Fix}(R) \times \mathbb{R} \to \text{Fix}(-R)$ defined by $F(x, \Omega) := f(x) - \Omega J_0 x$. Since $\text{Fix}(R)$ and $\text{Fix}(-R)$ have the same dimension we expect to see one-parameter families of symmetric relative equilibria.

Here is an example of a system satisfying our hypotheses. Take $n = 2$, identify $\mathbb{R}^4$ with $\mathbb{C}^2$ writing $x = (a, b)$, and consider the system

$$\text{(4)} \qquad \begin{cases} \dot{a} = ib, \\ \dot{b} = ia - i|b|^2 b. \end{cases}$$

The reversibility and S^1-equivariance are generated by respectively

$$\text{(5)} \qquad R(a, b) := (\bar{a}, \bar{b}) \quad \text{and} \quad J_0(a, b) := i(a, b), \qquad \forall (a, b) \in \mathbb{C}^2.$$

An extensive study of (4) can be found in [6]. There mainly the Hamiltonian structure of (4) was used; here we will mostly ignore this additional structure. The condition for a non-trivial symmetric relative equilibrium takes the form

$$\text{(6)} \qquad b = \Omega a \quad \text{and} \quad a - b^3 = \Omega b, \qquad (a, b, \Omega) \in \mathbb{R}^3, \ (a, b) \neq (0, 0).$$

It follows that $0 < b^2 = g(\Omega) := \Omega^{-1} - \Omega$ and hence either $\Omega < -1$ or $0 < \Omega < 1$. This gives us two branches of symmetric relative equilibria, both originating at the origin, and generated by the points

(7)

$$x_\Omega = \sqrt{g(\Omega)}(\Omega^{-1}, 1), \qquad \text{with } \Omega \in I_- :=]-\infty, -1[\text{ or } \Omega \in I_+ :=]0, 1[.$$

§2. The Poincaré map

In order to study the bifurcation of subharmonics at a symmetric relative equilibrium $\gamma_0 := \{e^{J_0 \Omega_0 t} x_0 \mid t \in \mathbb{R}\}$ of (1) we will use the Poincaré map at x_0 defined by the R-invariant section $x_0 + \Sigma$, where $\Sigma := (J_0 x_0)^\perp$. First we introduce an appropriate coordinate system in a neighborhood of γ_0.

Lemma 1. *There exists some $r_0 > 0$ such that the mapping*

$$\text{(8)} \qquad \begin{aligned} \Psi : \quad S^1 \times U_0 := (\mathbb{R}/2\pi\mathbb{Z}) \times \{y \in \Sigma \mid \|y\|^2 < r_0\} &\longrightarrow \mathbb{R}^{2n}, \\ (\theta, y) &\longmapsto e^{J_0 \theta}(x_0 + y) \end{aligned}$$

is a diffeomorphism onto the open S^1-invariant neighborhood $U := \Psi(S^1 \times U_0)$ of γ_0.

Proof. We have $\Psi(\theta,0)=e^{J_0\theta}x_0$ and $D\Psi(\theta,0)\cdot(\tilde\theta,\tilde y)=e^{J_0\theta}(\tilde\theta J_0x_0+\tilde y)$ for all $\theta\in S^1$, $\tilde\theta\in\mathbb{R}$ and $\tilde y\in\Sigma$. Since Σ is the orthogonal complement of J_0x_0 this shows that $D\Psi(\theta,0)$ is invertible (for each $\theta\in S^1$), and hence we can choose $r_0>0$ sufficiently small such that the restriction of Ψ to (a small neighborhood of θ in $S^1)\times U_0$ is a diffeomorphism onto an open neighborhood of $e^{J_0\theta}x_0$ in $\mathbb{R}^{2n}$. It follows that $U=\Psi(S^1\times U_0)$ is an open neighborhood of γ_0 (we call this a *tubular neighborhood*); U is obviously S^1-invariant. It remains to show that Ψ is injective on its full domain $S^1\times U_0$ if we choose r_0 small enough; this injectivity is equivalent to the statement that for all $\theta\in\mathbb{R}$ and all $y\in U_0$ we have

$$e^{J_0\theta}(x_0+y)\in x_0+U_0 \implies \theta\in 2\pi\mathbb{Z}. \tag{9}$$

Suppose we can not choose $r_0>0$ small enough such that (9) holds. Then there exist sequences $(\theta_n)_{n\in\mathbb{N}}$ and $(y_n)_{n\in\mathbb{N}}$ in respectively $\mathbb{R}$ and Σ such that

$$\theta_n\notin 2\pi\mathbb{Z},\quad y_n':=e^{J_0\theta_n}(x_0+y_n)-x_0\in\Sigma,$$

$$\lim_{n\to\infty}y_n=0\quad\text{and}\quad\lim_{n\to\infty}y_n'=0.$$

Adding multiples of 2π we can assume that the sequence (θ_n) is bounded and hence (taking subsequences) convergent: $\lim_{n\to\infty}\theta_n=\bar\theta$. Then $e^{J_0\bar\theta}x_0-x_0=\lim_{n\to\infty}y_n'=0$, and therefore $\bar\theta\in 2\pi\mathbb{Z}$. But then the equality $e^{J_0\theta_n}(x_0+y_n)=e^{J_0\bar\theta}(x_0+y_n')$ together with the fact that Ψ is a local diffeomorphism near $(\bar\theta,0)$ implies that $\theta_n=\bar\theta$ and $y_n=y_n'$ for n sufficiently large, giving us a contradiction since $\theta_n\notin 2\pi\mathbb{Z}$. Q.E.D.

Clearly $\mathbb{R}^{2n}=\operatorname{span}_{\mathbb{R}}\{J_0x_0\}\oplus\Sigma$; by continuity and by choosing r_0 sufficiently small we have then also $\mathbb{R}^{2n}=\operatorname{span}_{\mathbb{R}}\{J_0(x_0+y)\}\oplus\Sigma$ for all $y\in U_0$. So there exist unique mappings $\Omega:U_0\to\mathbb{R}$ and $h:U_0\to\Sigma$ such that

$$f(x_0+y)=\Omega(y)J_0(x_0+y)+h(y),\qquad\forall y\in U_0; \tag{10}$$

these mappings are smooth, with $\Omega(0)=\Omega_0$ and $h(0)=0$. Moreover, the uniqueness of Ω and h combined with the reversibility of f imply that

$$\Omega(Ry)=\Omega(y)\quad\text{and}\quad h(Ry)=-Rh(y),\qquad\forall y\in U_0. \tag{11}$$

In the tubular neighborhood U of γ_0 the vectorfield f is given by

$$f(e^{J_0\theta}(x_0+y))=e^{J_0\theta}\left(\Omega(y)J_0(x_0+y)+h(y)\right),\qquad\forall\theta\in\mathbb{R},\forall y\in U_0. \tag{12}$$

Next we consider the solution $\tilde{x}(t, x_0 + y)$ starting at a point $x_0 + y$ in the section $x_0 + \Sigma$ and close to x_0 (i.e. $y \in U_0$). By Lemma 1 we can write

$$\tilde{x}(t, x_0 + y) = e^{J_0 \tilde{\theta}(t,y)}(x_0 + \tilde{y}(t, y)) \tag{13}$$

for all $t \in \mathbb{R}$ such that $\tilde{x}(t, x_0 + y) \in U$, and with $\tilde{\theta}(t, y) \in \mathbb{R}$ and $\tilde{y}(t, y) \in U_0$. Expressing that this is a solution of (1) leads to

$$\begin{aligned} &e^{J_0 \tilde{\theta}(t,y)}(\dot{\tilde{\theta}}(t, y) J_0(x_0 + \tilde{y}(t, y)) + \dot{\tilde{y}}(t, y)) \\ &= e^{J_0 \tilde{\theta}(t,y)} \left(\Omega(\tilde{y}(t, y))(J_0(x_0 + \tilde{y}(t, y)) + h(\tilde{y}(t, y))\right)\end{aligned}$$

and hence

$$\dot{\tilde{\theta}}(t, y) = \Omega(\tilde{y}(t, y)) \quad \text{and} \quad \dot{\tilde{y}}(t, y) = h(\tilde{y}(t, y)).$$

Combined with the initial conditions $\tilde{y}(0, y) = y$ and $\tilde{\theta}(0, y) = 0$ we obtain the following.

Theorem 2. *In the tubular neighborhood U given by Lemma 1 the flow of (1) is given by*

$$\tilde{x}(t, x_0 + y) = e^{J_0 \Omega(y) t}(x_0 + \tilde{y}(t, y)), \tag{14}$$

where $\tilde{y}(t, y)$ is the flow of the reversible system

$$\dot{y} = h(y) \tag{15}$$

on the $(2n - 1)$-dimensional space Σ.

Proof. It follows from the foregoing that $\tilde{y}(t, y)$ as appearing in (13) must indeed be the flow of (15). The S^1-equivariance of (1) implies that $\tilde{x}(t, e^{J_0 \theta} x) = e^{J_0 \theta} \tilde{x}(t, x)$ for all $\theta \in \mathbb{R}$; we also have $\tilde{y}(t, \tilde{y}(\tau, y)) = \tilde{y}(t+\tau, y)$ and $\tilde{x}(t, \tilde{x}(\tau, x_0+y)) = \tilde{x}(t+\tau, x_0+y)$ for all $t, \tau \in \mathbb{R}$. Bringing (13) in this last equality and working out leads to

$$\tilde{\theta}(t + \tau, y) = \tilde{\theta}(t, y) + \tilde{\theta}(\tau, y), \qquad \forall t, \tau \in \mathbb{R}, \forall y \in U_0.$$

Differentiating in τ at $\tau = 0$ gives

$$\dot{\tilde{\theta}}(t, y) = \dot{\tilde{\theta}}(0, y) = \Omega(y) \quad \Longrightarrow \quad \tilde{\theta}(t, y) = \Omega(y) t, \qquad \forall (t, y) \in \mathbb{R} \times U_0.$$

Bringing this in (13) proves the theorem. Q.E.D.

The proof above shows that

$$\Omega(\tilde{y}(\tau,y)) = \Omega(y), \qquad \forall \tau \in \mathbb{R},\ \forall y \in U_0. \tag{16}$$

Also, for $t, \tau \in \mathbb{R}$ and $y \in U_0$ we have

$$\begin{aligned} \tilde{x}(t, x_0 + \tilde{y}(\tau,y)) &= e^{J_0\Omega(\tilde{y}(\tau,y))t}(x_0 + \tilde{y}(t, \tilde{y}(\tau,y))) \\ &= e^{J_0\Omega(y)t}(x_0 + \tilde{y}(t+\tau, y)) \\ &= e^{-J_0\Omega(y)\tau}\tilde{x}(t+\tau, x_0+y); \end{aligned}$$

this proves that the orbits $\{\tilde{x}(t, x_0 + \tilde{y}(\tau,y)) \mid t \in \mathbb{R}\}$ and $\{\tilde{x}(t, x_0+y) \mid t \in \mathbb{R}\}$ generated by respectively $x_0 + \tilde{y}(\tau,y)$ and $x_0 + y$ can be obtained one from the other by using the S^1-symmetry.

It follows from Theorem 2 that for each sufficiently small $y \in \Sigma$ the *first return time* — defined as the time $T = T(y) > 0$ close to $T_0 = 2\pi/\Omega_0$ such that $\tilde{x}(T, x_0 + y) \in x_0 + \Sigma$ — is given by $T(y) = 2\pi/\Omega(y)$. The corresponding Poincaré map $P : U_0 \to \Sigma$, defined by $P(y) := \tilde{x}(T(y), x_0 + y) - x_0$, is then given by

$$P(y) = \tilde{y}(T(y), y) = \tilde{y}(2\pi/\Omega(y), y), \qquad \forall y \in U_0. \tag{17}$$

We have $P(0) = 0$ and (using $\tilde{y}(t, 0) = 0$)

$$DP(0) = D_y\tilde{y}(T_0, 0) = e^{A_0 T_0}, \qquad \text{with } A_0 := Dh(0) \in \mathcal{L}(\Sigma).$$

Moreover, it follows from (11) that $T(Ry) = T(y)$, $\tilde{y}(t, Ry) = R\tilde{y}(-t, y)$ and $P(Ry) = \tilde{y}(T(Ry), Ry) = R\tilde{y}(-T(y), y) = P^{-1}(y)$, i.e. the Poincaré map P is a *reversible map*, in the sense that

$$R \circ P \circ R^{-1} = P^{-1}. \tag{18}$$

We finish this section by finding the relation between the eigenvalues of A_0 and $DP(0)$, and the multipliers of the relative periodic orbit γ_0. These multipliers are the eigenvalues of the monodromy matrix $M_0 := X(T_0)$, where $X(t)$ is the solution of the initial value problem

$$\dot{X}(t) = Df(e^{J_0\Omega_0 t}x_0) \cdot X(t), \quad X(0) = I, \qquad (X(t) \in \mathcal{L}(\mathbb{R}^{2n})). \tag{19}$$

It follows from the S^1-equivariance of f that $Df(e^{J_0\theta}x) = e^{J_0\theta}Df(x)e^{-J_0\theta}$ for all $(\theta, x) \in \mathbb{R} \times \mathbb{R}^{2n}$; bringing this in (19) and setting $X(t) = e^{J_0\Omega_0 t}Y(t)$ shows that $Y(t) = e^{\tilde{A}_0 t}$, with $\tilde{A}_0 := Df(x_0) - \Omega_0 J_0$. Hence $X(t) = e^{J_0\Omega_0 t}e^{\tilde{A}_0 t}$, and $M_0 = e^{\tilde{A}_0 T_0}$. Remember that x_0 is an equilibrium of the vectorfield $F(x) := f(x) - \Omega_0 J_0 x$, and observe that $\tilde{A}_0 = DF(x_0)$. Since F is reversible and $x_0 \in \mathrm{Fix}(R)$ also $\tilde{A}_0$ is reversible: $R\tilde{A}_0 = -\tilde{A}_0 R$. Consequently, if $\lambda \in \mathbb{C}$ is an eigenvalue of

$\tilde{A}_0$, then so is $-\lambda$, and both have the same algebraic and geometric multiplicities. In the same way M_0 is a reversible linear mapping, i.e. $R \circ M_0 \circ R = M_0^{-1}$, and if $\mu \in \mathbb{C}$ is a multiplier, then so is μ^{-1}. From this it is easy to see that 1 (which is always a multilier) must have an even (algebraic) multiplicity.

To see the relation between $\tilde{A}_0$ and A_0 we differentiate (12) in θ and y at the point $(\theta, y) = (0, 0)$; rearranging terms and using the definitions of $\tilde{A}_0$ and A_0 we find that

$$\begin{cases} \tilde{A}_0 \cdot J_0 x_0 &= 0, \\ \tilde{A}_0 \cdot y &= (D\Omega(0) \cdot y)\, J_0 x_0 + A_0 y, \qquad \forall y \in \Sigma. \end{cases}$$

This shows that $\tilde{A}_0$ is triangular with respect to the splitting $\mathbb{R}^{2n} = \mathrm{span}_{\mathbb{R}}\{J_0 x_0\} \oplus \Sigma$, and that A_0 and $\tilde{A}_0$ have the same eigenvalues with the same multiplicities, except for the eigenvalue 0 for which the multiplicity as an eigenvalue of A_0 is one lower than the multiplicity as an eigenvalue of $\tilde{A}_0$. Hence the eigenvalues of $DP(0)$ are exactly the multipliers of the periodic orbit γ_0, except that one has to lower the multiplicity of the trivial multiplier 1 by one.

§3. Subharmonic bifurcation

In order to study subharmonic bifurcation from a symmetric relative equilibrium γ_0 as described in the preceding sections we fix some $q \geq 1$ and make the following assumption:

(H1) The linear operator $\tilde{A}_0 = Df(x_0) - \Omega_0 J_0$ has:

- 0 as an eigenvalue with multiplicity 2;
- a pair of simple eigenvalues $\pm i\omega_0$, with $\omega_0 = \Omega_0 k q^{-1}$, $k > 0$ and $\gcd(k, q) = 1$;
- no other purely imaginary eigenvalues $\pm i\mu$ such that $q\mu \in \Omega_0 \mathbb{Z}$.

In the case $q \geq 3$ this means that next to the multiplier 1, which should have the lowest possible multiplicity (namely 2), the periodic orbit γ_0 has just one pair of simple multipliers which are q-th roots of unity. We should mention that the cases $q = 1$ (plain bifurcation of periodic orbits) and $q = 2$ (period-doubling) are allowed. Under the hypothesis (**H1**) we will study the existence of q-periodic points of the Poincaré map P near the fixpoint $y = 0$; such a q-periodic point $y \in U_0 \subset \Sigma$ generates a q-subharmonic solution $\tilde{x}(t, x_0 + y)$ of (1).

The small q-periodic points $y \in U_0$ of the Poincaré map P are determined by the equation

$$P^{(q)}(y) = y, \qquad \text{with } P^{(q)} := P \circ P \circ \cdots \circ P \ \ (q \text{ times});$$

since $P(y) = \tilde{y}(T(y), y)$ and $T(\tilde{y}(\tau, y)) = T(y)$ this takes the form

$$\tilde{y}(qT(y), y) = y, \qquad (y \in U_0). \tag{20}$$

From this we obtain immediately the following result.

Theorem 3. *A point $y \in U_0$ is a q-periodic point of P if and only if either*

(i) *y is an equilibrium of the vectorfield h, i.e. $h(y) = 0$, $\tilde{y}(t, y) = y$ for all $t \in \mathbb{R}$, and $\{\tilde{x}(t, x_0 + y) = e^{J_0 \Omega(y) t}(x_0 + y) \mid t \in \mathbb{R}\}$ is a relative periodic solution of (1),*

or

(ii) *$\tilde{y}(t, y)$ is a non-trivial periodic solution of (15), with some minimal period $\tilde{T}(y) > 0$, and*

$$\frac{qT(y)}{\tilde{T}(y)} \in \mathbb{N}. \tag{21}$$

In case (ii) *each point of the (periodic) orbit $\{\tilde{y}(\tau, y) \mid \tau \in \mathbb{R}\}$ will also be a q-periodic point of P; the corresponding subharmonic orbits $\{\tilde{x}(t, x_0 + \tilde{y}(\tau, y)) \mid t \in \mathbb{R}\}$ $(\tau \in \mathbb{R})$ of (1) are obtained one from the other by applying the S^1-symmetry and all lie on a 2-torus $\{\tilde{x}(t, x_0 + \tilde{y}(\tau, y)) \mid t \in \mathbb{R}\ \tau \in \mathbb{R}\}$ which is invariant both for the flow of (1) and for the S^1-action. We call such invariant torus filled with subharmonics a subharmonic torus.*

It follows that we have to determine all sufficiently small equilibria and periodic orbits of (15); for the periodic orbits also the condition (21) must be satisfied. We know from Hopf bifurcation theory that for y sufficiently small the possible minimal periods $\tilde{T}(y)$ which can appear will be close to $2\pi/\omega$, with $\pm i\omega$ a pair of purely imaginary eigenvalues of $Dh(0) = A_0$. Taking the "limit for $y \to 0$" in (21) and using the fact that $T(0) = 2\pi/\Omega_0$ gives us the condition $q\omega \in \mathbb{N}\Omega_0$ for the eigenvalues $\pm i\omega$ which we have to consider. It follows from our hypothesis (**H1**) that there is only one pair of eigenvalues which satisfies this condition, namely the pair $\pm i\omega_0$. Hence we can restrict our search for small periodic orbits to those which have a minimal period near $\tilde{T}_0 := 2\pi/\omega_0$; the resonance condition (21) then takes the form

$$qT(y) = k\tilde{T}(y). \tag{22}$$

For the further analysis we rely on the general reduction theory for Hopf bifurcation as worked out in [5]; for our particular problem this theory tells us the following. There is a smooth one-to-one relation between the small periodic orbits of (15) with a period near $\tilde{T}_0$ (this includes all small equilibria) and the small periodic orbits of a *reduced system* which lives on the subspace $V := \mathrm{Ker}(e^{A_0\tilde{T}_0} - I)$. This reduced system is reversible but also equivariant with respect to the S^1-action on V defined by $(\theta, v) \in S^1 \times V \mapsto e^{A_0\omega_0^{-1}\theta}v \in V$; also, its linearization at the origin is given by the restriction of A_0 to V. It follows from (**H1**) that 0 is a simple eigenvalue of A_0 and that the corresponding eigenvector belongs to $\mathrm{Fix}(R)$ (the eigenvector J_0x_0 of $\tilde{A}_0$ belongs to $\mathrm{Fix}(-R)$ and hence $\dim(\mathrm{Fix}(R) \cap \Sigma) = n$ and $\dim(\mathrm{Fix}(-R) \cap \Sigma) = n-1$). Hence $\dim(V) = 3$; one can easily show that by choosing an appropriate basis one can identify V with $\mathbb{R} \times \mathbb{C}$ in such a way that $A_0v = A_0(\alpha, z) = (\alpha, i\omega_0 z)$ and $Rv = R(\alpha, z) = (\alpha, \bar{z})$ for all $v = (\alpha, z) \in V \cong \mathbb{R} \times \mathbb{C}$. The reduced system must then have the form

$$\left\{ \begin{array}{rcl} \dot{\alpha} & = & 0, \\ \dot{z} & = & i\omega(\alpha, |z|^2)z; \end{array} \right. \tag{23}$$

here $\omega : \mathbb{R}^2 \to \mathbb{R}$ is a smooth function with $\omega(0,0) = \omega_0$. Since *all* solutions of (23) are periodic (or equilibria) the one-to-one relation given by the general reduction theory (which in principle must be restricted to periodic orbits) becomes a smooth conjugacy between the system (23) and the system (15) restricted to an appropriate invariant manifold. This allows us to formulate the following theorem.

Theorem 4. *Assume* (**H1**). *Then the set of equilibria and periodic orbits of (15) with period near $\tilde{T}_0 = 2\pi\omega_0^{-1}$ forms locally near the origin a smooth* 3*-dimensional invariant submanifold of Σ on which the flow is smoothly conjugate to the flow on $V \cong \mathbb{R} \times \mathbb{C}$ given by (23).*

We will denote the (locally defined) conjugacy by $\Phi : V \to \Phi(V) \subset \Sigma$; the theory in [5] shows that Φ commutes with R. The flow of (23) can be written out explicitly:

$$\tilde{v}(t, v) = \tilde{v}(t, \alpha, z) = (\alpha, e^{i\omega(\alpha,|z|^2)t}z), \qquad \forall v = (\alpha, z) \in V \cong \mathbb{R} \times \mathbb{C}.$$

For $z = 0$ we find a one-dimensional line $\mathbb{R} \times \{0\}$ of equilibria corresponding (locally near γ_0) to a one-parameter family $\{\gamma_\alpha \mid |\alpha| < \alpha_0\}$ of symmetric relative equilibria, given by

$$\begin{array}{l} \gamma_\alpha = \{\tilde{x}(t, x_\alpha) = e^{J_0\Omega_\alpha t}x_\alpha \mid t \in \mathbb{R}\} \\ \text{with } x_\alpha := x_0 + \Phi(\alpha, 0) \text{ and } \Omega_\alpha := \Omega(\Phi(\alpha, 0)). \end{array}$$

The periodic orbit γ_α has period $T_\alpha := 2\pi\Omega_\alpha^{-1}$; next to the trivial multiplier 1 (which has multiplicity 2) there is also a pair of simple multipliers of the form $e^{\pm 2\pi i\lambda(\alpha)}$, with $\lambda(\alpha) := \omega_\alpha\Omega_\alpha^{-1}$, $\omega_\alpha := \omega(\alpha,0)$. All other solutions $\tilde{v}(t,\alpha,z)$ (with $z \neq 0$) of (23) are periodic, corresponding to (non-equilibrium) periodic solutions $\tilde{y}(t,\Phi(\alpha,z)) = \Phi(\tilde{v}(t,\alpha,z))$ of (15) with minimal period $\tilde{T}(\Phi(\alpha,z)) = 2\pi\omega(\alpha,|z|^2)^{-1}$. The set of q-periodic points of the Poincaré map P is then given by those points $\Phi(\alpha,z) \in \Sigma$ for which (22) is satisfied. Setting $z = \rho e^{i\vartheta}$ and using $\Omega(\tilde{y}(t,y)) = \Omega(y)$ (see (16)) in combination with the fact that $\Phi(\alpha,\rho e^{i\vartheta}) \in \{\Phi(\alpha,\tilde{v}(t,\rho)) = \tilde{y}(t,\Phi(\alpha,\rho)) \mid t \in \mathbb{R}\}$ shows that $T(\Phi(\alpha,\rho e^{i\vartheta})) = T(\Phi(\alpha,\rho))$ for all $\vartheta \in \mathbb{R}$; hence we have $T(\Phi(\alpha,\rho e^{i\vartheta})) = 2\pi\Omega(\alpha,\rho^2)^{-1}$ for some smooth $\Omega : \mathbb{R}^2 \to \mathbb{R}$ with $\Omega(0,0) = \Omega_0$. Bringing this in (22) gives the equation

$$\text{(24)} \qquad \frac{\omega(\alpha,\rho^2)}{\Omega(\alpha,\rho^2)} = \frac{k}{q}.$$

For $\rho = 0$ this reduces to $\lambda(\alpha) = k/q$, which is satisfied for $\alpha = 0$ by (**H1**). Now we make a last hypothesis, namely we assume that the transversality condition

(**H2**) $\lambda'(0) \neq 0$

holds; this means that as we move along the family $\{\gamma_\alpha\}$ of symmetric relative equilibria (using α as a parameter) the multipliers $e^{\pm 2\pi i\lambda(\alpha)}$ move with non-zero speed through the q-th roots of unity $e^{\pm 2\pi ik/q}$. Under this transversality condition we can apply the implicit function theorem to solve (24) for $\alpha = \tilde{\alpha}(\rho^2)$, with $\tilde{\alpha} : \mathbb{R} \to \mathbb{R}$ a smooth function such that $\tilde{\alpha}(0) = 0$. Each of the points $\Phi(\tilde{\alpha}(\rho^2),\rho) \in \Sigma$ $(0 < \rho < \rho_0)$ satisfies the condition (ii) of Theorem 3 and hence generates a subharmonic 2-torus $\mathcal{T}_\rho := \{\tilde{x}(t, x_0 + \tilde{y}(\tau,\Phi(\tilde{\alpha}(\rho^2),\rho))) \mid t \in \mathbb{R},\ \tau \in \mathbb{R}\}$. Since $R(\tilde{\alpha}(\rho^2),\rho) = (\tilde{\alpha}(\rho^2),\rho)$ (on V) we have that $\Phi(\tilde{\alpha}(\rho^2),\rho) \in \mathrm{Fix}(R)$ and hence also $\mathcal{T}_\rho$ is invariant under R, i.e. $\mathcal{T}_\rho$ is a *symmetric* subharmonic torus. Hence we have reached the following conclusion.

Theorem 5. *Let $\gamma_0 = \{e^{J_0\Omega_0 t}x_0 \mid t \in \mathbb{R}\}$ be a symmetric relative equilibrium of the reversible and S^1-equivariant system (1). Fix some $q \geq 1$ and assume* (**H1**). *Then γ_0 belongs to a one-parameter family $\{\gamma_\alpha \mid |\alpha| < \alpha_0\}$ of symmetric relative equilibria. If also* (**H2**) *holds then we have at γ_0 the bifurcation of a single branch $\{\mathcal{T}_\rho \mid 0 < \rho < \rho_0\}$ of R-invariant q-subharmonic 2-tori. Locally near γ_0 there are no other relative equilibria or q-subharmonic tori.*

It should be noted that the first part of the conclusion (on the existence of the family $\{\gamma_\alpha\}$) already holds if only the first item of (**H1**) is satisfied (this follows by redoing the analysis above for $q = 1$ and with $\dim(V) = 2m + 1$, where m depends on the number of pairs of eigenvalues of $\tilde{A}_0$ of the form $\pm i\Omega_0 k$, with $k \in \mathbb{N}$, $k \geq 1$).

We finish this section with a remark on the transversality condition (**H2**). As such this hypothesis requires the use of the special parameter α when calculating the pair of critical multipliers $e^{\pm 2\pi i\lambda(\alpha)}$ of the branch $\{\gamma_\alpha\}$ of symmetric relative equilibria. However, it is possible to verify this hypothesis more directly, as we show now. Suppose that from direct calculation we have obtained a (smooth) local curve of points $\{\hat{x}(\sigma) \mid \sigma \in \mathbb{R}\} \subset \text{Fix}(R)$ which generate symmetric relative equilibria, i.e.

$$f(\hat{x}(\sigma)) = \hat{\Omega}(\sigma)J_0\hat{x}(\sigma), \qquad \forall \sigma \in \mathbb{R}.$$

Suppose also that $\hat{x}(0) = x_0$ (and hence $\hat{\Omega}(0) = \Omega_0$), and that for small σ the linear operator $\hat{A}(\sigma) := Df(\hat{x}(\sigma)) - \hat{\Omega}(\sigma)J_0$ has

- 0 as an eigenvalue with multiplicity 2;
- a pair of simple eigenvalues $\pm i\hat{\omega}(\sigma)$, with $\hat{\omega}(0) = \hat{\Omega}(0)kq^{-1}$, $k > 0$ and $\gcd(k, q) = 1$;
- no other purely imaginary eigenvalues $\pm i\mu$ such that $q\mu \in \Omega_0\mathbb{Z}$.

Then the symmetric relative equilibrium $\hat{\gamma}(\sigma) := \{e^{J_0\hat{\Omega}(\sigma)t}\hat{x}(\sigma) \mid t \in \mathbb{R}\}$ has next to the trivial multiplier 1 with multiplicity 2 also a pair of multipliers $e^{\pm 2\pi i\hat{\lambda}(\sigma)}$ with $\hat{\lambda}(\sigma) := \hat{\omega}(\sigma)\hat{\Omega}(\sigma)^{-1}$. Clearly our assumptions imply (**H1**), and hence also the first part of Theorem 3 holds. It follows that there exists a smooth function $\hat{\alpha}(\sigma)$ with $\hat{\alpha}(0) = 0$ and such that $\hat{x}(\sigma) = x_{\hat{\alpha}(\sigma)}$ and $\hat{\lambda}(\sigma) = \lambda(\hat{\alpha}(\sigma))$. Differentiating at $\sigma = 0$ gives

$$\hat{x}'(0) = \hat{\alpha}'(0)D_\alpha\Phi(0,0) \qquad \text{and} \qquad \hat{\lambda}'(0) = \lambda'(0)\hat{\alpha}'(0);$$

so, if $\hat{x}'(0) \neq 0$ then $\hat{\alpha}'(0) \neq 0$ and $\lambda'(0) \neq 0$ if and only if $\hat{\lambda}'(0) \neq 0$. Therefore, under the foregoing assumptions (**H2**) is satisfied if
(**H2***) $\hat{x}'(0) \neq 0$ and $\hat{\lambda}'(0) \neq 0$.

§4. Applications

Consider the system (4) already mentionned in the Introduction. Direct calculation shows that there are two branches of symmetric relative equilibria, generated by the points x_Ω (see (7)), with $\Omega \in I_- =]-\infty, 0[$ along one branch, and $\Omega \in I_+ =]0, 1[$ along the other; this is in agreement with the first part of Theorem 3. To find the bifurcation points of subharmonic tori we follow the approach described at the end of the last section, i.e. we calculate the eigenvalues of $Df(x_\Omega) - \Omega J_0$ (here $\sigma = \Omega$).

Explicitly this matrix is given by

$$\begin{pmatrix} 0 & \Omega & 0 & -1 \\ -\Omega & 0 & 1 & 0 \\ 0 & -1 & 0 & \Omega^{-1} \\ 1 & 0 & 2\Omega - 3\Omega^{-1} & 0 \end{pmatrix};$$

its eigenvalues are 0, with multiplicity 2, and $\pm i\hat{\omega}(\Omega)$, with $\hat{\omega}(\Omega) := \Omega\sqrt{1+3\Omega^{-4}}$. This last pair of eigenvalues is simple. It follows that $\hat{\lambda}(\Omega) = \Omega^{-1}\hat{\omega}(\Omega) = \sqrt{1+3\Omega^{-4}}$, and it is then straightforward to verify that

$$\frac{dx_\Omega}{d\Omega} \neq 0 \quad \text{and} \quad \hat{\lambda}'(\Omega) \neq 0, \qquad \forall \Omega \in I_- \cup I_+.$$

It follows that for each $\Omega \in I_- \cup I_+$ for which $\hat{\lambda}(\Omega)$ is rational we will have the bifurcation of a subharmonic 2-torus at the relative equilibrium $\gamma_\Omega = \{e^{i\Omega t}x_\Omega \mid t \in \mathbb{R}\}$. More precisely, using the fact that

$$\{\hat{\lambda}(\Omega) \mid \Omega \in I_-\} =]1,2[\quad \text{and} \quad \{\hat{\lambda}(\Omega) \mid \Omega \in I_+\} =]2,\infty[\,,$$

we can say the following:

- for each pair of integers (k,q) with $0 < q < k < 2q$ and $\gcd(k,q) = 1$ there is a unique $\Omega = \Omega(k,q) \in I_-$ for which $\hat{\lambda}(\Omega) = k/q$; at $\gamma_{\Omega(k,q)}$ a branch of q-subharmonic tori bifurcates from the branch $\Gamma_- := \{\gamma_\Omega \mid \Omega \in I_-\}$ of relative equilibria;
- for each pair of integers (k,q) with $0 < q < 2q < k$ and $\gcd(k,q) = 1$ there is a unique $\Omega = \Omega(k,q) \in I_+$ for which $\hat{\lambda}(\Omega) = k/q$; at $\gamma_{\Omega(k,q)}$ a branch of q-subharmonic tori bifurcates from the branch $\Gamma_+ := \{\gamma_\Omega \mid \Omega \in I_+\}$ of relative equilibria.

The paper [6] contains the result of a numerical study of these bifurcating branches of subharmonic tori. In that paper it is shown that each of the relative equilibria and each of the subharmonic tori correspond to a unique point in the *energy-momentum diagram* for (4); this diagram depicts the image of the *energy-momentum map* defined by $(a,b) \in \mathbb{C}^2 \mapsto (H(a,b), F(a,b))$, where $H(a,b) := a\bar{b} + \bar{a}b - \frac{1}{2}(b\bar{b})^2$ is the Hamiltonian for the system (4), and where $F(a,b) := a\bar{a} + b\bar{b}$ is the Hamiltoniam which generates the S^1-action on $\mathbb{C}^2$. In Figure 1 (taken from [6]) one can see the two branches of relative equilibria and some of the bifurcating branches of subharmonic tori. It appears that the branches bifurcating from Γ_- connect to branches bifurcating from Γ_+, forming "bridges" between the two families of relative equilibria; we refer to [6] for more details on the connection rules.

The example (4) is somewhat special, in the sense that it is not only an S^1-equivariant reversible system, but at the same time it is also an integrable Hamiltonian system with a well understood geometry — again see [6]. The same can be said about the spherical pendulum (see [4]) and the Furuta pendulum (see [7]), which are both more physical examples showing the bifurcation behaviour of subharmonic tori as described in this paper.

From the other side, our specific example (4) is just a special case of a larger class of systems which satisfy our basic hypotheses, but which are in general not Hamiltonian. These systems can briefly be described as follows. Identify $\mathbb{R}^{2n}$ with $\mathbb{C}^n$, writing $x = (z_1, z_2, \ldots, z_n)$. Let $A : \mathbb{C}^n \to \mathcal{L}(\mathbb{R}^n)$ be a smooth mapping (smooth when using $\mathbb{C}^n \cong \mathbb{R}^{2n}$) such that $A(e^{i\theta}x) = A(x)$ for all $\theta \in \mathbb{R}$ and all $x \in \mathbb{C}^n$. Then the system

$$\dot{x} = iA(x)x, \qquad (x \in \mathbb{C}^n) \tag{25}$$

is both reversible and S^1-equivariant, with $Rx = R(z_1, z_2, \ldots, z_n) := (\bar{z}_1, \bar{z}_2, \ldots, \bar{z}_n)$ and $J_0 x = ix$. Symmetric relative equilibria are determined by the equation

$$A(x)x = \Omega x, \qquad x = (x_1, x_2, \ldots, x_n) \in \mathbb{R}^n = \mathrm{Fix}(R); \tag{26}$$

this means that Ω must be a (real) eigenvalue of $L(x)$ such that $x \in \mathbb{R}^n$ is a corresponding eigenvector. From the preceding theory one expects such symmetric relative equilibria to appear in one-parameter families from which there bifurcate branches of subharmonic tori. Of course the

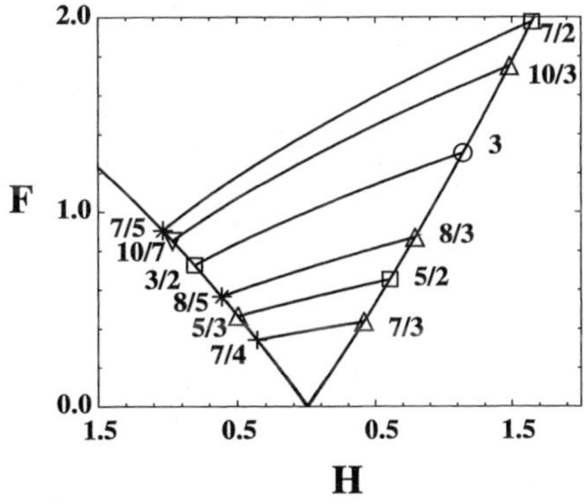

Fig. 1. The energy-momentum diagram for (4), with the two branches of relative equilibria, some subharmonic bifurcations and some connecting bridges. The numbers indicate the value of $\hat{\lambda}(\Omega)$. The figure is taken from [6].

explicit calculation of such branches and the verification of the hypotheses for subharmonic bifurcation will in general not be as easy as for our example (4).

§5. Final comment

The approach described in this note can be applied in much more general situations. For example, in order to avoid technical complications in the formulations, we have restricted to an S^1-action without non-trivial isotropies; it is straigthforward to adapt our results to general S^1-actions. For more general (continuous) symmetries relative equilibria are not necessarily periodic, but some of our arguments can still be used to study for example the bifurcation of relative periodic orbits from branches of relative equilibria.

Acknowledgments. The author wants to acknowledge many years of collaboration with Emilio Freire, Jorge Galán and Francisco Javier Muñoz-Almaraz from which the work presented in this paper originated. He also acknowledges the support of The Fund for Scientific Research—Flanders (Belgium) and from the Spanish Ministry of Education through the grant BFM2003-00336.

References

[1] M.-C. Ciocci and A. Vanderbauwhede, On the bifurcation and stability of periodic points in reversible and symplectic diffeomorphisms, In: Symmetry and Perturbation Theory, Proceedings SPT98, (eds. A. Degasparis and G. Gaeta), World Sci. Publ., 1999, pp. 159–166.

[2] M.-C. Ciocci and A. Vanderbauwhede, Bifurcation of periodic points in reversible diffeomorphisms, I: New Progress in Difference Equations, Proceedings ICDEA2001, (eds. S. Elaydi, G. Ladas and B. Aulbach), Chapman & Hall, CRC Press, 2004, pp. 75–93.

[3] M.-C. Ciocci, Subharmonic branching at a reversible 1 : 1 resonance, J. Difference Equ. Appl., **11** (2005), 1119–1135.

[4] R. Cushman and L. Bates, Global aspects of Classical Integrable Systems, Birkhäuser Verlag, Basel, 1997.

[5] J. Knobloch and A. Vanderbauwhede, A general reduction method for periodic solutions in conservative and reversible systems, J. Dynam. Differential Equations, **8** (1996), 71–102.

[6] F. J. Muñoz-Almaraz, E. Freire, J. Galán and A. Vanderbauwhede, Branches of invariant tori and rotation numbers in symmetric Hamiltonian systems: an example, In: Symmetry and Perturbation Theory,

Proceedings SPT2002, (eds. S. Abenda, G. Gaeta and S. Walcher), World Sci. Publl., 2003. pp. 267–276.

[7] F. J. Muñoz-Almaraz, E. Freire and J. Galán, Bifurcation behavior of the Furuta pendulum, Internat. J. Bifur. Chaos Appl. Sci. Engrg., **17** (2007), 2571–2578.

[8] A. Vanderbauwhede, Branching of periodic solutions in time-reversible systems, In: Geometry and Analysis in Non-Linear Dynamics, (eds. H. Broer and F. Takens), Pitman Research Notes in Math., **222** (1992), 97–113.

[9] A. Vanderbauwhede, Subharmonic bifurcation at multiple resonances, In: Proceedings of the Mathematics Conference, Birzeit, 1998, (eds. S. Elaydi et. al.), World Sci. Publl., 2000, pp. 254–276.

Department of Pure Mathematics and Computer Algebra
University of Ghent
Belgium

E-mail address: `avdb@cage.ugent.be`

Advanced Studies in Pure Mathematics 53, 2009
Advances in Discrete Dynamical Systems
pp. 387–398

A discrete-time rodent-hantavirus model structured by infection and developmental stages

Curtis L. Wesley, Linda J. S. Allen, Colleen B. Jonsson, Yong-Kyu Chu and Robert D. Owen

Abstract.

Hantaviruses are a group of viruses that infect wild rodents without causing any apparent illness or disease. New discrete-time models for the spread of hantavirus in a rodent population are formulated and analyzed. The models are structured by the stages of the infection, the stages of development, and the sex of the rodent. The basic reproduction number $\mathcal{R}_0$ is computed for the deterministic model and a condition is given for a simplified model with males only to be permanent. A stochastic model is also formulated. Numerical simulations illustrate the differences between the deterministic and stochastic models and the dynamics in the male and female rodents. It is shown, in the numerical examples, that a transcritical bifurcation occurs at $\mathcal{R}_0 = 1$ and a unique enzootic equilibrium exists when $\mathcal{R}_0 > 1$. The sensitivity of the equilibrium values to changes in the parameters is also investigated.

§1. Introduction

Hantaviruses are viruses that persistently infect wild rodents without causing any apparent illness or disease. However, they do produce disease in humans, either hantavirus pulmonary syndrome (HPS in the Americas) or hemorrhagic fever with renal syndrome (HFRS in Europe and Asia). Approximately 30 different hantaviruses are recognized throughout the world [22]. The widespread occurrence of HPS and its potential for severe human illness has identified it as an emerging disease with a significant public health threat.

Each hantavirus is generally associated with a single rodent host known as the reservoir host. Our goal is to formulate a realistic model for hantavirus in the reservoir host that will help us better understand how to prevent the spread of this disease to the human population. The new

Received October 12, 2007.

model is a system of difference equations, structured by the stages of the infection (susceptible and infected), the stages of development (juvenile, subadult, and adult) and the sex of the rodent (male and female). Other models for the rodent-hantavirus interaction have applied differential equations with stages for infection [1, 2, 3, 24] and males and females [4]. Our model is the first to consider the developmental stages of the rodent in a discrete-time formulation. Hantavirus spread is primarily through infections acquired during post-weaning intraspecific encounters [8]. Studies suggest that there are distinct differences in males and females in their duration of shedding and viremia [10, 14, 15, 21, 25], as well as differences in prevalence of hantavirus antibodies. In addition, males exhibit more aggressive behavior than females [5, 9, 10, 21, 25]. Inclusion of developmental stages and gender differences in our model will account for the differences in hantavirus seroprevalence seen in the field data.

In the next section, a susceptible-infected (SI) epidemic model with developmental stages and gender is formulated. The basic reproduction number $\mathcal{R}_0$ is computed for this model. It is shown for a simplified model with males only that the system is permanent. In section 3, a stochastic model is formulated. Then, in section 4, numerical simulations of the deterministic and stochastic models illustrate the differences in the models and of the persistent infection dynamics between males and females. It is shown that a transcritical bifurcation occurs at $\mathcal{R}_0 = 1$ and a unique enzootic equilibrium when $\mathcal{R}_0 > 1$. The sensitivity of the enzootic equilibrium values to changes in parameter values is investigated numerically.

§2. SI deterministic model

2.1. Description of model

Rodents are classified as susceptible or infected (and infectious) and further classified by gender, and stage of development. For male rodents, the variables $Y =$ juveniles, $C =$ subadults, and $M =$ adults. For female rodents, $J =$ juveniles, $D =$ subadults, $F =$ adults. We assume the juvenile and subadult stages are nonreproductive. The juvenile stage is when newborn rodents are in the nest. The subadult stage is after juveniles leave the nest but prior to reproduction. We assume that juvenile rodents have a small chance of contracting the virus in the nest and that subadults mature quickly to adulthood, so that the virus is only transmitted in the adult stage [5]. Data indicates that rodents infected with hantavirus may remain infected for life [6, 18], therefore there is no recovery stage in the model. A subscript S (susceptible) or I (infected and infectious) on the variables M, F denote the infection status of the adult classes. For example, M_S = susceptible adult males. The total male population is denoted as $N_M = Y + C + M_S + M_I$ and the total female population as $N_F = J + D + F_S + F_I$. The total population size is equal to $N = N_M + N_F$. Each developmental and infectious stage is followed over discrete time intervals $t = 1, 2, \ldots$, where $[t, t+1]$ equals one month. We assume rodents remain juveniles for one month and subadults for one month.

To model births, let

$$B(M, F) = \frac{2bMF}{M + F}$$

be the total number of births for $M = M_S + M_I$ reproductive males and $F = F_S + F_I$ reproductive females. The function B is known as a harmonic mean birth function, where b is the average litter size [7]. Juvenile rodents become reproductive adults in approximately three months. Therefore, the birth function is divided by three.

To model the infection process, we assume that the number of encounters between susceptible and infected rodents are distributed randomly among the rodent population. Then the number of encounters follows a Poisson distribution,

$$p(i) = \frac{\exp(-\lambda)\lambda^i}{i!}, \quad i = 0, 1, 2, \ldots,$$

where i is the number of infective encounters (that result in spread of infection from an infectious individual to a susceptible individual) and λ is the average number of infective encounters per susceptible individual in the population during the interval $[t, t+1]$. Since it only takes one successful encounter to become infected, the probability a susceptible rodent becomes infected is $1 - p(0)$. Female and male behavioral and immune system differences suggest that the interaction of the virus with its host may be sexually dimorphic [13, 16]. Aggressive encounters between males contribute to higher seroprevalence in males; therefore, it is reasonable to assume contacts among males are greater than among females [5, 9, 10, 21, 25]. The seroprevalence data in Paraguay also show a distinct difference between males and females (seroprevalence in *Akodon montensis* males $\approx$ 20% versus females $\approx$ 6% for years 2005 and 2006, unpublished data). Generally, male seroprevalence is approximately 3 to 5 times greater than female. If the average number of infective encounters by all susceptible individuals with infected males or females satisfies the law of mass action, then $\lambda S = (\beta_M M_I + \beta_F F_I)S$, where contact with infected males or females is differentiated by β_M or β_F, respectively. Hence,

$$1 - p(0) = 1 - \exp(-\beta_M M_I - \beta_F F_I),$$

where $\beta_M \gg \beta_F$. Encounters between infected and susceptible rodents may also depend on the particular developmental stage. Therefore, the parameters β_M and β_F may depend on C, M, D or F. We differentiate these infective encounters by assuming different values for the parameters, $\beta_{\nu M}$ or $\beta_{\nu F}$ for contacts between susceptible rodents in stages $\nu = C, M, D, F$, respectively.

Density-dependent mortality takes the form of the well-known Beverton-Holt growth equation. For example, in the simple case where the population is not structured by stage or sex, then

$$N(t+1) = \frac{(b/6+1)N(t)K}{K + (b/6)N(t)} = (b/6+1)N(t)L(N(t)),$$

where K is the carrying capacity. The constant $b/6$ appears because $(b/2)/3$ is the average litter size per rodent per month. In addition to the density-dependent mortality, every stage has probability μ_i, $0 < \mu_i \leq 1$, $i = Y, C, M, J, D, F$, of surviving to the next time interval. There are no deaths from hantavirus in the rodent population; hantavirus appears to have no effect on rodent survival [8]. Births, infections, survival, and transitions in time $[t, t+1]$ are followed by density-dependent deaths.

Based on the preceding assumptions, the SI difference equation model for the rodent population has the following form:

$$\begin{aligned}
Y(t+1) &= \frac{1}{3}\left[\frac{B(M(t),F(t))}{2}\right] L(N(t)) \\
C(t+1) &= [\mu_Y Y(t)]L(N(t)) \\
M_S(t+1) &= [\mu_C \exp(-\beta_{CF}F_I(t) - \beta_{CM}M_I(t))C(t) \\
\text{(1)} \quad & \quad + \mu_M \exp(-\beta_{MF}F_I(t) - \beta_{MM}M_I(t))M_S(t)]L(N(t)) \\
M_I(t+1) &= [\mu_C(1 - \exp(-\beta_{CF}F_I(t) - \beta_{CM}M_I(t)))C(t) \\
& \quad + \mu_M\{(1 - \exp(-\beta_{MF}F_I(t) - \beta_{MM}M_I(t)))M_S(t) + M_I(t)\}]L(N(t)) \\
J(t+1) &= \frac{1}{3}\left[\frac{B(M(t),F(t))}{2}\right] L(N(t)) \\
D(t+1) &= [\mu_J J(t)]L(N(t)) \\
F_S(t+1) &= [\mu_D \exp(-\beta_{DF}F_I(t) - \beta_{DM}M_I(t))D(t) \\
\text{(2)} \quad & \quad + \mu_F \exp(-\beta_{FF}F_I(t) - \beta_{FM}M_I(t))F_S(t)]L(N(t)) \\
F_I(t+1) &= [\mu_D(1 - \exp(-\beta_{DF}F_I(t) - \beta_{DM}M_I(t)))D(t) \\
& \quad + \mu_F\{(1 - \exp(-\beta_{FF}F_I(t) - \beta_{FM}M_I(t)))F_S(t) + F_I(t)\}]L(N(t)).
\end{aligned}$$

The initial conditions are positive and parameters are positive (except $\beta_{ij} \geq 0$). Therefore, it easily follows that solutions to (1)-(2) are positive for $t \geq 0$.

2.2. Basic reproduction number

The next generation approach [11, 19, 20] is used to compute the basic reproduction number, $\mathcal{R}_0$, the number of secondary infections caused by one infectious individual in an entirely susceptible population [12]. If $\mathcal{R}_0 < 1$, then the disease-free equilibrium (DFE) is locally asymptotically stable and if $\mathcal{R}_0 > 1$, the DFE is unstable.

To compute $\mathcal{R}_0$, first the DFE needs to be calculated. At the DFE, $\bar{M}_I = 0 = \bar{F}_I$. The remaining equilibrium values are

$$\bar{Y} = \frac{\bar{B}\bar{L}}{6}, \ \bar{C} = \frac{\mu_Y \bar{B}\bar{L}^2}{6}, \ \bar{M}_S = \frac{\mu_C \mu_Y \bar{B}\bar{L}^3}{6(1-\mu_M \bar{L})}, \ \bar{J} = \frac{\bar{B}\bar{L}}{6},$$

$$\bar{D} = \frac{\mu_J \bar{B}\bar{L}^2}{6}, \ \bar{F}_S = \frac{\mu_D \mu_J \bar{B}\bar{L}^3}{6(1-\mu_F \bar{L})}.$$

The value $\bar{B}$ is the birth function evaluated at the DFE and $\bar{L} = K/[K + (b/6)E]$, where E is the total population size at the DFE. For the case, $\mu_i \equiv \mu$ for all stages $i = Y, C, M, J, D, F$, we show that there exists a unique DFE under certain restrictions on μ. In this case, the male and female equilibrium values are equal. Thus, $\bar{B} = b\bar{M}$ and $E = 2(\bar{M} + \bar{C} + \bar{Y})$. A unique DFE exists if E is positive and unique. Applying the equation for $\bar{M}$, it can be shown that E is the unique positive real root of the following cubic equation:

$$b^3E^3 + 6Kb^2(3-\mu)E^2 + 36K^2b(3-2\mu)E - 36K^3(b\mu^2 + 6\mu - 6) = 0.$$

It follows from Descartes' rule of signs that there is a unique positive real root of this equation if the constant term in the preceding cubic equation satisfies

$$\mu > \frac{\sqrt{9+6b}-3}{b}. \tag{3}$$

This unique equilibrium value E can be computed explicitly. In general, $0 < E \leq K$.

To form the next generation matrix, the eight state variables are divided into infectious states, $\vec{\mathbf{X}} = (M_I, F_I)^T$, and noninfectious states, $(Y, C, M_S, J, D, F_S)^T$. Focusing on the infectious states, new infections $\vec{\mathbf{F}}$ are separated from other transitions $\vec{\mathbf{T}}$, $\vec{\mathbf{X}}(t+1) = \vec{\mathbf{F}}(t) + \vec{\mathbf{T}}(t)$, where

$\vec{\mathbf{F}}(t)$ is

(4)
$$\begin{pmatrix} \mu_C(1-\exp(-\beta_{CF}F_I(t)-\beta_{CM}M_I(t)))C(t) \\ \quad +\mu_M(1-\exp(-\beta_{MF}F_I(t)-\beta_{MM}M_I(t)))M_S(t) \\ \\ \mu_D(1-\exp(-\beta_{DF}F_I(t)-\beta_{DM}M_I(t)))D(t) \\ \quad +\mu_F(1-\exp(-\beta_{FF}F_I(t)-\beta_{FM}M_I(t)))F_S(t) \end{pmatrix} L(N(t))$$

and $\vec{\mathbf{T}}(t) = (\mu_M M_I(t), \mu_F F_I(t))^T$. Computing the Jacobian matrices of $\vec{\mathbf{F}}$ and $\vec{\mathbf{T}}$ with respect to the infectious states and evaluating at the DFE we obtain matrices $\mathcal{F}$ and $\mathcal{T}$,

$$(5)\qquad \begin{pmatrix} \mu_C\beta_{CM}\bar{C}+\mu_M\beta_{MM}\bar{M}_S & \mu_C\beta_{CF}\bar{C}+\mu_M\beta_{MF}\bar{M}_S \\ \mu_D\beta_{DM}\bar{D}+\mu_F\beta_{FM}\bar{F}_S & \mu_D\beta_{DF}\bar{D}+\mu_F\beta_{FF}\bar{F}_S \end{pmatrix}\bar{L}$$
$$\text{and} \quad \begin{pmatrix} \mu_M & 0 \\ 0 & \mu_F \end{pmatrix}\bar{L},$$

respectively. Matices $\mathcal{F}$ and $\mathcal{T}$ are nonnegative and $\rho(\mathcal{T}) < 1$. Thus $\mathcal{I}-\mathcal{T}$ is invertible. Matrix $\mathcal{F}(\mathcal{I}-\mathcal{T})^{-1}$ is the next generation matrix and the basic reproduction number, $\mathcal{R}_0 = \rho(\mathcal{F}(\mathcal{I}-\mathcal{T})^{-1})$, is the spectral radius of the next generation matrix [20], i.e.

$$\mathcal{R}_0 = \rho\begin{pmatrix} \dfrac{(\mu_C\beta_{CM}\bar{C}+\mu_M\beta_{MM}\bar{M}_S)\bar{L}}{1-\mu_M\bar{L}} & \dfrac{(\mu_C\beta_{CF}\bar{C}+\mu_M\beta_{MF}\bar{M}_S)\bar{L}}{1-\mu_F\bar{L}} \\ \dfrac{(\mu_D\beta_{DM}\bar{D}+\mu_F\beta_{FM}\bar{F}_S)\bar{L}}{1-\mu_M\bar{L}} & \dfrac{(\mu_D\beta_{DF}\bar{D}+\mu_F\beta_{FF}\bar{F}_S)\bar{L}}{1-\mu_F\bar{L}} \end{pmatrix}.$$

We have given the general form for $\mathcal{R}_0$, assuming a unique DFE exists. For the case $\mu_i \equiv \mu$, $i = Y, C, M, J, D, F$, we have shown that a unique DFE exists. In this case, $\bar{C} = \bar{D}$ and $\bar{M}_S = \bar{F}_S$. Due to male aggressive behavior, a reasonable assumption is that transmission is greater between two males than between males and females or two females. Hence, if we let $\mu_i \equiv \mu$, β_M be the male transmission coefficients and β be all other transmission coefficients, then the basic reproduction number has a relatively simple form,

$$(6)\qquad \mathcal{R}_0 = \left[\frac{\beta_M+\beta}{2} + \frac{1}{2}\sqrt{4\beta^2+(\beta_M-\beta)^2}\right]\frac{\mu\bar{L}(\bar{C}+\bar{M}_S)}{1-\mu\bar{L}}.$$

The expression (6) reduces to a more recognizable form if $\beta_M = \beta$,

$$\mathcal{R}_0 = \frac{\beta\mu\bar{L}}{1-\mu\bar{L}}2(\bar{C}+\bar{M}_S).$$

That is, $\mathcal{R}_0$ is the number of infective contacts by an infectious individual $\beta\bar{N} = 2\beta(\bar{C}+\bar{M}_S)$ during the individuals infectious period $\mu\bar{L}/(1-\mu\bar{L})$ (i.e., number of secondary infections caused by one infectious individual in an entirely susceptible population).

2.3. Permanence

Next, we focus on long-term survival of the population known as permanence or uniform persistence. Model (1) and (2) is permanent if there exist $\delta_1 > 0$ and $\delta_2 > 0$ such that $\delta_1 < \liminf_{t\to\infty} N(t) \leq \limsup_{t\to\infty} N(t) < \delta_2$ for all positive initial conditions. We simplify model (1) and (2), and consider only the males, assuming there are sufficient number of females, where the birth function is replaced by $B = bM$. The simplified model (1) can be expressed in the form $\vec{x}(t+1) = A_x\vec{x}(t)$, where $\vec{x} = (Y, C, M)^T$ and $M = M_S + M_I$. Matrix A_x depends continuously on $\vec{x}$ and the simplified model is dissipative. Let A_0 denote A_x evaluated at the extinction equilibrium, $\vec{x} = \vec{0}$,

$$A_0 = \begin{pmatrix} 0 & 0 & b/6 \\ \mu_Y & 0 & 0 \\ 0 & \mu_C & \mu_M \end{pmatrix}.$$

The form of A_0 and the fact that all parameters are positive implies that A_0 is irreducible. Hence, the simplified model is permanent if A_0 has a dominant eigenvalue greater than one (Theorem 3, pg. 519, [17]). The dominant eigenvalue of A_0 is greater than one if and only if the inherent net reproduction number

$$\mathcal{R} = \frac{b}{6}\mu_C\mu_Y + \mu_M > 1. \tag{7}$$

Inequality (7) is equivalent to (3) if the probability of survival between stages $\mu_i \equiv \mu$ for $i = Y, C, M$. In this case, the DFE exists and is feasible if and only if the simplified model is permanent.

§3. SI stochastic model

A discrete-time stochastic model is formulated based on (1) and (2). Random variables Y, $C, \ldots, F_I$ are assumed to be discrete with values in $\{0, 1, \ldots\}$. Similar assumptions apply to the stochastic model as in the deterministic model, except the births, deaths, and infections are expressed as probabilities. For example, a rodent in stage Y survives to stage C during the interval $[t, t+1]$ with probability μ_Y and does not survive with probability $1 - \mu_Y$. An adult male susceptible rodent becomes infected with probability $P_M = 1 - \exp(-\beta_{MF}F_I(t) - \beta_{MM}M_I(t))$ and does not become infected with probability $1 - P_M$.

For the birth process, we assume reproduction depends on the number of mating units at time t, $\lceil B(M(t), F(t))/3 \rceil$, where $\lceil u \rceil$ denotes the smallest integer greater than or equal to u. We assume each mating unit gives birth to r offspring with probability b_r, $r = 0, 1, \ldots$, where $\sum_{r=0}^{\infty} b_r = 1$. In addition, the mean of the offspring distribution, $\{b_r\}$, is $b = \sum_{r=0}^{\infty} r b_r$, where b is the average litter size. An offspring may be a male or female with probability 1/2.

§4. Numerical examples

Several numerical examples illustrate the dynamics of the SI deterministic and stochastic models. Baseline parameter values are given in Table 1 for the first simulation (Fig. 1). Because seroprevalence of hantavirus is approximately 3 to 5 times greater in males than females, the contact rates are adjusted. Surprisingly, the male contact rates need to be much greater than 3 to 5 times the females contact rates to obtain a seroprevalence of 3 to 5 times higher in males. Contact rates vary depending on hypothesized interactions among the rodent population. Some contact rates are set to zero indicating that some of the developmental stages have little or no interaction. The survival probability $\mu = 0.95$ is the same for all stages.

Param.	Value	Param.	Value
K	200 rodents	μ	$\frac{0.95}{\text{month}}$
β_{CM}	$\frac{0.02}{(\text{month})(\text{rodent})}$	β_{DM}	0
β_{CF}	0	β_{DF}	0
Param.	Value	Param.	Value
b	3 rodents	b_r	$\begin{cases} 1/7 & r = 0, 1, .., 6 \\ 0 & \text{otherwise} \end{cases}$
β_{MM}	$\frac{0.01}{(\text{month})(\text{rodent})}$	β_{FM}	$\frac{0.002}{(\text{month})(\text{rodent})}$
β_{MF}	$\frac{0.004}{(\text{month})(\text{rodent})}$	β_{FF}	$\frac{0.004}{(\text{month})(\text{rodent})}$

Table 1. Baseline parameter values.

In Fig. 1(a), the number of infected males at $t = 5$ years is approximately 4 times greater than females. The deterministic solution approaches an enzootic equilibrium. The total proportion of the population infected is $(\bar{M}_I + \bar{F}_I)/\bar{N} \approx 17\%$. In Fig. 1(b), one sample path of the stochastic model is graphed with the mean of 1000 stochastic sample paths. Notice that the mean is close to the deterministic solution. One sample path illustrates the variability in the number infected.

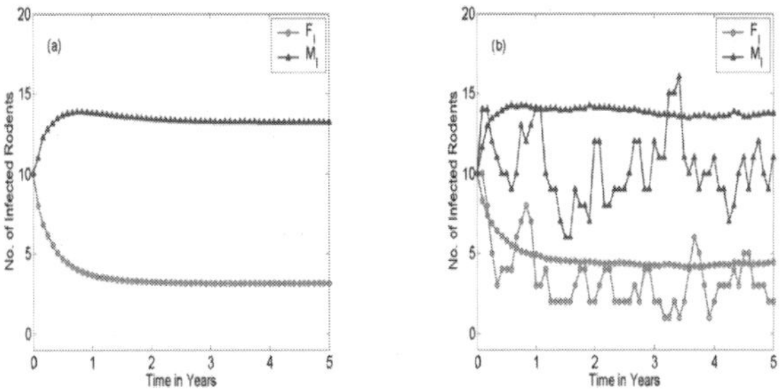

Fig. 1. (a) Deterministic solution, $\mathcal{R}_0 = 1.55$. (b) One stochastic sample path and the mean of 1000 sample paths.

In Fig. 2(a), the proportion of infected male and female rodents is graphed as a function of $\mathcal{R}_0$, where $\mathcal{R}_0$ is a function of the contact rates. At $\mathcal{R}_0 = 1$, there is a transcritical bifurcation. It is clear from the graph that the proportion of infected males is about four times that of females when $\mathcal{R}_0 \approx 2$. As $\mathcal{R}_0$ increases as a function of the contact rates, the gap between infected male and female rodents decreases.

In Figs. 2(b) and 2(c), the enzootic equilibrium $\bar{N}$ is graphed as a function of the male and female contact rates, carrying capacity, and litter size. In Fig. 2(b), the infected proportion quickly plateaus for relatively small values of the contact rates. Because juveniles and subadults are not infected, the entire population does not become infected. In Fig. 2(c), if carrying capacity K is fixed and litter size b increases, we see that the epidemic dies out. We can attribute this phenomenon to the fact that the population size is fixed but a large proportion of the offspring are not infected. A large number of births "flushes" out the population, increasing the number of susceptible juvenile and subadult rodents while decreasing the number of infected adult rodents. The size of the infected population is greatest when the carrying capacity is large, but the litter size is small.

§5. Discussion

In this investigation, we formulated new discrete-time, stage-structured SI epidemic models (deterministic and stochastic) for the spread of hantavirus in a rodent population. The stochastic model illustrates the variability in number infected over time (Fig. 1(b)) when births, deaths, and transitions are modeled via a discrete-time Markov process. The inclusion of juvenile and subadult stages in the model reduces the

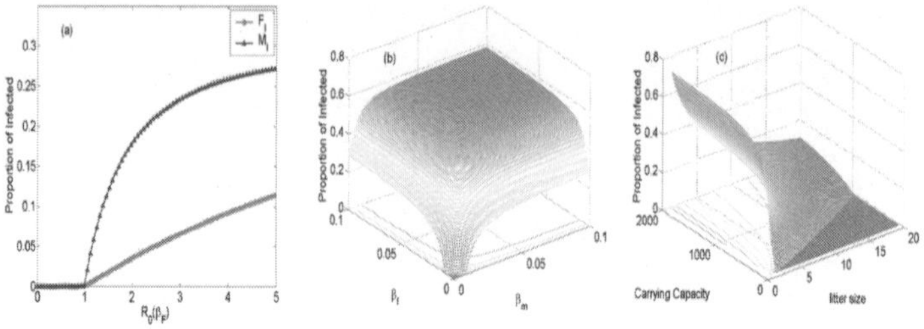

Fig. 2. (a) Graph of the proportion of infected rodents as a function of $\mathcal{R}_0$; $\beta_M = 10\beta_F$, where $\beta_M = \beta_{MF} = \beta_{MM} = \beta_{CM} = \beta_{CF}$ and $\beta_F = \beta_{FF} = \beta_{FM} = \beta_{DM} = \beta_{DF}$. Parameters are the same as in Table 1, except for the contact rates. (b) Proportion of infected rodents, $(M_I + F_I)/N$, at the enzootic equilibrium as a function of β_M and β_F. (c) Proportion of infected rodents at the enzootic equilibrium as a function of carrying capacity K and litter size b.

DFE of the population to a level below K, $E \leq K$. In addition, because the juvenile and subadult stages can not be infected, it is impossible for all rodents in the population to be infected. To control the spread of the infection, the most reasonable strategy is to reduce the population size. A population reduction invariably decreases the contact rates. The model simulations show that the proportion infected increases with K but decreases with litter size. However, this result depends on the independence of litter size b and carrying capacity K, which may not be true in general. Large population explosions (large number of births) often occur during periods of highly favorable environmental conditions [23]. Favorable environmental conditions are equivalent to a high K value in our models.

Acknowledgements. This research was supported by a grant from the Fogarty International Center #R01TW006986-02 under the NIH NSF Ecology of Infectious Diseases initiative.

References

[1] G. Abramson and V. M. Kenkre, Spatiotemporal patterns in hantavirus infection, Phys. Rev. E., **66**, 011912 (2002), 1–5.

[2] G. Abramson, V. M. Kenkre, T. L. Yates and R. R. Parmenter, Traveling waves of infection in the hantavirus epidemics, Bull. Math. Biol., **65** (2003), 519–534.

[3] L. J. S. Allen, M. Langlais and C. J. Phillips, The dynamics of two viral infections in a single host population with applications to hantavirus, Math. Biosci., **186** (2003), 191–217.

[4] L. J. S. Allen, R. K. McCormack and C. B. Jonsson, Mathematical models for hantavirus infection in rodents., Bull. Math. Biol., **68** (2006), 511–524.

[5] A. D. Bernshtein, N. S. Apekina, T. V. Mikhailova, Y. A. Myasnikov, L. A. Khlyap, Y. S. Korotkov and I. N. Gavrilovskaya, Dynamics of Puumala hantavirus infection in naturally infected bank voles (Clethrinomys glareolus), Arch. Virol., **144** (1999), 2415–2428.

[6] J. Botten, K. Mirowsky, D. Kusewitt, C. Ye, K. Gottlieb, J. Prescott and B. Hjelle, Persistent Sin Nombre virus infection in the deer mouse (Peromyscus maniculatus) model: sites of replication and strand-specific expression, J. Virol., **77** (2003), 1540–1550.

[7] H. Caswell, Matrix Population Models: Construction, Analysis, and Interpretation. 2nd ed., Sunderland, Mass., Sinauer Assoc. Inc, 2001.

[8] CDC, Hantavirus pulmonary syndrome–United States: updated recommendations for risk reduction, MMWR, **51** RR09 (2002), 1–12.

[9] J. E. Childs, T. G. Ksiazek, C. F. Spiropoulou, J. W. Krebs, S. Morzunov, G. O. Maupin, K. L. Gage, P. E. Rollin, J. Sarisky, R. E. Enscore, J. K. Frey, C. J. Peters and S. T. Nichol, Serologic and genetic indentification of *Peromyscus maniculatus* as the primary reservoir for a new hantavirus in the Southwestern United States, J. Infect. Dis., **169** (1994), 1271–1280.

[10] Y. K. Chu, R. D. Owen, L. Gonzalez and C. B. Jonsson, The complex ecology of hantavirus in Paraguay, Am. J. Trop. Med. Hyg., **69** (2003), 263–268.

[11] J. M. Cushing, An Introduction to Structured Population Dynamics, Soc. for Industrial and Applied Math., 1998.

[12] H. W. Hethcote, The mathematics of infectious disease, SIAM Rev., **42** (2000), 599–653.

[13] E. R. Hinson, S. M. Shone, M. C. Zink, G. E. Glass and S. L. Klein, Wounding: the primary mode of Seoul virus transmission among male Norway rats, Am. J. Trop. Med. Hyg., **70** (2004), 310–317.

[14] S. L. Klein, B. H. Bird and G. E. Glass, Sex differences in Seoul virus infection are not related to adult sex steroid concentrations in Norway rats, J. Virol., **74** (2000), 8213–8217.

[15] S. L. Klein, B. H. Bird and G. E. Glass, Sex differences in immune responses and viral shedding following Seoul virus infection in Norway rats, Am. J. Trop. Med. Hyg., **65** (2001), 57–63.

[16] S. L. Klein, A. Cernetich, S. Hilmer, E. P. Hoffman, A. L. Scott and G. E. Glass, Differential expression of immunoregulatory genes in male and female Norway rats following infection with Seoul virus, J. Med. Virol., **74** (2004), 180–190.

[17] R. Kon, Y. Saito and Y. Takeuchi, Permanence of single-species stage-structured models, J. Math. Biol., **48** (2004), 515–528.

[18] H. W. Lee and G. van der Groen, Hemorrhagic fever with renal syndrome, Prog. Med. Virol., **36** (1989), 62–102.

[19] M. A. Lewis, J. Renclawowicz, P. van den Driessche and M. Wonham, A comparison of continuous and discrete-time West Nile Virus models, Bull. Math. Biol., **68** (2006), 491–509.

[20] C. K. Li and H. Schneider, Applications of Perron-Frobenius theory to population dynamic, J. Math. Biol., **44** (2002), 450–562.

[21] N. E. McIntyre, Y. K. Chu, R. D. Owen, A. Abuzeineh, N. De La Sancha, C. W. Dick, T. Holsomback, R. A. Nisbet and C. Jonsson, A longitudinal study of Bayou virus, hosts, and habitat, Am. J. Trop. Med. Hyg., **73** (2005), 1043–1049.

[22] J. N. Mills, T. G. Ksiazek, B. A. Ellis, P. E. Rollin, S. T. Nichol, T.L. Yates, W. L. Gannon, C. E. Levy, D. M. Engelthaler, T. Davis, D. T. Tanda, J. W. Frampton, C. R. Nichols, C. J. Peters and J. E. Childs, Patterns of association with mammals in the major biotic communities of the southwestern United States, Am. J. Trop. Med. Hyg., **56** (1997), 273–284.

[23] J. N. Mills, T. G. Ksiazek, C. J. Peters and J. E. Childs, Long-term studies of hantavirus reservoir populations in the southwestern UnitedStates: A Synthesis, Emerg. Inf. Dis., **5** (1999), 135–142.

[24] F. Sauvage, M. Langlais, N. G. Yoccoz and D. Pontier, Modeling hantavirus in fluctuating populations of bank voles: the role of indirect transmission on virus persistence, J. Anim. Ecol., **72** (2003), 1–13.

[25] C. J. Yahnke, P. L. Meserve, T. G. Ksiazek and J. N. Mills, Patterns of infection with Laguna Negra virus in wild populations of *Calomys laucha* in the central Paraguayan chaco, Am. J. Trop. Med. Hyg., **65** (2001), 768–776.

Curtis L. Wesley and Linda J. S. Allen
Dept. of Mathematics and Statistics
Texas Tech University, Lubbock
TX 79409-1042
USA

Colleen B. Jonsson and Yong-Kyu Chu
Dept. of Biochemistry and Molecular Biology and Dept. of Emerging Pathogens
Southern Research Institute, Birmingham
AL 35206
USA

Robert D. Owen
Dept. of Biological Sciences
Texas Tech University, Lubbock
TX 79409-3131
USA

E-mail address: sizer@mnstate.edu

Advanced Studies in Pure Mathematics

A Series of Up-to-date Guides of Lasting Interest to Advanced Mathematics

Volume 1	Algebraic Varieties and Analytic Varieties. Edited by S. Iitaka. February, 1983	
Volume 2	Galois Groups and their Representations. Edited by Y. Ihara. December, 1983	
Volume 3	Geometry of Geodesics and Related Topics. Edited by K. Shiohama. June, 1984	
Volume 4	Group Representations and Systems of Differential Equations. Edited by K. Okamoto. March, 1985	
Volume 5	Foliations. Edited by I. Tamura. February, 1986	
Volume 6	Algebraic Groups and Related Topics. Edited by R. Hotta. March, 1985	
Volume 7	Automorphic Forms and Number Theory. Edited by I. Satake. February, 1986	
Volume 8	Complex Analytic Singularities. Edited by T. Suwa and P. Wagreich. February, 1987	
Volume 9	Homotopy Theory and Related Topics. Edited by H. Toda. February, 1987	
Volume 10	Algebraic Geometry, Sendai, 1985. Edited by T. Oda. July, 1987	
Volume 11	Commutative Algebra and Combinatorics. Edited by M. Nagata and H. Matsumura. October, 1987	
Volume 12	Galois Representations and Arithmetic Algebraic Geometry. Edited by Y. Ihara. November, 1987	
Volume 13	Investigations in Number Theory. Edited by T. Kubota. March, 1988	
Volume 14	Representations of Lie Groups, Kyoto, Hiroshima, 1986. Edited by K. Okamoto and T. Oshima. February, 1989	
Volume 15	Automorphic Forms and Geometry of Arithmetic Varieties. Edited by K. Hashimoto and Y. Namikawa. July, 1989	

Volume 16 Conformal Field Theory and Solvable Lattice Models.
Edited by M. Jimbo, T. Miwa and A. Tsuchiya.
June, 1988

Volume 17 Algebraic Number Theory—in honor of K. Iwasawa.
Edited by J. Coates, R. Greenberg, B. Mazur and I. Satake. August, 1989

Volume 18 I—Recent Topics in Differential and Analytic Geometry.
Edited by T. Ochiai. November, 1990
II—Kähler Metric and Moduli Spaces.
Edited by Edited by T. Ochiai. November, 1990

Volume 19 Integrable Systems in Quantum Field Theory and Statistical Mechanics.
Edited by M. Jimbo, T. Miwa and A. Tsuchiya.
November, 1989

Volume 20 Aspects of Low Dimensional Manifolds.
Edited by Y. Matsumoto and S. Morita.
December, 1992

Volume 21 Zeta Functions in Geometry.
Edited by N. Kurokawa and T. Sunada.
December, 1992

Volume 22 Progress in Differential Geometry.
Edited by K. Shiohama. September, 1993

Volume 23 Spectral and Scattering Theory and Applications.
Edited by K. Yajima. February, 1994

Volume 24 Progress in Algebraic Combinatorics.
Edited by E. Bannai and A. Munemasa. May, 1996

Volume 25 CR-Geometry and Overdetermined Systems.
Edited by T. Akahori, G. Komatsu, K. Miyajima, M. Namba and K. Yamaguchi. July, 1997

Volume 26 Analysis on Homogeneous Spaces and Representation Theory of Lie Groups, Okayama-Kyoto.
Edited by M. Kashiwara, T. Kobayashi, T. Matsuki, K. Nishiyama and T. Oshima. April, 2000

Volume 27 Arrangements – Tokyo 1998.
Edited by M. Falk and H. Terao. August, 2000

Volume 28 Combinatorial Methods in Representation Theory.
Edited by M. Kashiwara, K. Koike, S. Okada, I. Terada and H.-F. Yamada. December, 2000

Volume 29 Singularities – Sapporo 1998.
Edited by J.-P. Brasselet and T. Suwa.
December, 2000

Volume 30 Class Field Theory – Its Centenary and Prospect.
Edited by K. Miyake. April, 2001

Volume 31 Taniguchi Conference on Mathematics Nara '98.
Edited by M. Maruyama and T. Sunada. May, 2001

Volume 32 Groups and Combinatorics — in memory of Michio Suzuki.
Edited by E. Bannai, H. Suzuki, H. Yamaki and T. Yoshida. October, 2001

Volume 33 Computational Commutative Algebra and Combinatorics.
Edited by T. Hibi. February, 2002

Volume 34 Minimal Surfaces, Geometric Analysis and Symplectic Geometry.
Edited by K. Fukaya, S. Nishikawa and J. Spruck.
June, 2002

Volume 35 Higher Dimensional Birational Geometry.
Edited by S. Mori and Y. Miyaoka. July, 2002

Volume 36 Algebraic Geometry 2000, Azumino.
Edited by S. Usui, M. Green, L. Illusie, K. Kato, E. Looijenga, S. Mukai and S. Saito. November, 2002

Volume 37 Lie Groups, Geometric Structures and Differential Equations — One Hundred Years after Sophus Lie —.
Edited by T. Morimoto, H. Sato and K. Yamaguchi.
December, 2002

Volume 38 Operator Algebras and Applications.
Edited by H. Kosaki. January, 2004

Volume 39 Stochastic Analysis on Large Scale Interacting Systems.
Edited by T. Funaki and H. Osada. March, 2004

Volume 40 Representation Theory of Algebraic Groups and Quantum Groups.
Edited by T. Shoji, M. Kashiwara, N. Kawanaka, G. Lusztig and K. Shinoda. March, 2004

Volume 41 Stochastic Analysis and Related Topics in Kyoto
In honour of Kiyosi Itô.
Edited by H. Kunita, S. Watanabe and Y. Takahashi.
March, 2004

Volume 42	Complex Analysis in Several Variables – Memorial Conference of Kiyoshi Oka's Centennial Birthday, Kyoto/Nara 2001. Edited by K. Miyajima, M. Furushima, H. Kazama, A. Kodama, J. Noguchi, T. Ohsawa, H. Tsuji and T. Ueda. June, 2004
Volume 43	Singularity Theory and Its Applications. Edited by S. Izumiya, G. Ishikawa, H. Tokunaga, I. Shimada and T. Sano. January, 2007
Volume 44	Potential Theory in Matsue. Edited by H. Aikawa, T. Kumagai, Y. Mizuta and N. Suzuki. August, 2006
Volume 45	Moduli Spaces and Arithmetic Geometry (Kyoto, 2004). Edited by S. Mukai, Y. Miyaoka, S. Mori, A. Moriwaki and I. Nakamura. January, 2007
Volume 46	Singularities in Geometry and Topology 2004. Edited by J.-P. Brassele and T. Suwa. February, 2007
Volume 47	Asymptotic Analysis and Singularities. 1—Hyperbolic and dispersive PDEs and fluid mechanics. 2—Elliptic and parabolic PDEs and related problems. Edited by H. Kozono, T. Ogawa, K. Tanaka, Y. Tsutsumi and E. Yanagida. October, 2007
Volume 48	Finsler Geometry, Sapporo 2005 — In Memory of Makoto Matsumoto. Edited by S. V. Sabau and H. Shimada. November, 2007
Volume 49	Probability and Number Theory — Kanazawa 2005. Edited by S. Akiyama, K. Matsumoto, L. Murata and H. Sugita. December, 2007
Volume 50	Algebraic Geometry in East Asia — Hanoi 2005. Edited by K. Konno and V. Nguyen-Khac. February, 2008
Volume 51	Surveys on Geometry and Integrable Systems. Edited by M. Guest, R. Miyaoka and Y. Ohnita. November, 2008

Volume 52 Groups of Diffeomorphisms
in honor of Shigeyuki Morita
on the occasion of his 60th birthday.
Edited by R. Penner, D. Kotschick, T. Tsuboi,
N. Kawazumi, T. Kitano and Y. Mitsumatsu.
November, 2008

Volume 53 Advances in Discrete Dynamical Systems.
Edited by S. Elaydi, K. Nishimura, M. Shishikura and
N. Tose. June, 2009

Volume 54 Algebraic Analysis and Around.
Edited by T. Miwa, A. Matsuo, T. Nakashima and
Y. Saito. February, 2009

To be continued